THE PHYSICS OF FLOW THROUGH POROUS MEDIA

Third edition / Adrian E. Scheidegger

Here in one volume is summarized a vast amount of information on the physical principles of hydrodynamics in porous media, gathered from numerous publications. This new edition represents a substantial revision of the second, incorporating the most recent developments in the field. Alterations have been made throughout, new sections have been added, and the chapter on 'Miscible Displacement' has been rewritten. As in previous editions, the stress is on general physical aspects of phenomena rather than on particular cases, and theoretical rather than experimental aspects have been emphasized. Literature references have been chosen carefully to lead the reader to the most useful sources of additional information on particular aspects.

ADRIAN E. SCHEIDEGGER is head of the Institut für Geophysik, Technische Hochschule, Vienna.

ADRIAN E. SCHEIDEGGER

The physics of flow through porous media

Third edition

UNIVERSITY OF TORONTO PRESS

PHYSICS

To the memory of James W. Young

Preface

THE PRESENT MONOGRAPH is not the outcome of a project undertaken in order to produce a book. Rather, it had its beginning in a bibliography on the physical principles of hydrodynamics in porous media which the writer was asked to compile while working for Imperial Oil Limited in Canada.

Since the first edition of this book was published, a tremendous increase in interest in the subject matter of flow through porous media has occurred. The present edition is the third authorized one. In addition, an unauthorized translation into the Russian language has been published in an edition of two thousand copies.

The search for new techniques for the exploitation of oil reservoirs by petroleum engineers, as well as recent concern with groundwater pollution problems by hydrologists, has provided a great stimulus to studies of hydrodynamics in porous media. Correspondingly, there have been substantial advances in the understanding of the mechanics of flow through porous media, and wholly new approaches to the subject have been discovered. One need only recall the introduction of statistical-mechanical methods. The present edition of this book has been revised substantially to incorporate the most recent developments.

It is obviously impossible to present all of the pertinent information on the hydrodynamics of porous media in detail without writing a ten-volume encyclopaedia on the subject. A selection has therefore been made. Three guiding principles have been followed. (1) Emphasis has been laid on the general physical aspects of the phenomena rather than on particular cases applicable, say, to special engineering problems. (2) Of the many solutions available for some of the basic differential equations, only one has been chosen for presentation in each case. The theory of differential equations is a well-established part of mathematics and has been considered of interest in the present context only if pertinent physical concepts are revealed. (3) The theoretical aspects are given somewhat more stress than the experimental ones, although descriptions of procedures which enable one to determine theoretical 'constants' have always been supplied in order to establish the proper logical sequence.

These principles are admittedly somewhat arbitrary and naturally reflect the writer's own preference. Nevertheless, it is thought that sufficient care has been taken in choosing literature references from those collected in the original bibliography to enable anyone interested in any particular phase of the subject to obtain the required information from the original sources.

The presentation of material in this book often follows closely the presentation employed in the original sources. Wherever applicable, it is noted in the text that the exposition in a certain section is 'following a certain author.' The cooperation of the American Institute of Mining and Metallurgical Engineers, the American Chemical Society, the American Society of Mechanical Engineers, the Soil Science Society of America, the United States Bureau of Mines, the United States National Bureau of Standards, the Danish Academy of Technical Sciences, the Royal Society of London, and the Editor of the journal *Research* (London) in permitting the extensive use of material published in their journals, often more or less verbatim, is gratefully acknowledged. The writer is also indebted to the *Petroleum Engineer*, the *Producers Monthly*, the *Journal of Applied Physics*, *Geofisica Pura e Applicata*, and the *Handbuch der Physik* published by Springer in Berlin, for permission to use material from his own published articles.

The book is dedicated to the memory of Dr James W. Young, one of the most remarkable men the author has ever met. As manager of the Technical Service Department of Imperial Oil Limited in Calgary, Dr Young expended much energy in encouraging basic research in the Canadian oil industry. The truly visionary and congenial atmosphere in his department not only stimulated the birth of this book, but generally gave this author a start in his scientific career.

Vienna, Austria   A.E.S.
February 1972

Contents

THE PHYSICS OF FLOW THROUGH POROUS MEDIA

Introduction

THE STUDY of the physics of flow through porous media has become basic to many applied scientific and engineering fields, quite apart from the interest it holds for purely scientific reasons. Such diversified fields as soil mechanics, groundwater hydrology, petroleum engineering, water purification, industrial filtration, ceramic engineering, powder metallurgy, and the study of gas masks all rely heavily upon it as fundamental to their individual problems. All these branches of science and engineering have contributed a vast amount of literature on the subject.

The study of hydromechanics in porous media falls, quite naturally, into several subject-divisions. First, the properties of porous media and fluids, by themselves, must be discussed. The next step is to proceed to consider interactions between the two, first the statics and then the dynamics. Hydrodynamics in porous media, in turn, can be divided into hydrodynamics of single-phase fluids and hydrodynamics of several fluids. These divisions provide a natural structuring of this book into its various chapters.

Information on flow through porous media is scattered in a multitude of journals and books, which are generally concerned with specific engineering applications. Among the books that deal *generally* with this topic one that must be mentioned is a classic monograph by Muskat (1937), which, in spite of its age, still contains much useful information. A second monograph is a very concise book by Leĭbenzon (1947), which gives an excellent exposition of the physical principles of flow through porous media up to 1947. Unfortunately, it does not give many references to the original articles in which the various equations were developed although it mentions the names of their inventors. As a guide to the literature to which one should turn for further scientific study it is therefore of rather restricted value. Further books have been written by Von Engelhardt (1960); Collins (1961, on a rather elementary level); Childs (1969); Oroveanu (1963) – a volume which, apart from being in Rumanian (which may cause difficulty to many people), deals primarily with flow through inhomogeneous regions in porous media; and Aravin and Numerov (1965), on a quite theoretical level. Furthermore, Matheron (1967) published a highly sophisticated account of an abstract mathematical theory of porous media, and De Wiest (1969) edited a compilation of recent developments on the subject.

More specialized aspects of the physics of flow through porous media are contained in books on specific applications. Thus, on groundwater hydrology one must mention a classic monograph by Polubarinova-Kochina (1952), and books by Todd (1959), Harr (1962), Schoeller (1962), de Wiest (1965), Cedergren (1967), and Bear et al. (1968). On soil physics, there are books by Kirkham and Powers (1969) and Childs (1969). Carman (1956) wrote a book on the flow of gases through porous media, Muskat (1949) one on the physical principles of oil production, and Lykow (1958) one on transport phenomena in porous media.

Some books on general fluid flow also contain pertinent sections on flow through porous media, such as the monographs of Happel and Brenner (1965) and Yih (1965).

Apart from the textbooks mentioned above, some general reviews have appeared in journals and 'Handbooks' which cover part of the subject of flow through porous media. Thus, a lengthy article by Zimens (1944) discusses the characteristics of porous media, a similar work by Engelund (1953) reviews certain aspects of flow through porous media, a paper by Houpeurt (1957) discusses the elements of the mechanics of fluids in porous media, and finally some articles of my own (Scheidegger 1960, 1966) give brief reviews of the subject.

1
Porous media

1.1
DESCRIPTION AND GEOMETRICAL PROPERTIES OF POROUS MEDIA

1.1.1 *Definition of a porous medium*

In order to study the flow of fluids through porous media, it is first of all necessary to clarify what is understood by the terms that denote the two materials
involved: 'fluids' and 'porous media.'

Starting with the latter, one may be tempted to define 'porous media' as solid
bodies that contain 'pores,' it being assumed as intuitively quite clear what is
meant by a 'pore.' However, it is unfortunately much more difficult to give an
exact geometrical definition of what is meant by the notion of a 'pore' than may
appear at first glance. A special effort is therefore required to obtain a proper
description.

Intuitively, 'pores' are void spaces which must be distributed more or less
frequently through the material if it is to be called 'porous.' Extremely small
voids in a solid are called 'molecular interstices,' and very large ones are called
'caverns.' 'Pores' are void spaces intermediate in size between caverns and molecular interstices; the limitation of their size is therefore intuitive and rather
indefinite.

The pores in a porous system may be *interconnected* or *non-interconnected.*
Flow of interstitial fluid is possible only if at least part of the pore space is interconnected. The interconnected part of the pore system is called the *effective* pore
space of the porous medium.

According to the above description (the term definition would hardly be
applicable), the following are examples of porous media: towers packed with
pebbles, Berl saddles, Raschig rings, etc.; beds formed of sand, granules, lead
shot, etc.; porous rocks such as limestone, pumice, dolomite; fibrous aggregates
such as cloth, felt, filter paper; and finally catalytic particles containing extremely
fine 'micro'-pores.

It is thus seen that the term porous media encompasses a very wide variety of
substances. Because of this, it is desirable to arrange porous media into several
classes according to the types of pore spaces which they contain. Any one porous

medium is not, of course, restricted to having only one class of pore spaces, but may have pore spaces belonging to several classes. A suitable set of classes of pore spaces was devised by Manegold (1937 a, b, 1941) by dividing them into voids, capillaries, and force spaces: *voids* are characterized by the fact that their walls have only an insignificant effect upon hydrodynamic phenomena in their interior; in *capillaries*, the walls do have a significant effect upon hydrodynamic phenomena in their interior, but do not bring the molecular structure of the fluid into evidence; in *force spaces* the molecular structure of the fluid is brought into evidence.

In addition, pore spaces have also been classified according to whether they are *ordered* or *disordered*; and also according to whether they are *dispersed* (as in beds of particles) or *connected*. The meaning of these terms is self-evident.

1.1.2 *Geometrical quantities characterizing porous media*

A porous medium can be characterized by a variety of geometrical properties. First of all, the ratio of void to total volume is important. This quantity is called the *porosity* (denoted in this monograph by P) and is expressed either as a fraction of 1 or in per cent. If the calculation of porosity is based upon the interconnected pore space instead of on the total pore space, the resulting quantity is termed the *effective* porosity. In soil physics, one often uses, instead of the porosity, the void ratio e defined by

$$e = P/(1 - P). \tag{1.1.2.1}$$

Another well-defined geometrical quantity of a porous medium is its *specific*

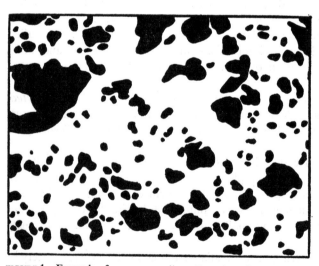

FIGURE 1   Example of a porous medium. Drawing of a cross-section of a porous medium with very large pores in slightly reduced size. The void spaces are shown in black. (Courtesy of Imperial Oil Limited)

internal area. This is the ratio of the internal area to the bulk volume and is therefore expressed as a reciprocal length. In this monograph, it will always be denoted
by S.

It would be most desirable to be able to define a geometrical quantity that
would characterize the pore system in any one porous medium. Unfortunately,
the pore system of a porous body forms a very complicated surface, one that is
difficult to describe geometrically (cf. fig. 1). Intuitively, one would like to talk
about the 'size' of pores, a convenient measure of the 'size' being the 'diameter.'
However, the term diameter makes sense geometrically only if all the pores are of
spherical shape – unless some further specifications are made. If one has the flow
of fluids through those pores in mind, however, it will not do to restrict oneself to
spherical pores only (these would have no effective porosity at all); the pores
must be visualized instead as rather tube-shaped things. One would then call the
diameter of such a tube the 'pore diameter.' Unfortunately, this visualization of
the term diameter is again geometrically quite meaningless unless the tubes are
circular. In general they will not be circular, and (which makes matters worse)
they will not even possess a normal cross-section since the walls will be irregularly
diverging and converging. Thus one cannot even speak about a 'biggest' or a
'smallest' diameter of the tube at any one point.

The writer is not aware of any citation in the literature where this geometrical
dilemma has been satisfactorily solved. Nevertheless, people still talk about the
'size' of pores, the 'pore size distribution,' etc., without defining accurately what
they mean. A possible way out of the dilemma might be to define the pore
diameter (denoted in this monograph by δ) at any one point within the pore
space as the diameter of the largest sphere which contains this point and remains
wholly within the pore space. Thus, to each point of the pore space a 'diameter'
could be attached rigorously and, if desired, a 'pore size distribution' could be
defined – simply by determining what fraction α of the total pore space has a
'pore diameter' between δ and $\delta+d\delta$. It is easily seen that the pore size distribution thus defined is normalized in the following way:

$$\int_\delta^\infty \alpha(\delta)d\delta = 1. \tag{1.1.2.2}$$

Instead of dealing with α, it will often be more convenient to deal with the
cumulative pore size distribution, which will be denoted here by f; this quantity
is defined as that fraction of the pore space which has a pore diameter larger than
δ:

$$f(\delta) = \int_\delta^\infty \alpha(\delta)d\delta, \qquad f(0) = 1. \tag{1.1.2.3}$$

Once the 'pore size distribution' is defined, one can try to characterize it by
certain parameters. This is best accomplished by fitting various standard distribution curves known from mathematical statistics to actual pore size distribution

curves; the parameters of the standard distribution curves will then serve as parameters for the pore size distribution curves. The simplest standard distribution curves suitable for this purpose are the various types of Gauss curves, which have been thus employed by Tollenaar and Blockhuis (1950).

In contrast to pore size distribution, it is easier to determine the *grain size* or grain size distribution of a porous medium, at least if the medium is unconsolidated (dispersed). Unfortunately, the grain size distribution as such does not mean too much as far as the properties of the corresponding pore space are concerned. In order to obtain correlations between grain size and pore size, one has first to investigate the theory of packing of grains (see sec. 1.5.2).

Finally, another geometrical property has been proposed, which has been called the *tortuosity* (denoted in this monograph by T). Originally this was introduced as a kinematical property equal to the relative average length of the flow path of a fluid particle from one side of the porous medium to the other. It is thus a dimensionless quantity.

1.1.3 *Statistical description of a porous medium*

Because of the great complexity of a porous medium it is difficult to describe it geometrically. Theoretically, a porous medium is defined by giving the analytical equation of the surface which bounds its pore space. For practical purposes this is, of course, impossible to accomplish, although one can overcome the difficulty by considering the problem from a statistical point of view. Instead of considering an actual porous medium, one considers an ensemble of porous media which are equivalent to each other within the limits of one's knowledge, and works statistically with this ensemble. In other words, one deals with a random porous medium whose geometrical properties one tries to relate to some easily measurable statistical properties of the medium.

A first attempt to work in this manner was made by Corte and co-workers (Kallmes and Corte 1960; Kallmes, Corte, and Bernier 1961; Corte and Kallmes 1961a, b; Corte 1962), who analysed random fibre networks such as paper. If one assumes that such a porous medium is built up of many layers of two-dimensional fibre networks, such geometrical properties as the total number of fibre crossings, the number of crossings per fibre, and the number, size, and shape of interfibre polygons can be expressed as a function of the independent fibre properties (mean length, width, and curl).

It is apparent that a knowledge of the above-mentioned geometrical properties constitutes an indispensable basis for the prediction of the physical behaviour of such media. The limits of this method of describing porous media lie in the fact that it is concerned with only very specialized media.

A more general statistical description of porous media has been proposed by Fara and Scheidegger (1961), who assume that a picture of a cross-section of a porous medium as represented by figure 1 can be obtained (for techniques for doing this, cf. sec. 1.2.1) and then draw arbitrary, rectifiable lines across it. Points on such lines are defined by giving their arc length s from an arbitrarily chosen origin. For certain values of s, the line will pass through solid space, and for other values of s, through void space. One then defines a function $f(s)$ which is $+1$ if the line is in void space and -1 if it is in solid space. It follows immediately that the mean $\bar{f}$ of f (taken over many values of s),

$$\bar{f} = \lim_{s \to \infty} \int_{-s}^{+s} f(s) ds \Big/ \int_{-s}^{+s} ds, \qquad (1.1.3.1)$$

is related to the porosity P of the medium by

$$\bar{f} = 2P - 1. \qquad (1.1.3.2)$$

Because the mean $\bar{f}$ of f is not zero, a new, normalized function f^* is defined by writing

$$f^*(s) = f(s) - (2P - 1). \qquad (1.1.3.3)$$

Then the porous medium can be characterized by the autocorrelation function R^*

$$R^*(\Delta) = \lim_{s \to \infty} \int_{-s}^{+s} f^*(s) f^*(s + \Delta) ds \Big/ \int_{-s}^{+s} ds. \qquad (1.1.3.4)$$

Instead of using this autocorrelation function, one can perform a spectral analysis directly on $f^*(s)$. However, because of the sharp corners, the spectrum will contain many high harmonics. This complication can be avoided by rounding the corners initially, but the procedure then becomes very artificial.

These techniques of Fara and Scheidegger (1961) were modified somewhat by Moran (1961), without, however, introducing anything fundamentally new. Later, the probabilistic function $f(s)$ introduced by Fara and Scheidegger was generalized by Matheron (1967). Accordingly, an actual porous medium consists of a set A of space occupied by solid matter and its complementary set A^0, which is the pore space. The two sets are defined by the function

$$f(x) = 1 \quad \text{for } x \in A, \qquad f(x) = 0 \quad \text{for } x \notin A, \qquad (1.1.3.5)$$

where x is a point. As is evident, this function is but a slight generalization of that introduced by Fara and Scheidegger (1961). It can be considered as a realization of a random function whose values are 0 and 1, characteristic of a random set A, i.e. of a random porous medium. The random set A is considered to be defined by its space law, i.e., by the ensemble of functions

$$P(x_1, x_2, ..., x_k; y_1, y_2, ..., y_{k'}) = P(x_1, x_2, ..., x_k \in A; y_1, y_2, ..., y_{k'} \notin A) \quad (1.1.3.6)$$

of points $x_1, x_2, ..., x_k, y_1, y_2, ..., y_{k'}$ for all the possible values of integers k and k'. Using this kind of description, Matheron showed how some geometrical properties of a porous medium, which in probabilistic terms are expressed as probabilities and expectation values, are related to other simple probabilities which can be determined experimentally.

Unfortunately, the above probabilistic techniques, although they are very satisfactory from a theoretical standpoint, suffer from the fact that they do not lead to easily applicable results.

1.2
THE MEASUREMENT OF POROSITY

1.2.1 *Methods*

The measurement of the geometrical quantity defined as 'porosity' in section 1.1.2 can be achieved by a great variety of methods. All these methods aim to measure somehow the void volume and the bulk volume of the material; the ratio of the two is then the porosity. The following is a review of the more common of the possible methods.

(*a*) *Direct method*   The most direct method of determining the porosity is to measure the bulk volume of a piece of porous material and then to compact the body so as to destroy all its voids, and to measure the difference in volumes. Unfortunately, this can be done only if the body is very soft. This method has therefore not been applied very extensively, although it is very suitable for the analysis of bread (Lÿnovskiĭ and Postnikova 1940).

(*b*) *Optical methods*   Another direct way to determine the porosity is simply to look at a section of the porous medium under a microscope. A value of porosity can be obtained in this manner since the plane porosity of a random section must be the same as that of the porous medium. It might frequently be advantageous first to photograph the section through the microscope and then to measure the area of the pores with a planimeter. Such visual methods of porosity determination have been applied by Zavodovskaya (1937) to porcelain, by Dallmann (1941) to bread, and by Verbeck (1947) to concrete. A stochastic method for the evaluation of porosity on a photomicrograph has been described by Chalkley, Cornfield, and Park (1949). A pin is thrown into the picture, and if its point is in a void area a 'hit' is scored. If the experiment is repeated often enough, the ratio of 'hits' to 'throws' is equal to the porosity of the specimen.

It is not always possible to make sections of a porous medium conveniently. Difficulties will be encountered especially if the porous medium is dispersed. Techniques have therefore been developed whereby the medium is first impregnated with wax, plastics, or Wood's metal. As a matter of convenience, a dye can be added to the impregnating material so as to make the voids more visible. Such methods have been described by, for example, Waldo and Yuster (1937), Imbt and Ellison (1946), Nuss and Whiting (1947), Ryder (1948), Locke and Bliss (1950), and Lockwood (1950). The method of injecting plastics or wax into porous media has the further advantage of differentiating between effective (i.e., interconnected) and total pore space: the plastic, of course, will reach only the former.

Instead of light rays, x-rays can be used to investigate the pore space of porous media. However, such methods are used to obtain the pore size distribution rather than the straight porosity of a porous medium. They will be discussed more fully in section 1.4.

(c) *Density methods*   If the density ρ_G of the material making up the porous medium is known, then the bulk density ρ_B of the medium is related to the fractional porosity P by

$$P = 1 - \rho_B/\rho_G. \tag{1.2.1.1}$$

The bulk density can be obtained in several ways. If measuring the outside dimensions and weighing the piece of material (Babcock 1945) do not give accurate enough results, a volumetric displacement method can be applied. For fluid to be displaced, a non-wetting fluid such as mercury has to be used, which (it is hoped) will not enter the porous medium – as long as the pore radii are not too large (Seil, Tucker, and Heiligman 1940). The displacement method can be improved upon if the porous medium is coated with a suitable material before immersion (Melcher 1921), in which case corrections have to be made to allow for the coating. Weighing under mercury is another method of obtaining the bulk density (Westman 1926; Athy 1930; King and Wilkins 1944). In this method, a heavy mass usually has to be attached to the sample in order to submerge it.

The density of the material of which the porous medium is composed can be obtained by crushing it and weighing the parts; determining their volume by a displacement method thereupon yields the effective porosity. An alternative means of porosity determination is provided by measuring the change in weight of a soaked porous medium outside the liquid and by measuring directly the volume of liquid that was soaked up. The soaking method is quite an old one; it has been applied by, for example, Gorodetskiǐ (1940), Schumann (1944), Lepingle (1945), Saunders and Tress (1945), Cross and Young (1948), Eichler (1950), Lakin (1951), Pollard and Reichertz (1952), and Russell (1957).

(d) *Gas expansion method*   The basic principle of the gas expansion method is direct measurement of the volume of air or gas contained in the pore space. This can be achieved either by continuously evacuating the air out of the specimen, or, in the more modern versions of the method, by enclosing the specimen of known bulk volume and a certain amount of gas (or air) in a container of known volume under pressure and then connecting this with an evacuated container of known volume. The new pressure of the system is read off and permits one to calculate immediately from the gas laws (Boyle-Mariotte) the volume of gas that was originally in the porous medium. It is obvious that this method gives effective porosities.

The gas expansion method was originally employed by Washburn and Bunting (1922). Equipment based on gas-expansion principles has been refined and described by McGee (1926), Stevens (1939), Page (1947), Hofsäss (1948), Kaye and Freeman (1949), Rall and Taliaferro (1949), and Beeson (1950).

Following Beeson (1950), a typical apparatus for porosimetry by the gas expansion method and its operation may be described as follows. As shown in figure 2, the apparatus consists of a gauge A connected to the sample chamber H through the needle valve K. Similar valves J and L are in the line to control the flow of a gas, such as helium, into the gauge and to exhaust the gas from the sample chamber. This is equivalent to two interconnected chambers, the gauge and the connecting lines representing one and the sample chamber and connecting lines the other. Valves M and N have been added to permit adjustment of the zero point by altering the amount of either valve stem included in the system.

In operation, helium is forced into the gauge chamber until the needle B on the dial reaches the 'start' position (100.00 psi), as seen through the illuminated eyepiece C mounted on a swivel attached to the gauge. This reading of the gauge, as well as all others, is made with the attached buzzer D in operation, to eliminate the effect of friction on the position of the needle. The gas is then expanded into the sample chamber and the position of the needle is noted. This reading, when

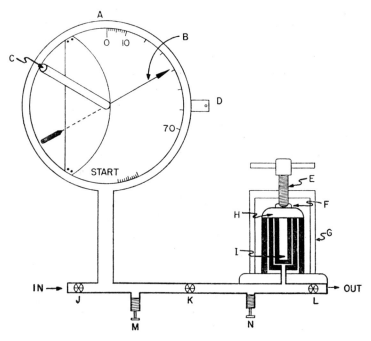

FIGURE 2   A typical apparatus for the measurement of porosity by the gas-expansion method (after Beeson 1950).

interpreted on a calibration table, yields the volume of the sample contained in the chamber.

The calibration table is obtained by applying the above procedure to samples of known volume, such as chrome-alloy ball-bearings that have been measured accurately with a micrometer. This method of calibration minimizes errors due to deviations of the gas from the perfect gas law.

The gas expansion method is probably most generally used at present. It is fast, relatively accurate, and leaves the sample in an undisturbed state so that other tests can be performed immediately.

(*e*) *Other methods*   Methods other than those outlined above have been proposed for porosity measurements. As stated at the beginning of this section (1.2.1), one has to determine, for a porosity measurement, both the bulk volume and the pore (or solid) volume of a piece of material. These 'other' methods are concerned with rather unorthodox measurements of the latter.

One such method is the adsorption of liquids on the internal surface; the surface area can then be determined, and, if the pore geometry is known, the porosity. These methods will be outlined in detail in section 3.2 on adsorption of fluids by porous media.

Another way to obtain the pore volume is to measure the resistance of the medium to flow through the pores. According to some formulas, to be developed in the chapters on fluid flow, the pore volume can be obtained if certain assumptions about pore geometry are made (cf. the Kozeny theory, sec. 6.3.2). Although

TABLE I

Representative values of porosity for various substances

Substance	Porosity range (porosity in %)	Literature reference
Berl saddles	68–83	Carman (1938)
Raschig rings	56–65	Ballard and Piret (1950)
Wire crimps	68–76	Carman (1938)
Black slate powder	57–66	Carman (1938)
Silica powder	37–49	Carman (1938)
Silica grains (grains only)	65.4	Shapiro and Kolthoff (1948)
Catalyst (Fischer-Tropsch, granules only)	44.8	Brötz and Spengler (1950)
Spherical packings, well shaken	36–43	Bernard and Wilhelm (1950)
Sand	37–50	Carman (1938)
Granular crushed rock	44–45	Bernard and Wilhelm (1950)
Soil	43–54	Peerlkamp (1948)
Sandstone ('oil sand')	8–38	Muskat (1937)
Limestone, dolomite	4–10	Locke and Bliss (1950)
Coal	2–12	Bond et al. (1950)
Brick	12–34	Stull and Johnson (1940)
Concrete (ordinary mixes)	2–7	Verbeck (1947)
Leather	56–59	Mitton (1945)
Fibreglass	88–93	Wiggins et al. (1939)
Cigarette filters	17–49	Corte (1955)
Hot-compacted copper powder	9–34	Arthur (1956)

this method is in general not well founded, Drotschmann (1943) claims that it works well when applied to plates in batteries.

The various methods of porosity determination have been compared and evaluated on several occasions (Lehmann and Nüdling 1954; Schmid 1955). However, no general statement as to which is the 'best' method can be made since this depends on the material whose porosity is to be determined.

1.2.2 *Specific cases*

The methods of porosity measurement have been applied and adapted to many cases, and lists of the range of porosity values have been obtained.

In table I, we supply a list of representative values of the porosity of a variety of substances which have been reported in the literature. The values listed under 'range' should not be regarded as *the* maximal and minimal values possible, but rather as values indicating the range in which the porosity of the respective substances is likely to be found. The literature references are the original papers from which the values shown were taken.

1.3
THE MEASUREMENT OF SPECIFIC SURFACE

1.3.1 *Methods*

 As outlined in section 1.1.2, the specific surface is defined as the surface of the pores per unit bulk volume of the porous medium. It is therefore expressed as a reciprocal length. There are several methods of measuring it. The following is a compilation of the more common methods found in the literature.

(*a*) *Optical methods*   In analogy with the optical methods of porosity measurement, the specific surface can also be determined from photomicrographs obtained according to the procedures outlined in section 1.2.1. If it is somehow possible to determine the ratio of the circumference of the pores to that of the total area of the section, then there is a simple relationship with the required specific area. Care must be taken, however, to allow for the over-all magnification of the photograph; since the dimension of the specific area is a reciprocal length, it will come out n times too small if the magnification of the picture was n-fold. The value of the specific area as determined from the picture must therefore be multiplied by the magnification factor employed in order to get the correct specific area of the sample.

The determination of the specific internal area from a photomicrograph is best accomplished by a statistical method due to Chalkley, Cornfield, and Park (1949). Let a needle of length r fall a great number of times upon the picture. Count the number of times the two end points of the needle fall in the interior of the pores,

and denote this by h for 'hits'; also count the number of times the needle inter-
sects the perimeter of the pores, and denote this by c for 'cuts.' Then, in a very
large number of throws, one will have

$$\frac{rh}{c} = 4 \frac{\text{volume of pores}}{\text{surface of pores}} \qquad (1.3.1.1)$$

if the magnification of the picture is 1. The bulk volume of the specimen is $1/P$
times the volume of the pores; therefore one obtains

$$S = 4Pc/(rh) \qquad (1.3.1.2)$$

and, allowing for the n-fold magnification of the photomicrograph, this yields

$$S = 4Pcn/(rh). \qquad (1.3.1.3)$$

(b) *Methods based on adsorption*   The adsorption of a vapour by the surface of
a solid is connected with the area of that surface. A fairly reliable determination
of the internal surface area can be obtained on this basis. There are several
theories of adsorption of fluids by porous media which will be discussed in a
separate section (3.2.1) of this study. The application of such methods to a
determination of the internal area will also be studied in detail, (see sec. 3.2.2).

(c) *Methods based on fluid flow*   Formulas have been developed which claim to
relate the rate of flow of fluids through porous media to their specific surface
area. These formulas are known as the Kozeny equation and the Kozeny-
Carman equation. Because of their importance in the theory of flow, they will be
discussed in detail later (see sec. 6.3.2). If extremely rarefied gases are used, a
modification of the Kozeny equation (Knudsen equation, see sec. 7.3.2) is
employed.

The Kozeny equation has been used extensively for the determination of the
specific surface of porous media; the particular applications of that equation for
this purpose, however, will be discussed in connection with the exposition of the
relevant theory (see sec. 6.4.1).

(d) *Other methods*   Some other methods of surface measurement have also
been proposed. One can, for instance, use the amount of heat conduction by a
gas in a porous medium for this purpose. This heat conduction depends on the
speed of self-diffusion within the gas, which, in turn, is determined by the pore
structure. In this way, a method of area determination is obtained (Kistler 1942).

A surface meter for unconsolidated porous media can be developed in a very
neat way if the media can be fluidized (Bell 1944). Such a meter is based upon
the assumption that the specific surface (per unit weight) is a linear function of
the reciprocal weight of particles that have to be added to a fluidizing agent (e.g.,
water) in order to obtain a certain optical opacity of the mixture.

In analogy with surface adsorption, there also exists an ionic adsorption
phenomenon by which ions from an electrolytic solution are adsorbed on a solid
in contact with it. Obviously, this provides another means by which the specific

surface of a porous medium can be determined (Schofield and Talibuddin 1948).

In addition, some further methods can be applied to determine the specific surface of a porous medium. Such methods are based upon the catalytic reaction and the direct chemical reaction of the surface with certain other substances. Similarly, the heat of wetting and certain exchange reactions of radioactive substances can also be used (Zimens 1944).

1.3.2 *Specific cases*

The above methods of area determination in porous media have been applied to many specific cases, and tables of representative values have been obtained. We supply, in table II, a list of some representative values for various cases. As previously, 'range' implies the representative range into which the area values of the substances concerned are likely to fall, and not the utmost extremes possible. Again, we also supply references to the literature whence the values shown were obtained.

1.4
THE DETERMINATION OF PORE SIZE DISTRIBUTION

The pore size distribution function $\alpha(\delta)$ was defined in section 1.1.2. Methods for determining it have been proposed in the literature, without, however, giving such an exact definition as given there. The investigators simply talk about 'pore diameters' without specifying in any way what is meant by this term. No evaluations exist, therefore, to determine whether the methods proposed actually do measure $\alpha(\delta)$ as defined in section 1.1.2 or something else.

The method most frequently used is that of injection of mercury into the pore system. In this manner, a capillary pressure curve is obtained which, in turn,

TABLE II

Representative values of specific surface for various substances

Substance	Specific surface range (specific surface in cm^{-1})	Literature reference
Berl saddles	3.9–7.7	Carman (1938)
Raschig rings	2.8–6.6	Ballard and Piret (1950)
Wire crimps	2.9×10–4.0×10	Carman (1938)
Black slate powder	7.0×10^3–8.9×10^3	Carman (1938)
Silica powder	6.8×10^3–8.9×10^3	Carman (1938)
Catalyst (Fischer-Tropsch, granules only)	5.6×10^5	Brötz and Spengler (1950)
Sand	1.5×10^2–2.2×10^2	Carman (1938)
Leather	1.2×10^4–1.6×10^4	Mitton (1945)
Fibreglass	5.6×10^2–7.7×10^2	Wiggins et al. (1939)

can be interpreted in terms of pore size distribution. A detailed account of the method and references to many applications will be given in connection with the discussion of capillarity (see sec. 3.4.3).

Surface adsorption has also been used to obtain indirectly a curve for pore size distribution. In a similar fashion capillary condensation in small spaces has been used to obtain pore size distribution curves. These methods will be discussed in the section dealing with these phenomena (see sec. 3.2.2).

If the size and frequency of the maximum pores that permit the passage of a fluid through the porous medium only is required, a method forcing gas bubbles through the porous medium and measuring threshold pressures can be employed (Knöll 1940).

More direct methods using the principles of x-ray scattering have been employed (Brusset 1948; Ritter and Erich 1948; Shull, Elkin, and Roess 1948; Avgul et al. 1951; Luk'yanovich and Radushkevich 1953; Clark and Liu 1957). Another method is to break down the porous medium by crushing it more and more finely. At each stage of crushing the porosity can be measured, which in turn permits evaluation of the pore size distribution as the larger pores are progressively destroyed (Gilchrist and Taylor 1951).

Direct optical methods such as the study of micrographs are, of course, also available. Procedures employing such methods were cited in connection with porosity measurements, where they give excellent results. Their application to pore size distribution studies, however, is somewhat problematic. The reason for this is, of course, that the void space shown in a section through the porous medium does not necessarily bear much relation to the void space in the medium. It is possible that the pores shown may all be rather large in one direction, but small in the other direction, normal to the section. Thus, at least some statistical considerations should be kept in mind if micrographic methods are to be used for pore studies.

The methods mentioned above yield the cumulative pore size distribution curve $f(\delta)$, from which the differential pore size distribution curve $\alpha(\delta)$ has to be calculated. A typical example is given in figure 3. This particular pore size distribution has been determined by the injection of mercury into a piece of limestone which showed very large (vuggy) and very small pores. The maximum point of $\alpha(\delta)$ has obviously not been obtained; the method breaks down for very small openings.

One can try to set up analytical expressions containing indetermined parameters for the characterization of cumulative and differential pore size distributions. Such an expression has been postulated by Milligan and Adams (1954) as follows, k being a normalization factor:

$$df(\delta)/d\delta = \tfrac{1}{4}k \operatorname{sech}^2[\tfrac{1}{2}\kappa(\delta - \delta')]. \qquad (1.4.1)$$

This expression contains two parameters, viz., δ', which is the most frequent pore diameter, and κ, which is a factor determining the sharpness of the distribution.

According to Milligan and Adams, the postulated expression fits experimental pore size distributions very closely so that, in most instances, it is possible to characterize them by just the two parameters κ and δ'.

1.5

THE CORRELATION BETWEEN GRAIN SIZE AND PORE STRUCTURE

1.5.1 *The measurement of grain size*

It has been stated in section 1.1.2 that the grain size distribution of unconsolidated porous media can serve as a basis for the investigation of pore properties. In principle, there are similar difficulties with the notion of 'grain size' as exist with the notion of 'pore size'. The grains are irregularly shaped and therefore

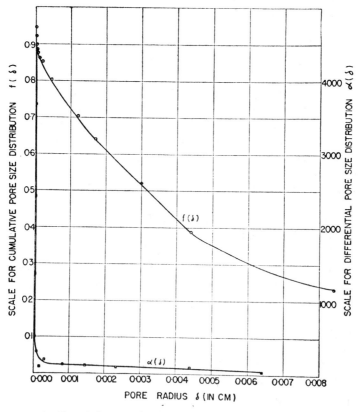

FIGURE 3   Cumulative pore size distribution curve and differential pore size distribution curve obtained on a piece of limestone by the mercury-injection method (courtesy of Imperial Oil Limited).

their 'size' is not defined a priori. However, at least the concept of 'largest diameter' of a grain makes sense geometrically.

The (largest) grain diameter has the dimension of a length, and could thus be given in cm, mm, or such like. Unfortunately, in the literature a different type of unit is usually found, the φ-unit (phi-unit). The number of phi units is related to the diameter δ by the following equation (cf. Krumbein and Pettijohn 1938):

$$\delta(\text{mm}) = (\tfrac{1}{2})^{\varphi}. \qquad\qquad (1.5.1.1)$$

The grain size or grain size distribution of an unconsolidated porous medium can be determined by a variety of methods. However, as the subject of grain size determination is only incidental to the study of the physical principles of hydromechanics in porous media, only a list of selected references will be given.

General discussions and reviews of particle size measurement have been given by Krumbein and Pettijohn (1938), Heywood (1938, 1947), Milner (1940), Uren (1943), Cadle (1955), and Hawksley (1951). The last article mentioned is a good general review listing 218 references. A book by Dalla Valle (1948) also contains much useful information. In addition to the general references given above, some of the methods applicable to particle size measurement have been described: hydrometric methods (such as sedimentation) by Steinberg (1946), Mills (1948), Rim (1952), Donoghue (1956), and Pramanik (1956); x-ray diffraction methods by DuMond (1947), Müller (1947), and Hirsch (1954); sieving by a multitude of authors, especially Dalla Valle (1948); light scattering methods by Gumprecht and Sliepcevich (1953) and Sedláček (1956); and the use of the electron microscope for the present purpose by Franklin et al. (1953). The various methods of particle size analysis have been evaluated and compared by Jarrett and Heywood (1954) and Joglekar and Marathé (1958).

1.5.2 *Theory of packing of spheres*

In order to establish a correlation between the grain size and pore size of an unconsolidated porous medium, one has to know something about the packing of the grains as well as about their shape. For even if the grain 'size' (i.e., the largest diameter) is known, the shape is still not determined. The grains that pass through a certain sieve-mesh and do not pass through another slightly smaller one are not necessarily all identical, owing to the irregularity of shape.

A qualitative visualization of the conditions involved can be obtained by the procedure of assuming 'model'-grains, usually spheres, and by studying their modes of packing. Thus, one arrives at a theory of models of porous media which are composed of spheres.

Such models are constructed in the following way. First, the average size of the grains is measured in one way or another and the spheres of the model are envisaged as being of the same diameter. Then the porosity of the original specimen is determined and the packing of the model-spheres is assumed to be arranged in such a fashion that the model has the same porosity as the original. Therefore, in order to be able to construct proper models, one has to investigate the packing of spheres.

The first study of the modes of packing of spheres and the porosity calculated therefrom appears to have been undertaken by Slichter (1899). Since then the theory has been reviewed, refined, and extended by Smith, Foote, and Busang (1929), Graton and Fraser (1935), Manegold (1937b), Manegold and Solf (1939), Hrubíšek (1941), Ackerman (1945), Foord (1945), and Busby (1950). Muskat (1937, p. 10) gives a short qualitative review of the theory in his book on flow through porous media, and Leĭbenzon, in his monograph (1947, pp. 18–24), gives a concise exposition of the theory of Slichter, including the mathematical deductions involved.

The article of Hrubíšek (1941) contains probably the most comprehensive investigation of the geometry of assemblies of spheres to date. It is based upon Minkowski's theory of the geometry of numbers and gives a very complete survey of the applications of this theory to the present problem. The most difficult part is, of course, an enumeration of the possible patterns of assemblies of spheres and the mathematical proofs their for maximal and minimal properties. Of practical importance are 'stable' packings only, i.e., packings where one sphere touches at least four others in such a manner that at least four of the contact-points are not contained in the same hemisphere.

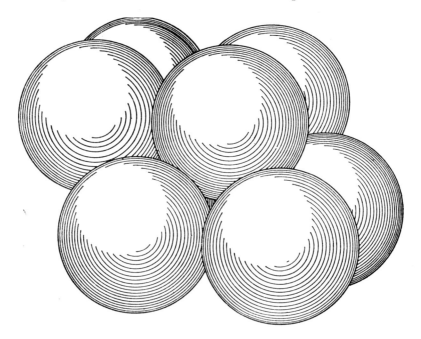

FIGURE 4   Rhombohedral packing of spheres (after Graton and Fraser 1935).

Without going into the mathematics of the theory, for which the reader is referred to the article of Hrubíšek (1941), one can state the principal results as follows:

(a) For any one mode of packing, the porosity of the bed is independent of the size of the spheres.

(b) The porosity of stable beds varies from 0.259 upwards. The system with a porosity of 0.259 represents 'rhombohedral' (fig. 4) or 'face-centred cubic' (fig. 5) packing. The 'thinnest' stable packing of spheres that is now known has been described by Heesch and Laves (cf. Hrubíšek 1941); it has a porosity of 0.875.

(c) A quantity, n, can be defined which gives the ratio of the area left void by the spheres in a plane cross-section through the centres of adjacent spheres to the area of that cross-section. It is a measure of the pore size of the array. For the different packings, it varies from 0.0931 upwards (cf. Leĭbenzon 1947, p. 23). It is again independent of the size of the spheres.

It is thus seen that a complete enumeration of all the stable packings of identical spheres has not yet been achieved. The stable packing with minimum porosity is rhombohedral or face-centred cubic, as mentioned above. A stable packing with maximum porosity has not yet been found. It is doubtful that one exists since larger and larger arches of spheres that are stable can be constructed; however, no mathematical proof of the non-existence of a stable packing with maximum porosity has been achieved.

In an assemblage of spherical particles there is a very neat method for calculating the size distribution of the spheres from the size distribution of the circles in a cross-section (Lenz 1954). The connection between the size distribution of the circles (in a section) and that of the spheres is given by the equation (cf. Fromm 1948)

$$g(r)dr = 2rdr \int_r^\infty \frac{\rho(R)dR}{\sqrt{R^2 - r^2}}. \tag{1.5.2.1}$$

In this equation, the functions $g(r)$ and $\rho(R)$ are defined in such a manner that $Ag(r)dr$ signifies the number of circles in the section of area A with radius between r and $r + dr$, and $V\rho(R)dR$ signifies the number of spheres in the volume

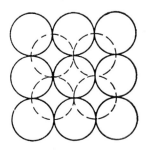

FIGURE 5   Face-centred cubic packing of spheres.

V with radius between *R* and *R* + *dR*. Equation (1.5.2.1) permits one to calculate *g* from ρ, but in the practical analysis of cross-sections, the problem is the inverse one. Thus one has to solve equation (1.5.2.1) for ρ, which leads to

$$\rho(R) = -\frac{1}{\pi R} \int_R^\infty \frac{r(dg/dr)dr}{\sqrt{r^2 - R^2}}. \tag{1.5.2.2}$$

This equation can be evaluated numerically, but, unfortunately, it is very sensitive to statistical fluctuations of *g* (since it is the derivative and not the function itself that enters). In practice, therefore, it is recommended that *dg/dr* be approximated by some analytical function, for instance by a series expansion of the form

$$dg/dr = \sum_{m,n} a_{mn} r^n e^{-mr}. \tag{1.5.2.3}$$

Then one has

$$\rho(R) = -\frac{1}{\pi} \sum_{m,n} (-1)^n a_{mn} \frac{d}{dR} \left(R^n \frac{d^n}{d(mR)^n} K_0(mR) \right), \tag{1.5.2.4}$$

where *K* is the usual modified Hankel function. This expression is a useful expression for the numerical evaluation of experimental data.

1.5.3 *Packing of natural materials*

Natural materials are composed of grains whose shape may deviate appreciably from that of spheres. Moreover, it will often be found that the grains are somewhat cemented together without being fully consolidated, but it would still be desirable to apply some correlations between the 'grain size' and pore size distribution and other characteristics of the porous medium, such as the specific area. Furthermore, the size of the grains will seldom be very uniform. Non-uniformity in size will, in general, permit the smaller particles to fill the spaces between the larger ones and thus reduce the porosity appreciably. Contrariwise, angularity of the particles permits bridging with a resulting increase in porosity.

It is only reasonable, therefore, to expect that spherical models as discussed above will not be entirely adequate to represent even the geometrical properties of porous media (not to speak of hydraulic ones). They will help us to understand some of the features better, but actual correlations between grain sizes and pore sizes, such as have been proposed, are based on experimental investigations rather than on theory. A true understanding of the effects, of course, cannot be obtained in this manner either.

Theoretical and experimental studies of the effect of grain size on pore size have been conducted by Tickell, Mechem, and McCurdy (1933), Nissan (1938), Cloud (1941), Hrubíšek (1941), Rosenfeld (1949), Griffiths (1952), and Gaither (1953). Kiesskalt and Matz (1952) and Wise (1954) discussed the dependence of specific area on the grain size distribution. These studies bear out some general facts such as those mentioned at the beginning of this sub-section, but the correlations obtained do not seem to be generally applicable without the introduction of sufficient 'cementation,' 'compaction,' and other 'factors' – which, actually, are only undetermined factors to allow adjustment of the actual findings to those one would like to expect. It should be noted that any correlation can be made to fit any set of findings if a sufficient number of indeterminate factors are used. A typical model of this type has, for instance, been proposed by Euler (1957).

Thus the correlations developed are at best valid for one type of material or rock strata. To expect that Raschig rings and spherical pebbles, which happen to have been screened out by the same set of sieves, should have identical geometrical pore structures is somewhat overoptimistic.

1.6

THE MEASUREMENT OF TORTUOSITY

The last geometrical quantity introduced in section 1.1.2 is the tortuosity. As noted, it is defined as the ratio of the length of a true flow path for a fluid and the straight-line distance between inflow and outflow. This is, in effect, a kinematical definition, although there is some hope that the tortuosity is also a geometrical quantity. If a suitable model of the porous medium is chosen, for example, one consisting of a network, then it is independent of the flow regime. It could be measured, for example, by an electrical resistivity measurement, as the current would have only one set of paths to flow through which would be identical with the flow paths of fluid particles, i.e., the paths prescribed by the capillaries making up the porous body. However, although the hydraulic paths are very similar to the electrical paths in this case, it must be noted that there are some differences. The hydrodynamic flow depends not only, like the electrical flow, on the total cross-section of the channels, but also on their shape. Furthermore, surface-flow phenomena in electrical double-layers become significant (Pfannkuch 1969). The discussion cannot be carried further at this stage, inasmuch as the particulars of the model employed are of great importance. The hydraulic tortuosity T_{hy} depends directly on the model chosen. Generally, one uses a formula of the type

$$T_{hy} = \frac{P}{k} \frac{A_{eff}}{8\pi},$$ *probably porosity*   (1.6.1)

where k is the permeability (expressing in some fashion the pore geometry; cf. chap. 6) and A_{eff} is an effective-cross section.

Regarding the electrical tortuosity T_{el}, it can be obtained from a measurement of the formation factor F,

$$F = \rho_0/\rho_w,$$ (1.6.2)

which is the ratio of the resistivity ρ_0 of the 100 per cent saturated porous medium to the resistivity ρ_w of the fluid. One can then deduce the tortuosity by the formula

$$T_{el} = PF. \tag{1.6.3}$$

The identity of T_{hy} and T_{el} is open to question, but is generally assumed (cf. Wyllie and Rose 1950). On this basis, the idea of determining the tortuosity from some sort of electrical measurements has been followed up by Fricke (1931), Stamm (1931), Archie (1942), Wyllie and Spangler (1952), Schopper (1966), and Bitterlich and Wöbking (1970).

1.7
RHEOLOGICAL PROPERTIES OF POROUS MEDIA

1.7.1 *Stresses in porous media*

Porous media have been treated so far as rigid, an assumption which cannot always be expected to be true. The geometric quantities describing porous media, as introduced, may themselves be functions of certain dynamic quantities, notably of the prevailing stresses.

The description of stresses in porous media meets with certain difficulties. It is obviously as impractical to give the actual stress-tensor microscopically at every point of the medium as it is to give the analytical equation of the surface constituting the pore-system.

Thus, some heuristic theory has to be developed which describes the stresses in porous media in macroscopic terms. Gassmann (1951) and Schiffman (1970) have developed a rather complete theory of such stresses, but for our purpose a more elementary theory proposed by Terzaghi (1951) is entirely sufficient. Without going into the details of Terzaghi's theory, one may state that the stresses in a porous medium are twofold: one part, termed the *neutral stress*, is the stress in the fluid; the second part, called the *effective stress*, is the difference between the total stress prevailing in the fluid-filled porous medium and the neutral stress. It is the effective stress which, according to Terzaghi, produces the deformation of the porous medium.

The macroscopic stress in the fluid is hydrostatic, and can therefore be denoted by a scalar p. It is the pressure of the pore-fluid. The total macroscopic stress of the porous medium (including the pore-fluid) may be denoted by a tensor $\mathcal{T}$ with components $\mathcal{T}_{ik}$. In most cases, however, the hydrostatic component of this stress will be the prevailing one, as it is equal to the 'overburden pressure' of the porous medium. We shall denote this scalar overburden pressure by $\mathcal{T}$, too, as it is quite clear from the context whether a tensor or a scalar is implied.

The effective stress in a porous medium can be defined in several ways. First, there is the microscopic effective stress $\mathcal{T}_{ik}{}^m$, which is the stress that actually acts on the average locally in the porous skeleton. Writing down the equilibrium condition yields (δ_{ik} is the Kronecker symbol):

$$\mathcal{T}_{ik} = \mathcal{T}_{ik}^{m}(1-P) + pP\delta_{ik}. \qquad (1.7.1.1)$$

Second, there is the macroscopic effective stress $\mathcal{T}_{ik}^{M}$ which acts on the average macroscopically in the porous medium. The equilibrium condition yields the relationship

$$\mathcal{T}_{ik} = \mathcal{T}_{ik}^{M} + pP\delta_{ik}. \qquad (1.7.1.2)$$

Finally, there is a fictitious stress $\mathcal{T}_{ik}^{F}$ introduced by Gersevanov (1933), which is the stress acting between grains in a porous medium if this is regarded as an assemblage of grains immersed in a fluid:

$$\mathcal{T}_{ik}^{F} = \mathcal{T}_{ik}^{M} - p(1-P)\delta_{ik}. \qquad (1.7.1.3)$$

It is Gersevanov's contention that it is the trace of the fictitious stress tensor $\mathcal{T}_{ik}^{F}$ which causes consolidation in a porous medium, at least if it is granular; thus (after Gersevanov):

$$P = P(\mathcal{T}_{jj}^{F}, p), \qquad (1.7.1.4)$$

where the summation convention has been applied. Gersevanov's 'fictitious stress' is the same as Terzaghi's 'effective pressure,' but in tensor notation.

It is clear from the above remarks that, in order to describe the stress state in a fluid-filled porous medium, one needs in all seven parameters: the six parameters of the (symmetric) total stress tensor $\mathcal{T}_{ik}$ and the fluid pressure p. If we have applications to underground strata in mind, it will be possible in most cases to regard the vertical direction as a principal stress direction; the reasons for this are outlined in, for example, Anderson's (1942) book. The other two principal directions then lie somewhere in the horizontal plane.

Using a Cartesian coordinate system x, y, z with z denoting the vertical and x, y principal axes, one has, under the present simplifying assumptions, three significant stress components: $\mathcal{T}_{xx}, \mathcal{T}_{yy}$, and $\mathcal{T}_{zz}$; the other stress components, viz., $\mathcal{T}_{xy}, \mathcal{T}_{xz}$, and $\mathcal{T}_{yz}$, are zero. The seven parameters therefore reduce to four (viz., $\mathcal{T}_{xx}, \mathcal{T}_{yy}, \mathcal{T}_{zz}$, and p). A further simplification is achieved if one assumes that the stress state is isotropic; one then has only two significant parameters (the overburden pressure and fluid pressure).

1.7.2 Deformation of porous media

In order to determine the dependence of the various geometrical quantities characterizing porous media on stress, it is necessary to develop some connection between the stresses and the corresponding strains of both the fluid and the solid. The fluids will be discussed at length in chapter 2 of this book, and, as for porous media, the following remarks can be made.

The simplest connection between geometrical quantities referring to the porous medium and the stresses will be obtained by assuming that the porous medium is elastic, in which case Hooke's law obtains. The effect of this will be that all the geometrical quantities relating to the pores are linear functions of the effective stress. By the boundary conditions, however, the fluid pressure p and the total stress $\mathcal{T}$ will generally be prescribed so that it appears that for a fixed total stress the pore geometry is a linear function of p. With reference to the porosity P, this leads to the equation

$$dP/dp = \beta_m(\mathcal{T}). \tag{1.7.2.1}$$

The above theory can be generalized somewhat if one undertakes to write down the equilibrium conditions of the theory of elasticity for all the stress and strain components involved in a porous medium. This has been done by Biot (1941). Assuming that all the displacements are small and subject to Hooke's law, the components ε_{ik} of the macroscopic strain tensor (i.e., of the bulk of the medium) at constant fluid pressure p are given in the usual fashion (cf. Jeffreys 1931):

$$\mathcal{T}_{ik} = \lambda' \delta_{ik} \varepsilon_{jj} + 2\mu' \varepsilon_{ik}, \tag{1.7.2.2}$$

where λ', μ' are Lamé's constants and the summation convention for tensors has been applied (note that the strains have been defined here in *tensor* form; hence the factor 2 in the above formula). If the fluid pressure is now changed at the rate $\dot{p}$, only the diagonal elements of the strain tensor can be affected. Hence Biot postulates the equation

$$\mathcal{T}_{ik} = \lambda' \delta_{ik} \left(\varepsilon_{jj} - \frac{\dot{p}}{\mathfrak{H}} \right) + 2\mu' \left(\varepsilon_{ik} - \frac{\dot{p}}{3\mathfrak{H}} \delta_{ik} \right), \tag{1.7.2.3}$$

where $\mathfrak{H}$ is a new constant of the porous medium. To this one has to add a relation for the rate of the porosity change

$$\dot{P} = \frac{1}{3\mathfrak{H}} \dot{\mathcal{T}}_{ii} + \frac{\dot{p}}{\mathfrak{R}}, \tag{1.7.2.4}$$

where $\mathfrak{R}$ is a further constant of the porous medium. The equations (1.7.2.3) and (1.7.2.4), in conjunction with the well-known compatibility relations (cf. Jeffreys 1931), represent the equilibrium conditions for a porous medium under elastic conditions. The constants $\mathfrak{H}$, $\mathfrak{R}$ may be termed *Biot's constants*.

In a more complicated case, the rheological equation of the porous medium may be assumed to be different from that of Hooke. The medium may be plastic; in fact it may itself behave externally almost like a 'fluid' (cf. Ferrandon 1950). Clay, for instance, will consolidate if any fluids contained in it are withdrawn. This will show up by the fact that its porosity, as well as the other geometrical quantities characterizing it, are not linear functions of p – they are not even reversible. Nevertheless, it will be possible to represent the porosity as some function of p, with the components of the total stress ($\mathcal{T}$) as parameters:

$$P = P_{\mathcal{T}}(p), \tag{1.7.2.5}$$

and similarly with the other geometrical quantities of interest. Following Grace (1953), Tiller (1953) showed that a widely applicable porosity-pressure relationship (especially with filter cakes) is given by the equation

$$P = P_0(\mathcal{T} - p)^{-c}, \tag{1.7.2.6}$$

where P_0 and c are constants; as usual $\mathcal{T}$ is the scalar total pressure on the porous system (i.e., the 'overburden' pressure), and p is the pressure in the fluid.

Generally, experimental investigations of 'rock compressibility' have been undertaken in such a manner as to determine the particular form of $P_{\mathcal{T}}(p)$ in equation (1.7.2.5). Thus, an 'overburden pressure' ($\mathcal{T}$) is chosen, and the porosities and strains of the porous body are measured for different fluid pressures p or vice versa. Such measurements have, for instance, been made on petroleum reservoir rock (consolidated sandstone) by Hall (1953), Hughes and Cooke (1953), Geertsma (1957), Fatt (1958), and McLatchie, Hemstock, and Young (1958); on soils by Nishida (1956) and Caquot and Kérisel (1956); and on filter cakes by Hutto (1957). A set of curves, demonstrating the dependence of porosity on overburden pressure and on fluid pressure, which was obtained by Tiller (1953) for kaolin, is shown in figure 6.

The above theory allows only for the existence of an overburden *pressure*. If allowance is to be made for an anisotropic overburden *stress*, the formulas have to be amplified somewhat (cf. Scheidegger 1959). To this end, we follow the general method in finite strain theory (cf. Scheidegger 1956) of introducing 'parameters' ξ_α. These parameters are the Cartesian coordinates of the porous medium in an arbitrary state (e.g. under zero overburden stress and zero fluid pressure). The parameters ξ_α will be the variables to which everything will ultimately be referred.

Let the coordinates of the point ξ_α in the stressed state be $x_i(\xi_\alpha)$. Then we have

$$x_i = x_i(\xi_\alpha, \mathcal{T}_{ik}, p, P), \tag{1.7.2.7}$$
$$\xi_\alpha = (x_i, \mathcal{T}_{ik}, p, P), \tag{1.7.2.8}$$

where it is implied that the rheological condition of the porous medium is such that its deformation is entirely given if a certain value is assumed for the total stress $\mathcal{T}_{ik}$, the fluid pressure p, and the fluid content (volume porosity P). This, of course, neglects all the inertial forces connected with the motion of the medium. However, we allow for the possibility of hysteresis by specifying single-valuedness for the xs only relative to the particular direction from which the stress state in question was attained.

The theory of deformation of porous media, as discussed so far, is entirely heuristic. The rheological equation (exemplified, for example, by the curves in fig. 6) is assumed to be experimentally determined. However, one would like to understand and explain the shape of these curves by a more detailed analysis of the microstructure of porous media. Because of the extreme complexity of the structure of such media one will, in general, have to have recourse to theoretical *models*. Reviews of attempts along these lines have been given, for instance, by Deresiewicz (1958a, b, and c); most models discussed in the literature do not appear to account for the fluid pressure in the pores since the medium is usually considered as empty.

Turning to a discussion of these models, we note that we have already mentioned the sphere-pack model of Gassmann in which all the spheres are assumed to be of like size in a rhombohedral array. Gassmann assumed that this array is subjected only to its own weight and used Hertz's (1895) theory (cf. Timoshenko and Goodier 1951, p. 372) to compute the stiffness at the individual contacts. In Hertz's theory only the normal components of the contact forces are accounted for. In a more refined treatment, local forces at the contact places which vary in direction and magnitude (slip!) as well as twisting couples should be introduced. The calculation of these forces may be based upon solutions of equations of elasticity theory provided by Cattaneo (1938), Mindlin (1949), and Mindlin and

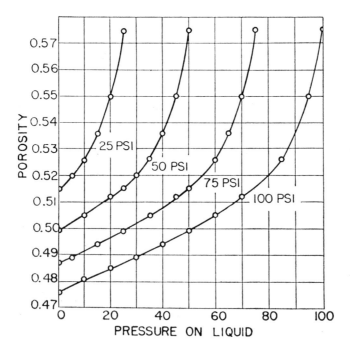

FIGURE 6   A set of curves demonstrating the dependence of porosity on overburden pressure and on fluid pressure, in the case of kaolin (after Tiller 1953).

Deresiewicz (1953). These solutions are not only non-linear, but inelastic as well owing to the dissipation of energy during slip. Thus, if the corresponding forces are introduced into a theoretical model of a porous medium, the resulting stress-strain relations will depend on load history, and will have to be expressed in terms of increments of stress and strain. In spite of the mathematical difficulties involved, such relations have been deduced by Deresiewicz (1958c) for a simple cubic lattice of spheres, by Duffy and Mindlin (1957) for a face-centred cubic arrangement (see fig. 5), and by Duffy (1957) for a rhombohedral (fig. 4) array. The relations obtained are in the form of non-linear differential equations; as has been outlined above, the finite (integral) relationships depend on the total previous stress history. As an example we present here the solution of Thurston and Deresiewicz (1958) for a face-centred cubic arrangement that has been brought homothetically to the isotropic compressive stress $\mathcal{T}_0^{(0)}$ and is then subjected to a further isotropic compressive stress $\mathcal{T}_0$. This solution for the additional strains ε_{xx}, ε_{yy}, ε_{zz} is

$$\varepsilon_{xx} = \varepsilon_{yy} = \varepsilon_{zz} = \frac{3K}{4}(\mathcal{T}_0^{(0)})^{\frac{2}{3}}\left[\left(1 + \frac{\mathcal{T}_0}{\mathcal{T}_0^{(0)}}\right)^{\frac{2}{3}} - 1\right] \qquad (1.7.2.9)$$

with

$$K = 2\left(\frac{1-m}{3G^2}\right)^{\frac{1}{3}}, \qquad (1.7.2.10)$$

where G is the shear modulus and m is Poisson's ratio for the spheres.

The theory for the face-centred cubic arrangement has been tested experimentally in a compression test by Thurston and Deresiewicz (1958), who claim to have obtained an excellent confirmation thereof.

The above models all employ regular packings of spheres to represent a porous medium. A more complicated model has been chosen by Brandt (1955) in which several sets of like spheres are considered, each set fitting into the interstices left by the preceding set. The ratio of the radii of the spheres in consecutive sets was taken as very small so that each set could be assumed to be in a random packing. Thus Brandt evolved the following equations for one set of uniform spheres:

$$V_{\text{bulk}} = \frac{4\pi NR^3}{3(1-P)} - \frac{4\pi NR^3}{1-P}\left(\frac{1.75(1-m^2)}{E}\right)^{\frac{2}{3}}p^{\frac{2}{3}}, \qquad (1.7.2.11)$$

$$V_{\text{void}} = \frac{4\pi PNR^3}{3(1-P)} - \frac{4\pi NR^3}{1-P}\left(\frac{1.75(1-m^2)}{E}\right)^{\frac{2}{3}}p^{\frac{2}{3}}, \qquad (1.7.2.12)$$

where E is the modulus of elasticity of the spheres, N is the number of spheres, p is the pressure on the pack (overburden pressure with zero fluid pressure), R is the radius of the spheres, P is the porosity, and m is Poisson's ratio of the spheres.

It stands to reason that Brandt's model is a much better approximation of a porous medium than a regular packing of spheres. However, the calculation of the contact stresses between spheres is based on Hertz's theory and thus does not allow for the influence of slip, etc., which was introduced in the work of Deresiewicz. This may be the reason why Fatt (1957), in a test of Brandt's equation, did not find very good agreement with experimental values.

1.7.3 Compaction

A special type of deformation of porous media is the vertical compaction of sedimentary layers in situ due to the (generally increasing) weight of the overburden (Hedberg 1936). This, in effect, is a process which is of great importance in the genesis of sedimentary rocks and has, therefore, been studied extensively by sedimentoligsts (see Von Engelhardt 1960; Skempton 1970). A full mathematical treatment of the problem would have to take the flow of the fluid during the deformation of the porous medium into account; this has been tried in the theories of consolidation (cf. sec. 4.3.3). However, for the problem of the compaction of sediments a much simpler approach is usually taken, inasmuch as the dynamics of the fluid flow can be neglected entirely. For a constant fluid pressure p, equation (1.7.2.5) can be written differentially as

$$dP/d\mathscr{T} = -\kappa P, \tag{1.7.3.1}$$

where κ is a coefficient of compaction which now depends only on the porosity P and the overburden pressure $\mathscr{T}$.

On this basis it is possible to set up a model of the sediment-compaction process which occurs in three stages (Marsal and Philipp 1970). In the *first stage*, fresh material is deposited in water, leading to a loose assemblage of grains with high porosities (35–80 %). In the *second stage* the overburden builds up and the material begins to compact. In the *third stage* a reduction in the porosity is caused by crushing of the individual grains.

All these stages can be represented heuristically by giving a relationship between κ, $\mathscr{T}$, and P. If this relationship is known, then, assuming that sediment compaction is solely a gravitational process, porosity-depth curves can be calculated by a simple numerical integration based on equation (1.7.3.1), taking account of the increasing overburden. A relation between κ, $\mathscr{T}$, and P had been proposed by Terzaghi for clays; it can be represented as follows (Marsal and Philipp 1970):

$$\kappa = \frac{\ln(P_0/P)}{\mathscr{T}-1}. \tag{1.7.3.2}$$

Inserting this into (1.7.3.1) and integrating numerically yields κ as a function of P. Marsal and Philipp have compared this value of κ with that obtained from known porosity-depth curves and obtained excellent agreement.

The remarks of this section also explain why fluid withdrawal from a formation can cause land subsidence above (Poland 1969).

1.7.4 *Failure in porous media*

If the stresses in a porous medium exceed a certain limit, the medium will fail just like any other solid body. The strength limits, however, must be represented in terms of *effective pressure* or *effective stress* as defined in section 1.7.1. The modes of failure will be identical with those observed in solid substances. The general types of failure which may occur in solids are discussed in standard treatises on fracture (see Orowan 1949; Liebowitz 1969).

In addition, the influence of the pore fluid on failure is of particular interest in a porous medium. Murrell (1963) has investigated the modifications necessary in the Griffith theory of fractuie due to pore pressure. The pore fluid has also the effect that a mode of failure may occur which is in addition to the modes otherwise observed in solids. This mode has been termed *splitting failure* by Terzaghi (1945). It consists of a tensile breakdown of the porous structure due to the fluid pressure from within. Consequently, the medium 'splits.' The fluid pressure causing splitting failure depends on the total stress on the medium; conditions can be written down easily. Generally, one assumes that the porous medium has no internal strength at all, so that splitting failure occurs as soon as the effective pressure falls below zero.

The breakdown of a porous medium due to fluid pressure has an interesting application in the fracturing of oil wells. Fracturing treatments consist of pumping fluid into a well until the pressure is raised to such a value that the formation fractures. As soon as this occurs, fluid enters the formation rapidly at a pressure which is just sufficient to keep the fractures open. Various theories explaining this process have been advanced, e.g., by Van Poollen (1957), Maksimovich (1957), Cleary (1958), and Morgenstern (1962). The most satisfactory explanation is probably that due to Hubbert and Willis (1957), according to whom the fracturing in a well is due to tensile failure of the formation. According to the prevailing geological stiess state, the resulting fractures will be either vertical or horizontal. Conversely, from the bottom hole pressures in a well which are observed during a fracturing operation, one can in turn determine the prevailing regional geological stress state (Scheidegger 1960; Pulpan and Scheidegger 1965). However, the details of these discussions are beyond the scope of the present monograph and the reader is referred to the papers cited above for further details.

2
Fluids

2.1
GENERAL REMARKS

The subject of hydromechanics in porous media is concerned with the pore space of a porous medium as filled with various fluids. It is therefore also necessary to investigate the mechanics of fluids that may fill the pore spaces, i.e., of liquids and gases.

There are many good reference books on the mechanics of liquids and gases, and thus the present chapter is but a very incomplete review of the subject for the convenience of the reader. Only those aspects of the theory that will be of importance later will be stressed.

From the outset, there are two possible aspects of the mechanics of fluids: macroscopic and microscopic. The macroscopic aspect is manifested in what one might call the 'continuous matter theory,' meaning that the fluid is treated as a continuous medium, its motion being determined if the motion of every material point of the fluid is given by mathematical equations. The microscopic aspect is obtained if the molecular structure of the fluid is taken into account. Usually the microscopic aspect will effect but minor corrections to the equations of continuous matter theory.

Of particular importance to the flow through porous media is the interaction of fluids with surfaces, and the final sections of this chapter will therefore be devoted to that subject.

2.2
CONTINUOUS MATTER THEORY

2.2.1 *The fundamental equations*

The motion of a fluid, if the fluid is regarded as a continuum, is described geometrically if the position of every material point of the fluid is known at every time-instant. There are three kinds of physical conditions which determine such motion. The first is the continuity condition, the second the rheological equation of state, and the third Newton's law of motion.

These physical conditions are expressed mathematically as a system of differential equations. An additional set of initial or boundary conditions is needed to make the problem fully determined. Such boundary conditions, for example, specify whether or not the particular fluid is sticking to the walls of a container. Different combinations of rheological equations and boundary conditions will determine whether the fluid is a liquid or a gas, whether it is viscous or non-viscous, etc.

The *continuity condition* and *Newton's law* of motion are well known; for example, Lamb (1932) expressed them in a representation suitable to the description of a continuous medium.

The *rheological condition* is the connection between the stresses and the strains in the fluid (and their time derivatives). For an 'ideal' fluid it is assumed that there are no shearing-stresses possible and that the fluid is incompressible. In general, however, we shall assume that the fluids are viscous and compressible. The conditions applied to account for viscosity (introducing a constant, μ, termed the 'viscosity') are well known (see, for example, Lamb 1932), and for the compressibility we shall assume the following general equation (Muskat 1937, p. 131):

$$\rho = \rho_0 \, (p/p_0)^m \, e^{\beta_f(p-p_0)}, \tag{2.2.1.1}$$

where ρ is the density and p the hydrostatic pressure. The quantities m and β_f are constants. The particular fluids of signficance may now be classified as follows (after Muskat, loc. cit.):

Liquids: $m = 0$
   Incompressible liquids: $\beta_f = 0$
   Compressible liquids: $\beta_f \neq 0$
Gases: $\beta_f = 0$
   Isothermal process: $m = 1$
   Adiabatic process: $m = C_V/C_p$,

where C_V denotes the specific heat at constant volume and C_p the specific heat at constant pressure.

Finally the *initial* and *boundary conditions* will determine, first, the shape of the walls of the container of the fluid; second, the external conditions (such as the throughput or the pressure drop); and, third, the interaction between the fluid and the walls. If the fluid is viscous, it is generally assumed that it sticks to the walls.

2.2.2 *Special cases of viscous fluid flow*

(a) *The Navier-Stokes equation*   The set of conditions outlined in section 2.2.1

can be combined to form various differential equations which are applicable to different kinds of fluids. The best known of these equations is that of Navier and Stokes (see Lamb 1932), which is applicable to incompressible viscous fluids. Because of its fundamental importance it is restated here:

$$\mathbf{v} \operatorname{grad} \mathbf{v} + \partial \mathbf{v}/\partial t = \mathbf{F} - (1/\rho) \operatorname{grad} p - (\mu/\rho) \operatorname{curl} \operatorname{curl} \mathbf{v}. \qquad (2.2.2.1)$$

Here $\mathbf{v}$ is the local velocity-vector of a point of the fluid, t the time, $\mathbf{F}$ the volume-force per unit mass, and, as before, p, μ, and ρ are respectively the pressure, viscosity, and density of the fluid. The boundary conditions prescribed that $\mathbf{v} = 0$ at the walls of the container.

The structure of the Navier-Stokes equation and the boundary conditions make the analytical solution of (2.2.2.1) very difficult. One is therefore led to look for some approximation which could be treated analytically with more ease. It may be observed that the Navier-Stokes equation is simplified considerably if μ is set equal to zero – in fact, one then obtains the well-known equation of Euler for non-viscous flow where solutions may be obtained from a flow-potential. However, the boundary condition ($\mathbf{v} = 0$ at the walls) does not contain the viscosity and thus remains unaffected by the assumption $\mu = 0$. One can, there-fore, state that, apart from a thin boundary layer at the walls, a slightly viscous fluid behaves like a non-viscous one. By introducing a sufficient amount of epsilontics, the order of magnitude of the boundary layer can be estimated in terms of the order of magnitude of the viscosity, and an approximate equation (which is analytically relatively simple) for the viscous fluid inside the boundary layer can be deduced from equation (2.2.2.1). One thus arrives at a 'boundary layer theory.' This theory was originated by Prandtl. The boundary layer has actually been made visible experimentally by ingenious experiments (Koncar-Djurdevic 1953).

(*b*) *The Hagen-Poiseuille equation*   The Navier-Stokes equation can be solved *exactly* for a *straight, circular* tube. In this case, the term $\mathbf{v} \operatorname{grad} \mathbf{v}$ is zero because of the orthogonality of $\mathbf{v}$ and the gradient of all the components of $\mathbf{v}$. The rest of the equation can be integrated quite easily for the prevailing boundary condi-tions. In terms of total volume throughput Q through a circular tube of radius a and length h, the pressure drop from end to end being Δp, the solution which is called the 'Hagen-Poiseuille equation,' becomes (Lamb 1932):

$$Q = \frac{\pi}{8} \frac{\Delta p}{h} \frac{a^4}{\mu}. \qquad (2.2.2.2)$$

As noted, the Hagen-Poiseuille equation was deduced under the assumption that the term $\mathbf{v} \operatorname{grad} \mathbf{v}$ is zero. This is only true for *straight* tubes. In curved tubes, equation (2.2.2.2) is no longer true; one then has the conditions of non-linear laminar flow (cf. (d) below).

(*c*) *Turbulence and Reynolds number*   Theory and experiment show that for high flow velocities the flow pattern becomes transient although the boundary conditions remain steady: eddies are formed which proceed into the fluid at intervals. For any one system, there seems to be a 'transition point' below which steady flow is stable. Above the 'transition point' the steady flow is more and more likely to become unsteady and to form eddies upon the slightest disturbance. The steady flow is often termed *laminar*, and the flow containing eddies *turbulent*. In turbulent flow, the law of Hagen-Poiseuille is no longer valid – the resistance (pressure drop) becomes a quadratic function of the throughput.

Although the transition point has been calculated from the Navier-Stokes equation for certain simple systems, it is obvious that such a calculation is a very difficult undertaking. One has therefore to have recourse to experiments to determine when turbulence will set in. If some systems can be shown to be dynamically similar, then the transition point in one system will have a corresponding point in the dynamically similar system, which can then be calculated.

It has been shown by Reynolds that circular straight tubes are dynamically similar, as far as the Hagen-Poiseuille equation is concerned, if the following 'Reynolds number' (denoted by Re) is the same:

$$\mathrm{Re} = 2\rho a v/\mu, \tag{2.2.2.3}$$

where all the constants have the same meaning as before, and v is now the average flow velocity in the tube.

It must be expected, therefore, that turbulence will occur in any straight tube if a certain Reynolds number is reached. This critical Reynolds number has been determined to be in the neighbourhood of 2200 (Joos 1947, p. 204).

Formula (2.2.2.3) contains only the radius a of the tube, apart from constants referring to the fluid. It has, therefore, been indiscriminately applied to *any* tube, straight or not, with rather devastating results, for the calculation of flow through a straight tube is based on the neglect of the term $\mathbf{v}$ grad $\mathbf{v}$ in the Navier-Stokes equation. The Reynolds number is, therefore, a valid criterion of dynamical similarity *only* if $\mathbf{v}$ grad $\mathbf{v}$ vanishes in the two systems to be compared. This term, however, is anything but zero in a curved tube.

This statement has been confirmed experimentally by Comolet (1949), who has shown that the critical Reynolds number at which water flowing in a tube becomes turbulent is changed greatly by a slight curvature of the tube. This was proved by making a quantitative study using a flexible tube projecting from a reservoir. Topakoglu (1951) took a theoretical approach to the problem and showed explicity that the flux is a function of both the Reynolds number *and the curvature of the tube.*

We should therefore like to emphasize once more that identity of Reynolds numbers is *not* a sufficient condition to ensure dynamical similarity between two systems consisting of flow channels; in this instance the Reynolds number has been greatly abused, as *it is also necessary that the channels be straight.*

As mentioned above, turbulent flow is characterized by the fact that the flow pattern is transient. The velocity fluctuations that occur are essentially random so that a statistical theory is advised. Such a theory consists in introducing a velocity correlation tensor R_{ik} (the bar denoting the average):

$$R_{ik}(\mathbf{r}) = \overline{v_i(\mathbf{x})v_k(\mathbf{x} + \mathbf{r})}. \tag{2.2.2.4}$$

Based on the Navier-Stokes equations, one then sets up dynamical relations for this tensor, or, more conveniently, for its Fourier transform. A review of the statistical theory of turbulence may be found in a book by Batchelor (1953).

(*d*) *Non-linear laminar flow* The preceding remarks about the Reynolds number also entail certain consequences with regard to the description of laminar flow. As a special solution of the laminar flow problem, we have mentioned the Hagen-Poiseuille equation for tubes, which is valid for Reynolds numbers smaller than 2200. However, just as turbulent flows may not be dynamically similar in spite of the equality of their Reynolds numbers, the same holds true also for laminar flows. The solution of the Navier-Stokes equations represented by the Hagen-Poiseuille formula neglects inertia terms; it does not apply, therefore, to cases where these terms become important. This occurs in curved tubes or in very short tubes where the end-effects are appreciable. The net result of introducing the inertia terms is to produce a relationship between the pressure drop Δp and the flow rate Q that is no longer linear. Laminar flow, therefore, is not necessarily characterized by a proportionality between these two quantities.

Formally, the inertia effects in laminar flow are expressed in the same fashion as in turbulent flow, so that one has

$$\Delta p = k_1 Q + k_2 Q^2. \tag{2.2.2.5}$$

Experiments to study relationships of this type for very short tubes have been reported, for instance by Kreith and Eisenstadt (1956). In some instances it is also convenient to represent the connection between Δp and Q by assuming an exponent of Q intermediate between 1 and 2.

2.2.3 *Non-Newtonian flow*

The flow formulas discussed above are concerned with purely viscous fluids for which the Navier-Stokes equation (2.2.2.1) is applicable. There are, however, substances capable of flow whose rheological equation is considerably different from that of viscous ('Newtonian') materials: the corresponding flow patterns have commonly been termed 'non-Newtonian.' A survey of the various possibilities for constructing rheological equations for continuous matter has been given by Reiner (1949), and another survey of their application to flow problems has been given by Metzner (1956) and Oldroyd (1956).

One of the most commonly considered modifications of the flow equation for viscous flow is that introduced by Bingham in which the material is assumed to have a yield strength. For the simplest case in which a fluid flows parallel to a wall, the 'Bingham condition' can be expressed as follows:

$$\tau - \tau_y = \mu \, dv/dr, \tag{2.2.3.1}$$

where τ is the shear stress, τ_y is the yield strength, v is the local flow velocity (parallel to the wall), and r is the distance from the wall to the point under consideration. The Bingham equation has been integrated (by Buckingham 1921) for a tube; the result is

$$\frac{8\bar{v}}{D} = \frac{\tau_w}{\mu} \left[1 - \frac{4}{3} \frac{\tau_y}{\tau_w} + \frac{1}{3} \left(\frac{\tau_y}{\tau_w} \right)^4 \right], \tag{2.2.3.2}$$

where $\bar{v}$ is the average flow velocity over the cross-section of the tube, D is the tube diameter, and τ_w is the shear stress at the wall of the tube.

The Bingham relation for flow parallel to a wall can be represented in a diagram showing the connection between the shear stress and the shear rate dv/dr (see fig. 7), by a straight line that does not pass through the origin. In such a diagram, viscous flow is represented by a straight line passing through the origin. An intermediate case, called 'pseudoplastic' flow, is represented by a curve that is concave downward (see fig. 7). No theoretical justification for assuming any particular such curve seems to exist, but Powell and Eyring (1944) showed that the relation

$$\tau = \mu \frac{dv}{dr} + \frac{1}{B} \sinh^{-1} \left(\frac{1}{A} \frac{dv}{dr} \right) \tag{2.2.3.3}$$

can be made to fit most observational data by an appropriate choice of the constants μ, A, B. The integration of this equation for a tube has been achieved numerically by Stevens (1953), who presented charts for various cases.

Other rheological equations will be mentioned in section 7.5 when their significance in connection with flow through porous media is discussed.

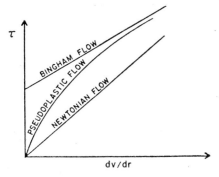

FIGURE 7   Relation between shear stress and shear rate in Newtonian (viscous) flow, Bingham-type flow, and pseudoplastic flow,

At high rates of flow, Bingham-type and pseudoplastic flows also show the phenomenon of *turbulence*. However, the description of turbulence in this case becomes very complicated (cf. Dodge and Metzner 1959).

The above flow equations are applicable to flow parallel to a wall. The inertia terms can then be neglected. A formulation of some flow equations in tensor notation which is applicable to the general three-dimensional case has been given by Oldroyd (1956). However, the mathematical difficulties in the integration of such flow equations become very great so that no solution for practical cases seems to be available.

In addition to the Bingham-type and pseudoplastic behaviour, other peculiar phenomena may occur. These include time-dependent behaviour such as thixotropy, in which the flow resistance decreases if a certain shear rate is being maintained, or rheopecty, in which the resistance increases. However, the theory of such phenomena is only very poorly understood.

2.3

MOLECULAR FLOW

It has been observed experimentally that the theorems obtained from the continuous matter theory are in need of correction if the distance between the walls confining the fluid is of the same order of magnitude as the free molecular path length in the fluid. It has been found that the essential correction needed is in the boundary condition, which stated that $\mathbf{v} = 0$ at the walls. In fact, it has been observed that the molecular nature of the fluid essentially effects a 'slip' at the walls such that $\mathbf{v} \neq 0$ at the (fixed) walls.* It is to the credit of Kundt and Warburg (1875) that they observed this fact first. As a consequence, the quantity of gas flowing through a capillary is larger than would be expected from Poiseuille's formula. Warburg (see Klose 1931) applied this concept to obtain a 'slip correction' of Poiseuille's formula by adding a constant term. This means that under a zero pressure differential there is a finite 'slip flow' through a capillary. At other pressure differentials, the flow is always bigger by this finite value than would be calculated from Poiseuille's formula.

Experimental investigations prompted Knudsen (1909) to propose yet another condition for the total volume rate of gas flow Q (measured at pressure p) through a capillary of radius a and length h. He postulated that

$$Q = \frac{4}{3}\sqrt{\frac{2\pi R\mathfrak{T}}{M}}\frac{a^3}{h}\frac{\Delta p}{p}, \tag{2.3.1}$$

where R is the gas constant, $\mathfrak{T}$ the (absolute) temperature, and M the molecular

---

* Reviews of problems connected with molecular flow of gases may be found in a monograph by Patterson (1956) and an article by Schaaf (1956).

weight of the gas. According to the kinetic theory of gases, the mean free path
length λ is given by

$$\lambda = c\frac{\mu}{\bar{p}}\sqrt{\frac{R\mathfrak{T}}{M}}, \tag{2.3.2}$$

where c is a constant approximately equal to 2, μ the viscosity, and $\bar{p}$ the mean
pressure. Thus Knudsen's equation can be rewritten as follows:

$$Q = \frac{4}{3}\sqrt{2\pi}\,\frac{\lambda}{c\mu}\,\frac{a^3}{h}\,\bar{p}\,\frac{\Delta p}{p}, \tag{2.3.3}$$

where λ is to be measured at the mean pressure.

It soon became established that Knudsen's equation described the flow
correctly if the mean free path is very large compared with the tube radius a (see,
for example, Klose 1931), but that Poiseuille's equation has to be used if the
mean free path is very small. It is therefore a ready conjecture that, for intermediate cases, the two equations have to be combined. This has been proposed,
for example, by Adzumi (1937), who thus arrived at the equation

$$Q = \frac{\pi a^4 \bar{p}\Delta p}{8\mu hp} + \varepsilon\frac{4}{3}\sqrt{2\pi\frac{R\mathfrak{T}}{M}}\,\frac{a^3}{h}\,\frac{\Delta p}{p}, \tag{2.3.4}$$

where ε is a dimensionless proportionality factor which has generally values of
about 0.9 for single gases and about 0.66 for gaseous mixtures. It may thus be
assumed safely that ε has some constant value for any flow phenomenon of
interest. It is called the 'Adzumi constant.'

Later on, the Adzumi equation was modified by Deryagin, Fridlyand, and
Krȳlova (1948) and by Arnell (1946) with the intention of applying it to model-
porous media. These modifications will be discussed in detail later, in connection
with the models of porous media for which they were intended.

2.4
THE INTERACTION OF FLUIDS WITH SURFACES

2.4.1 *Adsorption*

The boundary conditions at the walls of a container of a fluid are not always as
simple as outlined in either section 2.2 or 2.3. Especially if the fluid is a vapour
near its condensation point, further remarkable effects may occur.

Thus, the molecules of the fluid may be *adsorbed* by the walls of the container.
This signifies that within a few molecular distances from the wall there is a strong
attractive potential between it and the molecules of the fluid. Thus, a much
greater concentration of molecules of the fluid may be found at the walls of the

container than in its interior. The general aspects of adsportion may be found in any textbook on colloid chemistry (e.g., Hartman 1947) or in the fundamental monograph of Brunauer (1943).

During the process of adsorption, energy is liberated at the walls of the container. This energy is called the heat of adsorption. Thus, if any pressure (or other) measurements are made during the process of adsorption, the temperature of the system has to be carefully watched. In general, conditions are arranged in such a manner that the liberated heat is conducted away so that the process is isothermal. One therefore refers to curves obtained in such a manner as adsorption *isotherms*. In particular, it is understood that an 'adsorption isotherm' is a plot of (molecular) pressure (in the centre of the container) against the number of molecules of the fluid on the walls of the container, and care must be taken that the process of adsorption is isothermic. Examples of some isotherms are given in figure 8.

Because of the large potential present in the adsorbed layers, the rheological state of the fluid inside the container may be different from that on the walls. In particular, if the fluid is a vapour near its condensation point, the adsorbed molecules may well form a film of *liquid* along the walls.

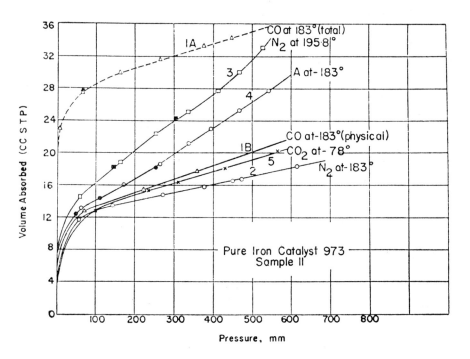

FIGURE 8   Examples of adsorption isotherms on iron catalyst for various gases near their boiling points (after Emmett 1948).

The quantitative aspects of adsorption will be dealt with in chapter 3 on 'Hydrostatics in Porous Media,' since this is practically the only instance where large surfaces come into contact with fluids in such a manner that adsorption may become significant.

2.4.2 Interfacial tension

If a liquid is in contact with another substance (gas or solid), there is free interfacial energy present between the two (see, for example, Adam 1941). This means that a certain amount of work has to be performed in order to separate a liquid from, say, a solid. The interfacial energy arises from the inward attraction of the molecules in the interior of a substance upon those at the surface. Since a surface possessing free energy contracts if it can do so, the free interfacial energy manifests itself as *interfacial tension*. The interfacial energy between a substance and the vacuum is called *surface tension*. Surface tensions will be denoted in this monograph by γ, with subscripts S, L, A, etc., for solid, liquid, air, etc., respectively. Correspondingly, interfacial tension will be denoted by the same symbol γ, but with two indices indicating to which substances it refers. It should be noted, however, that the terms 'surface tension' and 'interfacial tension' are often used indiscriminately.

The work W_{SL} required to separate a unit area of, for example, a solid from a liquid is related to the corresponding surface and interfacial tensions by an equation established by Dupré:

$$W_{SL} = \gamma_S + \gamma_L - \gamma_{SL}. \tag{2.4.2.1}$$

If we assume now that two fluid phases (for instance, liquid and air) are at one point in contact with a solid, then the surface tensions between the various phases can be calculated from (2.4.2.1). It can be shown that equilibrium is possible only if the interface between the air and the liquid forms a definite angle with the solid. This angle has been termed the 'contact angle' and will be denoted here by θ. It is determined by the condition

$$\cos \theta = (\gamma_{SA} - \gamma_{SL})/\gamma_{LA}. \tag{2.4.2.2}$$

This equation shows that no equilibrium is possible if $\gamma_{SA} - \gamma_{SL}$ is larger than γ_{LA}. In this case, the liquid will spread indefinitely over the solid.

According to equation (2.4.2.2), the cosine of the contact angle is given by the ratio of the energy released in forming a unit area of interface between solid and liquid to the energy expended in forming a unit area of interface between liquid and air. The contact angle thus depends on a proper determination of the areas of these two interfaces. The area of the interface between liquid and air can always safely be assumed to be equal to the apparent surface of contact, but the area of interface between solid and liquid may be greater than the apparent or 'macroscopic' area of contact. If this is the case, the surface of the solid is said to be 'rough.' It depends, of course, solely upon the accuracy of measurement how much of the 'roughness' of the surface is incorporated in the 'macroscopic' area in the first place. Nevertheless, it is often convenient to express 'surface roughness' by stating that the 'true' surface is σ (where σ is a measure of roughness) times bigger than the 'apparent' area. The apparent contact angle θ' on such a rough surface is then given by

$$\cos \theta' = \sigma(\gamma_{SA} - \gamma_{SL})/\gamma_{LA} = \sigma \cos \theta \qquad (2.4.2.3)$$

because the surface energy released in forming the solid-liquid interface is σ times that which would have been released if the surface had not been rough.

Care should be taken not to carry the above argument ad absurdum. For, relating to molecular dimensions, any surface would have to be considered as rough and thus any contact angle is only an 'apparent' one.

The theoretical aspects of surface tension and capillarity have been discussed and reviewed by Brown (1947), Shuttleworth and Bailey (1948), Burden (1949), Gurney (1949), Koenig (1950, 1953), and Elton (1951). Prigogine and his co-workers (Prigogine 1950; Prigogine and Maréchal 1952; Prigogine and Saroléa 1952) developed a molecular theory of the surface tension of solutions; Mac-Lellan (1952) advanced a statistical-mechanical theory of the same phenomenon. However, these theories are beyond the scope of this study.

Interfacial tension may cause a fluid to penetrate into small spaces that are filled with another liquid. The rate of flow of liquids into capillaries under the action of surface forces has been studied by Calderwood and Mardles (1955); the rise of liquids into a wedge-shaped crevice has been investigated by Tollenaar (1954).

With interfacial tension, there is the complication that it may not be the same when a liquid is advancing as when it is receding on a solid. One thus observes a possible hysteresis effect in interfacial tension, and so also in contact angles. This phenomenon has been reviewed and discussed by Cassie (1948), Shuttleworth and Bailey (1948), and Bartell and co-workers (Ray and Bartell 1953; Bartell and Shepard 1953). For our purposes it is sufficient to realize the possibility of its occurrence.

The quantities γ_{LA}, etc., are specific to the two phases which the indices signify. There is no general expression for the γs which would be applicable to any fluid or solid. Thus, the contact angle is specific to all *three* phases involved. In particular, if the fluid phases are not changed at all, this angle depends on the condition of the solid. Measurements of interfacial tensions have been made for many combinations of substances and are available in textbooks on physical chemistry.

If the properties of the solid are such that it makes the contact angle towards one of the fluid phases less than 90°, one says that the solid has a preferential *wettability* to that phase. The term 'wettability' has therefore a relative meaning only; moreover, it is also subject to hysteresis in the same manner as the surface tensions. Thus, of two fluids one may preferentially wet a solid if it is receding from the solid, but not if it is advancing.

2.5
MULTIPLE PHASE FLOW IN CAPILLARIES

2.5.1 *Immiscible fluids*

In a discussion of simultaneous flow of various fluids ('phases') in a flow channel, the distinction as to whether these phases are miscible or not is of prime import-ance.

The flow of immiscible phases has been studied extensively in connection with the design of pipelines. However, if one has applications to the flow through porous media in mind, the study of flow in capillaries would be of more interest. In this connection, it appears that, on a microscopic scale, in a capillary one has usually single phase flow with different capillaries being occupied by the various components of an immiscible mixture, if such a mixture is forced through an array of capillaries. Immiscible multiple phase flow in small capillaries does not seem to have been studied very much, and no applications to the flow of porous media are known to the writer.

2.5.2 *Miscible fluids*

The situation is different for miscible fluid flow in capillaries. Particular phenom-ena are encountered if two fluid phases, which are completely miscible, are in contact with a bounding surface. In this case the discussion in terms of capillarity and interfacial tension is no longer applicable and new concepts have to be developed.

Miscibility of two fluid phases implies that there cannot be any equilibrium other than that achieved by total mixing so that the concentration of one phase within the other is constant throughout the whole system under consideration. Therefore, no distinct interface between the two phases can exist and all the phenomena of interest are transient.

It turns out that the investigation of the dynamics of miscible fluids is much more difficult than the investigation of immiscible fluids, chiefly because of their fundamentally transient character. The speed of mixing of the two fluids is, on the one hand, conditioned by the speed of the internal molecular diffusion, and, on the other, by the mechanical convection imposed on the mixture. Thus, consider-ing the simplest case, that of a soluble fluid being introduced into a fluid flowing slowly through a small-bore tube, the soluble substance spreads out under the combined action of molecular diffusion and the variation of velocity over the cross-section. Thus, the invading soluble substance at the centre of the tube moves much more rapidly than near the edge. If radial diffusion were absent, an ever-lengthening 'needle' of invading fluid would extend down the tube. However, in actuality the fluids interdiffuse radially, 'washing out' the needle. An approximate treatment of this simple case has been given by Taylor (1953); Taylor's theory has been reviewed by Von Rosenberg (1956). Experiments to test this 'diffusion-type' theory have been discussed by Levenspiel and co-workers (Levenspiel and Smith 1957; Levenspiel 1958).

Following Taylor (1953), the dispersion of the invading fluid has, thus, to be separated into two contributions originating in convection and in radial diffusion. Considering first convection, we may note that in a circular tube of radius a the velocity v at distance r from the axis is, according to the analysis of Hagen-Poiseuille,

$$v = v_0(1 - r^2/a^2), \tag{2.5.2.1}$$

where v_0 is the maximum velocity at the axis. If the solute is distributed symmetrically, the concentration C is

$$C = f(x, r), \tag{2.5.2.2}$$

where $f(x, r)$ denotes 'a function of' x and r.

After the elapse of time t, the concentration will be

$$C = f(x - vt, r).$$

Accordingly, the mean value C_m of the concentration over a cross-section of the tube is defined by

$$C_m = \frac{2}{a^2} \int_0^a Cr dr. \tag{2.5.2.3}$$

Taylor now considered two cases. First, assuming that the space between two planes $x = 0$ and $x = X$ (X/a being small) is filled initially with solute of concentration C_0, it is seen that the amount which lies between r and $r + \delta r$ is constant during the flow and equal to $2\pi r C_0 X \delta r$. The solute will be distorted into a paraboloid defined by the equation

$$x = v_0 t(1 - r^2/a^2). \tag{2.5.2.4}$$

The total amount of solute between x and $x + \delta x$ is therefore given by $2\pi r C_0 X(dr/dx)\delta x$, and, inserting dr/dx from equation (2.5.2.4), one obtains

$$C_m = C_0 X/(v_0 t). \tag{2.5.2.5}$$

The mean concentration C_m therefore has the constant value $C_0 X/(v_0 t)$ in the interval $0 < x < v_0 t$ and is zero when $x < 0$ and when $x > v_0 t$.

Second, if solute of constant concentration enters a tube which at time $t = 0$ contains only solvent, one has

$$\left.\begin{array}{ll} C = C_0, & x < 0 \\ C = 0, & x > 0 \end{array}\right\} \text{ at time } t = 0. \tag{2.5.2.6}$$

This case can be treated by imagining that the constant initial concentration for $x < 0$ consists of a number of thin sections of the type considered above, leading to equation (2.5.5). In this manner one finds:

$$\left.\begin{array}{ll} C_m = C_0, & x < 0, \\ C_m = C_0[1 - x/(v_0 t)], & 0 < x < vt, \\ C_m = 0, & x > v_0 t. \end{array}\right\} \tag{2.5.2.7}$$

After considering dispersion by convection alone, Taylor considered the effect of molecular diffusion. It is natural to assume that the concentration C is symmetrical about the axis of the tube so that C is a function of r, x, and t only. The equation of diffusion is then

$$D\left(\frac{\partial^2 C}{\partial r^2} + \frac{1}{r}\frac{\partial C}{\partial r} + \frac{\partial^2 C}{\partial x^2}\right) = \frac{\partial C}{\partial t} + v_0\left(1 - \frac{r^2}{a^2}\right)\frac{\partial C}{\partial x}. \tag{2.5.2.8}$$

Here D, the coefficient of molecular diffusion, is assumed to be, in approximation, independent of C.

In all the cases which Taylor considered, $\partial^2 C/\partial x^2$ is much less than $\partial^2 C/\partial r^2 + (\partial C/\partial r)/r$, so that, writing

$$z = r/a, \tag{2.5.2.9}$$

the diffusivity equation (2.5.2.8) becomes

$$\frac{\partial^2 C}{\partial z^2} + \frac{1}{z}\frac{\partial C}{\partial z} = \frac{a^2}{D}\frac{\partial C}{\partial t} + \frac{a^2 v_0}{D}(1 - z^2)\frac{\partial C}{\partial x}. \tag{2.5.2.10}$$

The boundary condition, which indicates that the walls of the tube are impermeable, is

$$\partial C/\partial z = 0 \quad \text{at } z = 1. \tag{2.5.2.11}$$

The solution of the diffusivity equation, even in the simplified form (2.5.2.10) under the boundary condition (2.5.2.11), is difficult to achieve; however, Taylor showed that it is relatively easy to obtain an approximate solution under the assumption that the time necessary for appreciable effects to appear, owing to convective transport, is long compared with the 'time of decay' during which radial variations of concentration are reduced to a fraction of their initial value through the action of molecular diffusion. This assumption is applicable in 'slow' flow.

Since molecular diffusion in the longitudinal direction has been neglected, the whole of the longitudinal transfer of concentration is due to convection. Considering the convection across a plane which moves at constant speed $\frac{1}{2}v_0$, and writing

$$x_1 = x - \tfrac{1}{2}v_0 t \tag{2.5.2.12}$$

(note that $\frac{1}{2}v_0$ is the mean speed of flow), equation (2.5.2.10) becomes

$$\frac{\partial^2 C}{\partial z^2} + \frac{1}{z}\frac{\partial C}{\partial z} = \frac{a^2}{D}\frac{\partial C}{\partial t} + \frac{a^2 v_0}{D}(\tfrac{1}{2} - z^2)\frac{\partial C}{\partial x_1}. \tag{2.5.2.13}$$

Since the mean velocity across planes for which x_1 is constant is zero, the transfer of C across such planes depends only on the radial variation of C. Under the above assumptions, this radial variation is small and can therefore be calculated from the equation

$$\frac{\partial^2 C}{\partial z^2} + \frac{1}{z}\frac{\partial C}{\partial z} = \frac{a^2 v_0}{D}(\tfrac{1}{2} - z^2)\frac{\partial C}{\partial x_1}, \tag{2.5.2.14}$$

where $\partial C/\partial x_1$ may be taken as independent of z.

A solution of (2.5.2.14) satisfying the boundary condition is

$$C = C_{x_1} + A(z^2 - \tfrac{1}{2}z^4), \tag{2.5.2.15}$$

where C_{x_1} is the value of C at $z = 0$ and A is a constant.

Substituting (2.5.2.15) into (2.5.2.14), Taylor found that

$$A = \frac{a^2 v_0}{8D}\frac{\partial C}{\partial x_1}. \tag{2.5.2.16}$$

The rate of transfer Q of C across the section at x_1 is

$$Q = -2\pi a^2 \int_0^1 v_0(\tfrac{1}{2} - z^2)Cz\,dz. \tag{2.5.2.17}$$

Inserting the value of C from above, Taylor obtained the equation

$$Q = -\frac{\pi a^4 v_0^2}{192D}\frac{\partial C_{x_1}}{\partial x_1}. \tag{2.5.2.18}$$

If (2.5.2.18) is expressed in terms of the mean concentration C_m, one obtains, using the assumption that radial variations of C are small,

$$Q = -\frac{\pi a^4 v_0^2}{192D}\frac{\partial C_m}{\partial x_1}. \tag{2.5.2.19}$$

It is therefore seen that C_m is dispersed relative to a plane which moves with velocity $\frac{1}{2}v_0$ as though it were being diffused by a process obeying the same law as molecular diffusion, but with a diffusivity coefficient D_1 where

$$D_1 = \frac{a^2 v_0^2}{192D}. \tag{2.5.2.20}$$

The fact that no material is lost in the process is expressed by the continuity equation for C_m, viz.:

$$\partial Q/\partial x_1 = - \pi a^2 \, \partial C_m/\partial t. \tag{2.5.2.21}$$

Substituting the value for Q, the equation governing longitudinal dispersion becomes

$$D_1 \, \partial^2 C_m/\partial x_1{}^2 = \partial C_m/\partial t. \tag{2.5.2.22}$$

This equation can be applied to the case where dissolved material of uniform concentration C_0 is allowed to enter the tube at uniform rate at $x = 0$ starting at time $t = 0$. Initially the tube is assumed to be filled with pure solvent. One obtains (Von Rosenberg 1956):

$$\frac{C}{C_0} = \frac{1}{2} + \frac{1}{2} \operatorname{erf}\left[x_1 \left(\frac{48D}{a^2 v_0{}^2 t} \right)^{\frac{1}{2}} \right] \quad \text{for } x_1 < 0,$$

$$\tag{2.5.2.23}$$

$$\frac{C}{C_0} = \frac{1}{2} - \frac{1}{2} \operatorname{erf}\left[x_1 \left(\frac{48D}{a^2 v_0{}^2 t} \right)^{\frac{1}{2}} \right] \quad \text{for } x_1 > 0,$$

where

$$\operatorname{erf} z = 2\pi^{-\frac{1}{2}} \int_0^z e^{-z^2} \, dz. \tag{2.5.2.24}$$

Equation (2.5.2.23) signifies (Von Rosenberg 1956) that the length of the concentration front – i.e., the length over which a certain fraction, say 80 per cent, of the total concentration change takes place – at any distance traversed is directly proportional to the square root of the velocity and inversely proportional to the square root of the diffusion coefficient. Also, at any given velocity, the length of the front increases as the square root of the distance traversed.

The above-outlined theory of Taylor (1953) is, as is evident from the exposition, an approximate one. Nevertheless, it seems to describe experimental results quite well (Taylor 1953). There is no doubt, therefore, that it is at least qualitatively correct.

3
Hydrostatics in porous media

3.1
PRINCIPLES OF HYDROSTATICS

The statics of fluids within porous media is governed by the same principles which obtain when fluids are confined in other types of vessels. Thus the basic theorems, such as that of Torricelli, will still be valid. There are, however, certain effects that are peculiar to fluids which are confined within porous media as a result of the proximity of the walls to practically all molecules of the fluid. These effects will be grouped for purposes of discussion into several classes, as follows:

(i) The statics of one fluid phase. The all-important phenomenon is that of adsorption of fluid particles at the surface of the pores. Such adsorption grossly alters the relationship between the pressure and volume observed in bulk quantities of the fluids.

(ii) The statics of two phases of one fluid. The influence of the walls of the pores may cause some of the fluid to condense to form another phase. This second phase will cover the walls of the pores as a film of variable thickness. If it merges to form menisci, the phenomenon is known as 'capillary condensation.'

(iii) The statics of two different immiscible fluids. This subject is treated by the theory of capillaric forces acting at the interface between the two fluids. This is also the theory of the quasistatic displacement of one fluid by another from a porous medium. 'Quasistatic' means that the displacement is taking place through a series of equilibrium-conditions without any proper dynamic effects occurring.

(iv) In a final section, the subject of wettability will be treated. The concept of wettability is actually nothing but a straightforward consequence of the displacement theory; nevertheless it merits special attention because of its widespread importance.

Hydrostatics in porous media has numerous applications to measurements of geometrical properties of porous media, which were mentioned in earlier sections of this monograph. These applications will be treated here in greater detail in connection with the basic underlying theories.

3.2
ADSORPTION OF FLUIDS BY POROUS MEDIA

3.2.1 *Theory of isotherms*

If an adsorbable fluid is confined within a porous solid, then the relationship
between the pressure and the volume of the fluid is not the same as if the fluid
were confined in a sphere of the same volume as the pore space. A static interaction between the porous medium and the fluid is thus taking place. For a
particular fluid and porous solid, one can find experimentally the relationship
between the fluid pressure and fluid density at a given temperature inside the
porous medium. Such curves are referred to as adsorption (or desorption)
isotherms.

Of the large number of papers on theories of adsorption, some of the most
important will be reviewed here. A general presentation of adsorption phenomena
has been given in the fundamental monographs of Brunauer (1943) and de Boer
(1953); less detailed reviews may be found in any textbook on colloid chemistry
(see, for example, Hartman 1947), as well as in articles by Cremer (1950) and
Everett (1950).

Gibbs (Hartman 1947) was probably the first to investigate the problem. Using
general principles of thermodynamics he derived a relationship between the
excess number ω of adsorbed molecules per unit area, the concentration C of
molecules in the interior of the fluid, and the surface tension γ. This relationship,
which is known as the Gibbs adsorption equation (for a write-up of the derivation
of this equation see, for example, Glasstone 1946, p. 1206), is as follows:

$$\omega = (C/R\mathfrak{T})\partial\gamma/\partial C. \tag{3.2.1.1}$$

As usual, R signifies the gas constant and $\mathfrak{T}$ the absolute temperature.

The Gibbs equation reduces the phenomenon of adsorption to one of surface
tension. The actual mechanics of the process is understood only if the dependence
of γ on C is understood. This dependence has to be determined experimentally,
which is equivalent to making an adsorption experiment and thus does not help
the understanding of the mechanism. Nevertheless, the Gibbs equation restricts
somewhat the general aspects of the adsorption process by admitting only possibilities that are thermodynamically sound. The Gibbs equation is usually
applied to the analysis of the adsorption of a solute at the surface of a solution.
It is, however, also applicable to the problem of adsorption on solids and in this
respect it has been scrutinized by Bangham (1937).

Polanyi (1920) made a different attempt to understand the mechanism of
adsorption. He developed a theory based on the assumption that the density of
the adsorbed layer varies continuously with the distance from the surface of the
adsorbent, owing to the action of an 'adsorption potential.' The adsorption
potential is assumed to depend on the nature of the adsorbent as well as of the

adsorbate, but not on the temperature and on the number of adsorbed molecules present. It should thus be possible to determine the adsorption potential from an isotherm; and, if the quantity of adsorbed matter is determined at a certain temperature as a function of the pressure, it should be possible to predict the quantity of adsorbed matter at any other temperature. Experimental checks of the Polanyi theory yielded discrepancies in many instances. Polyakov, Kuleshina, and Neĭmark (1937), for example, stated that in the case of gels the Polanyi theory is applicable only for low adsorbent concentration, and yields incorrect results at higher concentrations.

Another theory of adsorption was initiated by Langmuir (1916), who developed some consequences of the kinetic theory of gases. The latter theory provides a means of calculating the number of molecules of gas hitting a unit surface area in terms of the gas pressure, its molecular weight, and its temperature. Langmuir assumed that any molecules arriving at the surface of an adsorbent would stay there for a time long enough for a film of monomolecular thickness to develop. The molecules in the film, in turn, would re-emanate by a certain finite probability. The net effect of this process would be that a finite number of molecules would always be present at the surface of the adsorbent. This represents the phenomenon of adsorption.

Starting from these considerations, Langmuir (for an account of the mathematics see, for example, Hartman 1947, p. 54) obtained an expression, in terms of the gas pressure p, for the quantity of gas ω which is adsorbed:

$$\omega = abp/(1 + ap), \tag{3.2.1.2}$$

or

$$\frac{p}{\omega} = \frac{p}{b} + \frac{1}{ab}. \tag{3.2.1.3}$$

The last equation shows that one should obtain a straight line if p/ω is plotted against p. The quantities a and b are constants. They are constants because of the particular assumptions that Langmuir made, namely that adsorption could occur only in a single layer and that the energy of adsorption would be the same everywhere on the adsorbing surface. It follows therefore that the Langmuir equation should be valid only as long as there is free surface available to be taken up by molecules. After all free surface has been taken up, the adsorption process should come to an abrupt halt. This should manifest itself in a break in the adsorption isotherm.

Experiments showing that adsorption does not come to an abrupt halt in correspondence with the Langmuir assumptions prompted Brunauer, Emmett, and Teller (1938) to postulate the existence of adsorbed molecules in more than one layer. By a kinetic approach similar to that of Langmuir, they arrived at the following equation (the 'BET equation'):

$$\frac{\beta}{V(1 - \beta)} = \frac{1}{V_m G} + \frac{\beta(G - 1)}{V_m G}, \tag{3.2.1.4}$$

where $\beta = p/p_0$ is the 'relative' pressure (p is the pressure and p_0 the vapour pressure of the bulk liquid) at which a particular volume V of gas (expressed in some standard measure) is adsorbed by the porous medium; V_m is a constant, namely the volume of gas required to cover the entire internal surface of the porous medium; and G is another constant. The two constants (V_m and G) can be determined for any particular system from a set of adsorption data by plotting $\beta/[V(1 - \beta)]$ against β.

The BET equation has been extended and scrutinized by many authors. Notably, the form (3.2.1.4) assumes the possibility of an infinite number of adsorbed layers. One can derive a similar equation that assumes only a finite number of such layers. The consequences of such a restriction have been discussed by Joyner, Weinberger, and Montgomery (1945) and by Carman and Raal (1951). The basic assumptions of the multilayer adsorption theory have been analysed by McMillan and Teller (1951); the relationship of the BET equation to the Langmuir theory and earlier multilayer adsorption theories has been described by Keenan (1948).

A totally different approach to the theory of adsorption was initiated by Harkins and Jura (1943, 1944c). These authors assumed that the same condition is applicable to the adsorption of gases on solids which had been found to govern the correlation between the surface spreading force of adsorbed films and the area occupied per adsorbed molecule. Thus they were led to postulate that adsorption data should fit the equation

$$\log \beta = B - A/V^2, \tag{3.2.1.5}$$

where A and B are constants and the other symbols have the same meaning as in equation (3.2.1.4). Furthermore, the constant A is expressible in terms of the internal surface area S, by the equation

$$S = cA^{\frac{1}{2}}, \tag{3.2.1.6}$$

where c is a constant.

It is quite obvious that the adsorption data of one experiment cannot fit both the BET *and* Harkins-Jura equations. Nevertheless, Emmett (1946), discussing this difficulty, has shown that for quite a wide range of constants A, B, G, the two equations can be made to fit identical data very nearly. For the greater part of adsorption phenomena, the two equations seem therefore to be equally applicable.

3.2.2 *Applications of isotherms to area measurements*

The theories of adsorption isotherms described above can be used to determine experimentally the internal area of porous media. A general review of the methods available has been given, for instance, by Emmett (1948).

It has been pointed out in section 3.2.1 that, according to the Langmuir theory, one should expect adsorption to cease after all surface available for occupation in a monomolecular layer by molecules of the adsorbate has been taken up. If the area occupied by one single molecule were known, one could then calculate the surface of the adsorbent knowing the amount of gas that can be totally adsorbed.

Unfortunately, adsorption does not come to a halt abruptly as expected by Langmuir since adsorbed molecules may exist in more than a single layer. It is, however, reasonable to expect that adsorption isotherms should contain a point that might correspond to the completion of the first layer of adsorbed molecules. This hypothetical point has been called 'point B.'

The method of selection of 'point B' has been described by Emmett and Brunauer (1937). The method makes use of the fact that most isotherms are characterized by a long linear part extending over a considerable portion of the pressure range, which may be thought to represent the building up of the second layer of adsorbed molecules. Therefore, the beginning of the long linear part would have to be identified with 'point B.' However, the selection of 'point B' in this fashion is obviously somewhat arbitrary.

A better way to determine the volume of gas that makes up a single layer of adsorbate on the surface of the adsorbent is therefore obtained by the application of the BET equation. As outlined in section 3.2.1, the constant V_m in this equation represents precisely the sought-after quantity. If, in addition, the volume V_0 required to cover a unit area of adsorbent with a monomolecular layer of adsorbate is known, the internal area S of adsorbent is given by

$$S = V_m/V_0. \tag{3.2.2.1}$$

The BET method is enjoying great popularity. It gives good results for 'smooth' porous media, i.e., media without too narrow interstices. If the interstices are very narrow, the modification of the BET equation for a finite number of layers has to be used, and this greatly complicates matters.

Most of the area measurements reported earlier (in table II) have been obtained by the BET method. Among the gases that have been used for surface measurements are nitrogen, oxygen, argon, hydrogen. The experiments have usually been performed at low temperature, around $-190°C$, i.e., at the temperature of liquefaction of the gases.

The BET method still hinges on a determination of V_0, which cannot always be obtained very accurately. The theory of Harkins and Jura permits one to get around this difficulty. Their equations (3.2.1.5–6) can be applied immediately to surface-measurements if the constant c can be evaluated. In some ingenious experiments, Harkins and Jura (1944a, 1944b, 1945) were able to provide an independent means of evaluating c. They were able to measure the area of fine titanium oxide powder directly from measurements of the heat of wetting. Using this surface area, c was evaluated. For other solids, then, equation (3.2.1.6) could be applied directly by assuming the constant c as independent of the type of surface. In the cited papers, Harkins and Jura give several applications of their method to specific area determinations. A comparison of the Harkins-Jura with the BET method has been made by Emmett (1946), who found satisfactory agreement.

Apparatus for the determination of surface areas by the adsorption method has been described in the literature on many occasions. Apart from the references already given, the reader may consult papers by Halasz and co-workers (Halasz 1954; Halasz and Schay 1956; Halasz, Schay, and Wencke 1956); Daneš (1956), Rubinshteïn and Afanas'ev (1956), Deryagin (1957a), Mathews (1957), and Starkweather and Palumbo (1957) for further details.

3.3
CAPILLARY CONDENSATION OF FLUIDS IN POROUS MEDIA

3.3.1 *Theory of sorption hysteresis*

The theories of adsorption discussed so far assume that adsorption is an equilibrium process. However, it has been observed that under certain conditions phenomena of hysteresis occur. Such phenomena are usually explained in terms of capillarity effects or by more or less unspecific reference to phase transitions. The latter possibility is often referred to as the 'capillary condensation theory.'

The various versions of the theory of capillary condensation are all based on the assumption of Zsigmondy (1911) that vapours adsorbed on any porous solid are in equilibrium with a certain vapour pressure which is uniquely determined by the curvature of the menisci formed in the pores. One has thus two phases present in the fluid whose behaviour in relation to the porous medium is described by the relationships governing interfacial tension (see sec. 2.4.2). Since the interfacial tensions are subject to hysteresis, the behaviour of the capillary-condensed liquid will exhibit the same phenomenon. The original theory of Zsigmondy has been modified and improved by Foster (1932) and Broad and Foster (1945). These authors suggested that hysteresis could also be due to a delay in the formation of the capillaric menisci. This implies that the equilibrium postulated by Zsigmondy might not be reached instantaneously, but only after a certain time-lag.

In any case, the capillary condensation theory is founded upon the assumption that, in a capillary, the vapour pressure of a liquid is reduced from, say, p_0 to, say, p. The ratio p_0/p is termed the 'relative pressure.' Assuming zero contact angle and a circular capillary of radius a, the relative pressure is given by the following equation established by Kelvin:

$$\ln(p_0/p) = 2\gamma M/(\rho a R \mathfrak{T}). \tag{3.3.1.1}$$

Here γ is, as usual, the surface tension, M the molecular weight, ρ the density, $\mathfrak{T}$ the absolute temperature, and R the gas constant. The Kelvin equation describes matters correctly if it is assumed that (i) all 'adsorption' is entirely due to capillary condensation, (ii) 'adsorbate' densities equal bulk densities, (iii) differences of pore shape from circular can be ignored, and (iv) the validity of the Kelvin equation, including constancy of γ and p, is unimpaired at low values of a (Carman 1951).

Hirst (1947) has shown that the capillary condensation theory may account for sorption hysteresis without the assumption of hysteresis in interfacial tension – for a reason other than an assumed time-lag in the formation of the menisci. If condensation takes place in a cylindrical cavity with thin isotropic walls, the radius of the capillary will be altered owing to the pressure of the liquid film. As a result of the change of radius, the equilibrium pressure over the liquid film is altered and hysteresis is observed.

In this connection, investigations by Shull (1948), Barrett, Joyner, and Halenda (1951), and Carman (1951) are concerned with the fact that the first condition basic for the validity of the Kelvin equation, viz., assuming that 'adsorption' is due *only* to capillary condensation, is really quite unreasonable, since the formation of truly adsorbed layers certainly begins before capillary condensation is manifested and must continue at the surface of the unfilled capillaries. To obtain a realistic theory, a correction must therefore be introduced for the adsorbed layers. The details of this are, however, beyond the scope of this monograph.

The sorption hysteresis has also been attributed to an 'ink bottle effect' instead of to hysteresis in γ (Katz 1949). This is a blockage effect of the smaller capillaries which may stay filled during desorption and prevent the connecting larger ones from being properly emptied. It appears thus that in any interpretation the concept of 'capillary condensation' leads to sorption hysteresis.

Most of the above-mentioned investigations discuss some experimental tests. In addition, Carman and Raal (1951) provided direct evidence of capillary condensation and of blockage of capillaries with adsorbed layers by comparing the adsorption on a given surface both as a free surface and as the internal surface of a porous medium. Polyakov, Kuleshina, and Neĭmark (1937) made some experimental tests and showed that at high adsorbent concentration (i.e., for narrow pore spaces) the capillary condensation theory may give satisfactory results. However, Bond, Griffith, and Maggs (1948) conducted a set of experiments to determine the actual state of aggregation of the 'capillary-condensed' film of 'liquid.' They pointed out that certain aspects of the behaviour of the adsorbed 'film' stand in very sharp contrast to the behaviour of bulk quantities of such liquid. It thus appears that there exist various types of sorption hysteresis and that the explanation is different in every case (de Boer 1958). In view of these remarks it must be conceded that sorption hysteresis is not yet completely understood.

3.3.2 Determination of pore size distribution

The various theories of 'adsorption' (including capillary condensation) can be used to determine the pore size distribution. Naturally, as was outlined in section 1.1.2, the meaning of the term pore size distribution is not very well defined in most publications and it is therefore usually a little vague. Reported determinations of the pore size distribution are therefore not always entirely satisfactory from a quantitative standpoint. However, a good qualitative indication of the nature of the pore space involved is usually obtained.

There are various methods for determining the pore size distribution based on adsorption isotherms. Foster (1948) discerns three classes:

(a) Methods based on the capillary condensation theory. It follows from the Kelvin equation that the radius a of the smallest pores that may be invaded by a given liquid at constant temperature is solely a function of the pressure p, provided that the surface tension γ and the density ρ are assumed to be unaffected by a. It therefore also follows that, after an 'adsorption' isotherm for a porous medium has been determined, it is merely necessary to convert values of p to corresponding values of a to obtain the pore size distribution of the porous medium. Actually, corrections should be made for the simultaneous occurrence of multilayer adsorption with the capillary condensation, as was stated above. The Kelvin equation by itself, without such a correction, is somewhat restricted in its accuracy for the purpose of determining the pore size distribution, although it has been applied to such measurements with good qualitative success.

(b) Methods based upon the determination of the internal surface. The radius a of a straight circular cylinder can be calculated from the ratio of its volume V to its wall-surface S by the equation

$$a = 2V/S. \tag{3.3.2.1}$$

If this ratio is assumed to apply approximately to the pores of a porous medium (Foster 1934; Emmett and de Witt 1943), then internal surface values of porous media can be converted to yield 'average' pore radii.

(c) Methods yielding approximate estimates based on a consideration of the shape of adsorption isotherms. If adsorption practically ceases after what may be assumed to be the formation of a monolayer, one may infer that the pore radius is of the order of 1–2 molecular diameters. It can therefore be estimated (Foster 1945).

The various methods of determining the pore size distribution outlined above have been applied to a great variety of substances (see, e.g., Dubinin 1956; Dubinin and Zhukovskaya 1956; Deryagin 1957b; Imelik and François-Rosetti 1957; Innes 1957; Tovbin and Savinova 1957). These methods have been compared (Joyner et al. 1951) with the more common mercury-injection method (see sec. 3.4.3). It was found that there is, in view of the general vagueness of the notion of the pore size distribution, satisfactory qualitative agreement.

3.4
THE QUASISTATIC DISPLACEMENT OF ONE LIQUID BY ANOTHER IN A POROUS MEDIUM

3.4.1 *Theory of capillary pressure*

Let us now consider the hydrostatics of two immiscible fluids or phases that may exist simultaneously in a porous medium. In general, one phase will wet the solid. Experimental investigations have shown (Versluys 1917, 1931) that there are three general types of occurrence of one of the two phases, or regimes of saturation with that phase.

The saturation s is defined as the fraction of the pore space which is filled by the fluid in question. In soil physics, instead of the saturation s thus defined, one often uses the 'volumetric liquid content' θ (the second phase being assumed to be a gas)

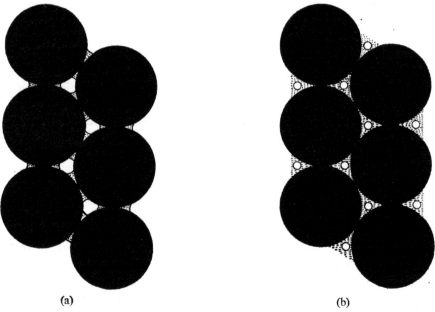

(a)   (b)

FIGURE 9   Drawing of pendular (a) and funicular (b) saturation regimes for an idealized porous medium consisting of packed spheres (after Versluys 1931).

$$\theta = V_{\text{liquid}}/V_{\text{bulk}} = sP \tag{3.4.1.1}$$

or the liquid ratio ϑ

$$\vartheta = (1 + e)\,\theta = sP/(1 - P), \tag{3.4.1.2}$$

where e is the void ratio as defined in section 1.1.

The three saturation regimes that are possible in a porous medium are then:

(a) *Complete saturation regime*   The porous medium is completely saturated with one phase.

(b) *Pendular regime*   The porous medium has the lowest possible saturation with one phase. This phase occurs in the form of pendular bodies throughout the porous medium. These pendular bodies do not touch each other so that there is no possiblity of flow for that phase. A drawing of the pendular regime for an idealized porous medium (consisting of spheres) is shown in figure 9(a).

(c) *Funicular regime*   The porous medium exhibits an intermediate saturation with both phases. If the pendular bodies of the pendular regime expand through addition of the corresponding fluid, they eventually become so large that they touch each other and merge. The result is a continuous network of both phases across the porous medium. It is thus possible that simultaneous flow of both phases occurs along what must be very tortuous (funicular) paths. A drawing of the funicular saturation regime for an idealized porous medium (consisting of spheres) is shown in figure 9(b).

The entrance of fluid into a small pore against another liquid is opposed (or helped) by the surface tension (Washburn 1921). The combined actions of all these forces are such that, at any given relative saturation, a certain pressure differential between the displacing phase and the displaced phase will have to be maintained to create equilibrium. This pressure is called the 'capillary pressure' (p_c); it is a function of the saturation s:

$$p_c = p_c(s) = p(\text{phase 1}) - p(\text{phase 2}). \tag{3.4.1.3}$$

Capillary pressure, like surface tension, exhibits the phenomenon of hysteresis.

As defined above, the capillary pressure $p_c(s)$ is the negative work done against the forces arising from the interaction of the fluid with the porous medium. If one neglects hysteresis effects, it can be treated as a potential. In soil physics, a related quantity Ψ defined by

$$\Psi = - p_c/\rho g \tag{3.4.1.4}$$

is used, ρ being the water density and g the gravitational acceleration (both these quantities are usually treated as constants). The quantity Ψ has then the dimension of a length; it is variously called the 'capillary potential,' 'pressure potential,' 'matric potential,' and 'moisture potential' (cf. Philip 1969a).

One can obtain an idea about the origin of capillary pressure by noting that the interfacial area between the fluids will change. The interfacial tension γ_{12}

between the two fluids is, by definition, the free surface energy F per unit interfacial area:

$$\gamma_{12} = dF/d\Sigma, \tag{3.4.1.5}$$

where Σ is the two-fluid interfacial area per unit pore volume. Following Leverett (1941), we *assume* for the time being the existence of capillary pressure. Then the difference in free surface energy between two saturation states (denoted by s_A and s_B) in the same porous medium is

$$\Delta F = -\int_{s_A}^{s_B} p_c ds; \tag{3.4.1.6}$$

and therefore one has

$$\Sigma_B - \Sigma_A = \frac{1}{\gamma_{12}} \int_{s_A}^{s_B} p_c ds. \tag{3.4.1.7}$$

Thus, if there occurs a change in interfacial area, the capillary pressure on the right-hand side of equation (3.4.1.7) *must actually exist* and be different from zero.

A further qualitative explanation, as discussed for example by Adam (1948), of the existence of capillary pressure is obtained by visualizing the porous medium as consisting of an assembly of capillaries of diameter δ_c and frequency $\alpha(\delta)$ (see chapter 1). In a single capillary, the curvature r of the interfacial surface gives rise to a pressure differential across the interface equal to

$$p_c = 2\gamma/r. \tag{3.4.1.8}$$

The radius of curvature of the meniscus is equal to $\delta_c/(2 \cos \theta)$ so that one has for a single circular capillary (θ is as usual the contact angle):

$$p_c = (4\gamma \cos \theta)/\delta_c. \tag{3.4.1.9}$$

The combined pressure in all the capillaries will give rise to the capillary pressure. The hysteresis between advancing and receding contact angles explains, then, why capillary pressure will also exhibit the effect of hysteresis.

Capillary pressure is specific to the nature of the two fluids involved. However, if no further specification is made, it is usually understood that the displaced 'fluid' is the vacuum.

If the displaced 'fluid' is the vacuum, and an external pressure p_c is applied in a non-wetting fluid, then all capillaries with a diameter larger than δ_c will be totally filled. The connection between saturation and this capillary pressure is thus given by the equation

$$s = \int_{\delta = (4\gamma \cos \theta)/p}^{\infty} \alpha(\delta)d\delta. \tag{3.4.1.10}$$

If the capillaries are not circular, the equation for the capillary pressure has to be generalized by replacing $2/r$ by $1/r_1 + 1/r_2$

$$p_c = \gamma \left(\frac{1}{r_1} + \frac{1}{r_2} \right), \tag{3.4.1.11}$$

where r_1 and r_2 are the principal radii of curvature of the meniscus. Schultze (1925a, b) has shown that the capillary pressures for such capillaries under the assumption of zero contact angle are given approximately by the equation

$$m = \frac{\gamma}{p_c} \quad \text{or} \quad \frac{1}{m} = \frac{1}{r_1} + \frac{1}{r_2}, \tag{3.4.1.12}$$

where m is the ratio of volume to surface of the capillary. The quantity m is sometimes also called the 'hydraulic radius' of the capillary. A list of comparative values to test equation (3.4.1.12) is given in table III (after Carman 1941).

An alternative possibility for calculating the volume of fluid contained in a porous medium is to represent the medium as a regular assemblage of spheres rather than as an assemblage of capillaries. This has been done by Fisher (1926) and later by Von Engelhardt (1955). The calculation in this case has to start with the capillary pressure equation (3.4.1.12). If this equation is assumed to be valid for every surface element of the meniscus, one can obtain the shape of the meniscus by an integration. Writing equation (3.4.1.12) in cylindrical coordinates, one has

$$\frac{1}{r} \frac{d}{dr} \frac{r\,dz/dr}{\sqrt{1 + (dz/dr)^2}} = \frac{p_c}{\gamma}, \tag{3.4.1.13}$$

where the meaning of the symbols is as explained in figure 10. This differential equation has to be integrated for different values of p_c/γ, which can be achieved

TABLE III

List of comparative values to show equivalence of the reciprocal hydraulic radius $(1/m)$ and the reciprocal mean radius of curvature $(1/r_1 + 1/r_2)$ in a capillary (r_i is the radius of the inscribed circle) (after Carman 1941)

Cross-section		$1/r_1 + 1/r_2$	$1/m$
Circle		$2/r$	$2/r$
Parallel plates		$1/b$	$1/b$
	$a{:}b = 2{:}1$	$1.50/b$	$1.54/b$
Ellipse	$a{:}b = 5{:}1$	$1.20/b$	$1.34/b$
	$a{:}b = 10{:}1$	$1.10/b$	$1.30/b$
Rectangle		$1/a + 1/b$	$1/a + 1/b$
Equilateral triangle		$2/r$	$2/r$
Square		$2/r$	$2/r_i$

in closed form only if p_c/γ is equal to zero. One obtains in this case the equation of a catenoid

$$r = r_0 \cosh(3/r_0). \tag{3.4.1.14}$$

If, in addition, the contact angle is zero, one obtains the following values for the parameters describing the catenoid:

$$r_0 = 0.683R, \tag{3.4.1.15}$$

$$\rho = 0.827R, \tag{3.4.1.16}$$

and, correspondingly, for the volume V of fluid in the pendular ring

$$V = 0.0521 \tfrac{4}{3}\pi R^3. \tag{3.4.1.17}$$

For $p_c/\gamma > 0$, equation (3.4.1.13) cannot be solved in closed analytical form and a numerical approximation method has to be applied. A series of results, calculated by Von Engelhardt (1955), is shown in table IV.

The various formulas obtained for the fluid contained in a porous medium consisting of a regular assemblage of spheres can be simplified by assuming that the cross-sections of the menisci are circular or straight (Rose 1958) rather than

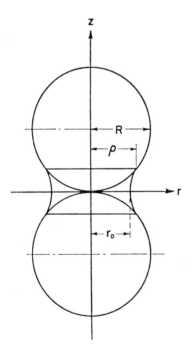

FIGURE 10 Pendular ring in an idealized porous medium consisting of a regular assemblage of spheres (after Von Engelhardt 1955).

given by the integration of equation (3.4.1.13). In many applications, particularly in cementation studies of sandstones, the error committed by this approximation can be shown to be small.

If the pore-openings are not of a simple geometric form, formula (3.4.1.11) is still a valid expression for the capillary pressure. In order to obtain a theoretical relationship between the saturation and capillary pressure for a porous medium, the crucial point is obviously to find an analytical expression for the average interfacial curvature as a function of saturation. This is a very difficult proposition.

As equation (3.4.1.12) can give a reasonably accurate correlation of capillary pressure in non-circular capilliaries, it will likely also be applicable to the capillary channels in a porous medium. One thus tries to generalize the relationship (3.4.1.12), which was obtained from a particular model of straight capillaries, to arbitrary porous media (Carman 1941). Thus, denoting by P the fractional porosity of the porous medium and by S its specific surface area, one sets

$$p_c = \gamma S/P, \tag{3.4.1.18}$$

If the porous medium is unconsolidated, and if S_0 is the particle surface per unit *solid* volume, then $S = (1 - P)S_0$. For spherical particles of uniform diameter δ, S_0 is equal to $6/\delta$, and therefore one has

$$p_c = 6(1 - P)\gamma/(P\delta). \tag{3.4.1.19}$$

This last equation may be used as a general relationship for arbitrary porous media although it was derived using a specific model. Data substantiating it can be taken from the experimental work of Atterberg (1918), Smith, Foote, and Busang (1931) and Hackett and Strettan (1928) (cf. Carman 1941).

The theory of Carman is built on an intuitive generalization of a simple model. A consistent theory of capillary pressure in porous solids should provide an

TABLE IV

Calculated volumes of pendular rings as a function of $p_c R/\gamma$ (after Von Engelhardt 1955)

$p_c R/\gamma$	$V/(\tfrac{4}{3}\pi R^3)$
1.90	1.74×10^{-2}
3.23	1.123×10^{-2}
4.67	0.753×10^{-2}
6.5	0.513×10^{-2}
8.5	0.337×10^{-2}
13.0	0.192×10^{-2}
18.5	0.114×10^{-2}
24.8	0.075×10^{-2}
32.2	0.0478×10^{-2}
59.2	0.0166×10^{-2}

explanation of the fundamental relationship between saturation and capillary pressure (or interfacial curvature). To date, this does not seem to have been obtained. Therefore, Leverett (1941) chose a semi-empirical approach by showing that a dimensionless expression, viz.:

$$J = \frac{p_c}{\gamma}\left(\frac{k}{P}\right)^{\frac{1}{2}} = J(s),$$

(3.4.1.20)

which is termed the 'Leverett function,' can be plotted against the saturation s of the wetting fluid and that the data for a number of unconsolidated porous media

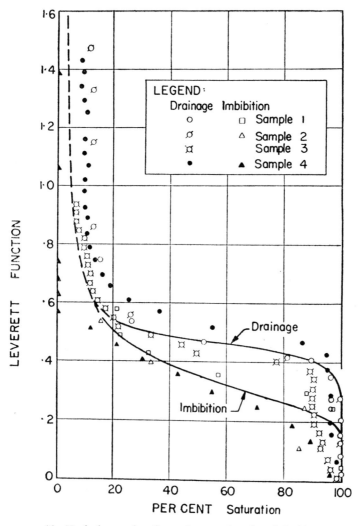

FIGURE 11   Typical examples of some Leverett functions (after Leverett 1941).

fall satisfactorily near two curves, one for imbibition of the wetting fluid and the other for drainage. In (3.4.1.20) k is the permeability (see sec. 4.2) of the medium. Typical examples of some Leverett functions are given in figure 11.

The form of the Leverett correlation may be derived from either of two assumptions (Leverett 1941): (i) that the pressure at which a definite wetting fluid saturation is found at equilibrium is inversely proportional to an 'equivalent circular diameter' of the pores calculated from the porosity and permeability of the medium (cf. chapter 4), inversely proportional to the density difference, and directly proportional to the interfacial tension; or (ii) that the interfacial surface area between the two fluids is, at a given water saturation, a definite fraction of the total surface of the porous medium itself. Leverett does not shed any light, however, upon the particular form of the curves $J = J(s)$, which, from a physical standpoint, is unaccounted for. Nevertheless, the assumption of certain shapes of J-curves for certain groups of porous materials serves well as a working assumption for applications.

3.4.2 *Measurement of capillary pressure*

A capillary pressure (moisture potential) curve requires the measurement of the equilibrium pressures at various saturations of a porous medium during a displacement process. This is usually achieved by enclosing the porous medium in a cell and displacing one fluid with another slowly by changing the pressure in the displacing phase. Unless the displaced phase is a vacuum or a highly compressible fluid, an outlet has to be provided for it; this can be done by applying a semipermeable medium which is impermeable to the displacing phase (Peck and Rabbidge 1966). Various applications of the above principle of measurement have been used in the oil industry and in soil physics.

The *capillary pressure* is then measured directly by a gauge or by an open-end manometer; the latter procedure is customary in soil physics as one obtains directly the length of the 'moisture potential.'

The *saturation* at any pressure can be determined by measuring the amount of displacing fluid that entered the porous medium, by electrical measurement (Martin et al. 1938), by a magnetic susceptibility method (Whalen 1954), or by x-ray techniques (Boyer, Morgan, and Muskat 1947; Morgan, McDowell, and Doty 1950; Laird and Putnam 1951).

Capillary pressure curves have been obtained in large numbers by workers in the oil industry, with oil as displaced and water or gas as displacing fluid. Typical experiments have been reported by Hassler and Brunner (1945), Calhoun, Lewis, and Newman (1949), Powers and Botset (1949), Rose and Bruce (1949), Stahl and Nielsen (1950), Slobod, Chambers, and Prehn (1951), and Dunning et al. (1954). The general shape of the curves bears out the fact that often no displacement occurs until a certain minimum 'displacement pressure' is reached in the displacing phase, and that always a certain residual saturation of the displaced fluid remains no matter how great a pressure is applied in the displacing fluid. A typical example of a capillary pressure curve, obtained by displacing water by oil from a porous medium, is shown in figure 12.

Similarly, moisture potential curves (corresponding to capillary pressure curves in the oil industry) have also been obtained by workers in soil physics; here the experiments involve water and air, the latter at atmospheric pressure. One also typically finds hysteresis (Philip 1964; Carey and Taylor 1967).

Capillary pressure curves are very sensitive to the state of the internal surface of the porous medium owing to the dependence of p_c on γ and the contact angle. Apart from the fact of their showing much hysteresis, a phenomenon which is aggravated by the possibility of physical blockage during a depletion experi-

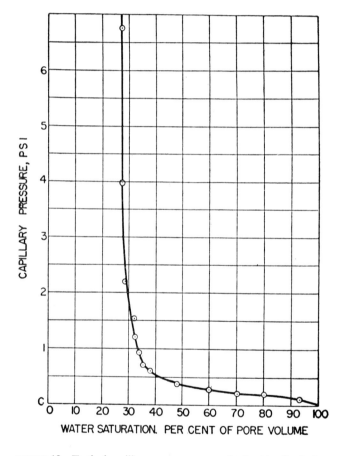

FIGURE 12   Typical capillary pressure curve, obtained by displacing water by oil from a porous medium. Note the residual ('connate') water saturation in spite of the application of very high capillary pressures. (After Bruce and Welge 1947)

ment, it is often difficult even to repeat experiments to obtain the same curves: the preferential wettability (i.e., the contact angle) may have been changed by the previous experiment. Thus, for petroleum-engineering purposes, it has been proposed that one should duplicate the process performed by nature, which consists of displacing water by oil in certain rock strata down to a residual water saturation (which is always found in oil-bearing formations), in the laboratory (Bruce and Welge 1947; Thornton and Marshall 1947), which would give a means of determining the 'connate' water saturation of an oil field a posteriori. It soon became evident, however, that this method does not work (Yuster and Stahl 1948) simply because in the laboratory experiment the contact angle is different from that encountered in the original natural process. Moreover, it is also impossible to correlate capillary pressure curves that have been obtained with different fluids (Brown 1951), although one might think from the formal appearance of p_c (equation 3.4.1.11) that this should be possible (Purcell 1949).

3.4.3 *Capillary pressure and pore size distribution*

Equation (3.4.1.10) states a connection between the void size distribution $\alpha(\delta)$ of a certain type of capillaric model and the capillary pressure curve where the displacing fluid does not wet the medium and is entering it against vacuum. One might propose, therefore, to use this equation to obtain $\alpha(\delta)$ from a capillary pressure curve for a given porous medium.

It should be noted, however, that the capillaric model on which equation (3.4.1.10) is based is a far cry from an actual porous medium. It is therefore not to be expected that, if $\alpha(\delta)$ is calculated from such capillary pressure curves, one will actually obtain the pore size distribution as defined in chapter 1. In this instance Meyer (1953) at least attempts a correction for those large pores that are connected to the 'outside' only by small ones and thus will not be filled at the corresponding pressure. However, most of these 'pore size distribution determinations' have been made without such corrections, which, at best, gives a good qualitative indication of the nature of the pore spaces involved.

The capillary pressure method of determining the pore size distribution usually takes the form of evacuating the porous medium in a cell, and then of injecting non-wetting fluid – usually mercury – at various pressures. The method goes back to Washburn (1921) and was later applied on various occasions by Henderson, Ridgway, and Ross (1940) and Loisy (1941). It was finally greatly popularized by Ritter and Drake (1945, also Drake 1949), after whom a great number of pore size distribution determinations were performed by the mercury injection method. Other versions of the mercury apparatus have been described, for instance, by Plachenov, Aleksandrov, and Belotserkovskiĭ (1953), by Bucker, Felsenthal, and Conley (1956), and by Watson, May, and Butterworth (1957).

A typical apparatus for routine determination of the pore size distribution by the mercury injection method is shown in figure 13. It consists, essentially, of a cell F to take up the sample, and a pumping mechanism A to inject the mercury. In detail, the experiment proceeds as follows. Prior to the use of the apparatus, the internal volume of the cell F is measured by the use of the calibrated pump A and measured steel blanks. The sample is then placed in the cell F and the lid tightened down. With the cell at atmospheric pressure, the mercury level is brought up to the lower reference point in the lucite window E and the pump reading recorded. The vacuum pump L is turned on and a vacuum of less than 0.1 mm as read on the closed arm manometer I is obtained. Pumping is continued for one hour to remove absorbed vapours, and the tightness of fittings checked. The mercury is drawn back into the pump A during evacuations to remove any air in the pump lines. The mercury is run back to the lower reference point and the pump reading taken. The mercury is then raised by the pump to the upper reference line in the top window E and the pump reading recorded. The sample is now ready for mercury injection. The vacuum pump is isolated and the valve G is opened to allow mercury in the manometer J to show a pressure of about 10 cm. The mercury is readjusted to the reference point in the upper window and allowed to stand until no further drop is recorded. In order to establish equilibrium, the cell B is tapped vigorously with a plastic hammer. The procedure of

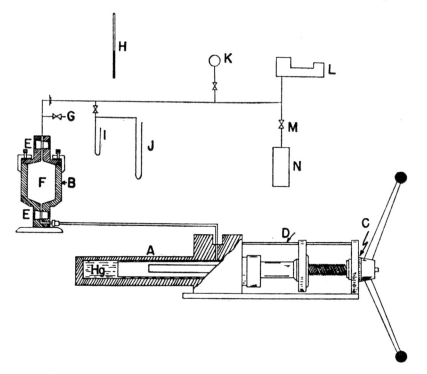

FIGURE 13   A typical apparatus for the determination of the pore size distribution by the mercury-injection method (courtesy of Imperial Oil Limited).

adjusting pressure and volume is repeated until atmospheric pressure is reached, using the manometer J for reference. Valve G and the valves connecting the manometer and vacuum pump to the system are then closed, the valve to pressure gauge K is opened and nitrogen from a pressure container N is admitted to the system through valve M. This allows more mercury to be injected into the sample. Readings of volumes injected are taken at various pressures, in each case tapping and allowing time for mercury to enter the pores.

The pore size distribution curves shown earlier in figure 3 were obtained by the above method.

3.4.4 Hydrostatic equilibrium of fluids in vertical columns of porous media

A special case of quasistatic displacement is the establishment of an equilibrium by a fluid against air in a vertical porous column. Because of the capillary pressure, a saturation equilibrium will develop. As noted in section 3.4.1, the capillary pressure can be treated as a potential if hysteresis is neglected. There must then be a balance between this potential and the gravity potential Φ, for only then is no energy gained or lost by moving a fluid particle up or down. The gravity potential is given by

$$\Phi = \rho g y, \qquad (3.4.4.1)$$

where ρ is the density, g the gravitational acceleration, and y the vertical coordinate upward. Thus, the equilibrium condition is simply

$$- p_c(s) + \rho g y = \text{const.} \qquad (3.4.4.2)$$

If $p_c(s)$ is known, the saturation profile $s(y)$ in the column can be calculated.

In soil physics, the vertical coordinate z is usually measured downward, i.e., $y = -z$. Furthermore, the fluid is standard, viz. water. Therefore, one can divide by ρg and introduce the matric potential Ψ instead of p_c (cf. eg. 3.4.1.4); the latter is always defined so as to be a negative quantity and normalized so that it is $-\infty$ for $s_{\text{water}} = 0$. Then the equilibrium condition becomes

$$\Psi - z = \text{const.} = -Z, \qquad (3.4.4.3)$$

where the constant Z can be identified with the water depth over soil. The last equation leads to

$$z(s) = Z + \Psi(s). \qquad (3.4.4.4)$$

An interesting extension of the above theory can be obtained if one assumes that the porous medium may *swell* by the addition of fluid (this occurs frequently in soils when water is added). Then an overburden potential Ω has to be added to the above potentials (Philip 1969a, b, 1970),

$$\Omega = \frac{de}{d\vartheta} p(z), \qquad (3.4.4.5)$$

where as usual $e = P/(1 - P)$, $\vartheta = sP/(1 - P)$, and $p(z)$ is the total vertical load carried by the porous medium at z expressed in centimetres of water:

$$p(z) = p(z_0) + \int_{z_0}^{z} \frac{\vartheta + \rho_c/\rho_w}{1 + e} \, dz. \tag{3.4.4.6}$$

Then the equilibrium condition becomes

$$\Psi - z + \Omega = \text{const.} \tag{3.4.4.7}$$

This is an integral equation for ϑ which has two classes of non-singular solutions, separated by a singular one:

$$\vartheta(z) = \vartheta_p. \tag{3.4.4.8}$$

Profiles where ϑ slowly approaches ϑ_p with depth so that $\vartheta < \vartheta_p$ are called *xeric*; profiles where ϑ slowly approaches ϑ_p with depth so that $\vartheta > \vartheta_p$ are called *hydric*. The value ϑ_p represents the *pycnotatic point*.

The above theory becomes considerably more complicated if further changes in the porous medium are also taken into account (Philip 1969c). However, this is beyond the scope of the present book.

3.4.5 *Limitations of the capillary pressure concept*

We have already mentioned on several occasions that the capillary pressure curves are different for drainage and imbibition experiments; in other words, they show the phenomenon of hysteresis. This fact invalidates in effect considerations based on the existence of an energy potential. However the pertinent equations may still be used for the calculation of saturation profiles, etc., as long as the direction of the saturation change is not reversed and the appropriate curve is taken.

The reason for capillary hysteresis seems to lie in problems of instability of interface configurations; changes seem to occur often spontaneously in 'Haines (1930) jumps' which produce an irreversibility in the process. When they occur, they evidently involve viscous flow of the fluids involved with an attendant energy dissipation. This problem has been studied by Miller and Miller (1956) and Heller (1959, 1968).

3.5
RELATIVE WETTABILITY

3.5.1 *The concept of wettability*

As has been stated before (sec. 2.4.2), the concept of relative wettability is another consequence of the presence of capillary forces. Depending on the contact angle between two fluids, one is said to wet the solid more preferentially than the other.

Because of possible hysteresis in contact angles, the relative wettability of a solid by two fluids may also be hysteretic.

In the case of a porous medium, it is difficult to observe contact angles. Nevertheless, the concept of wettability still has a meaning: of two fluids, one will generally wet the porous medium more preferentially than the other; in different words, one fluid (the preferentially wetting one) will penetrate into a porous medium against a less wetting one. Of any two fluids it should therefore be possible to say which wets a given porous medium more preferentially than the other.

Thus, the term *relative wettability* has a definite intuitive appeal and the task is to find a proper quantitative characterization for it. Such a quantitative characterization of 'wettability' would be obtained if a suitable measurement could be devised that would define it. However, when introducing the concept of 'wettability,' it is difficult to refer to a numerical value, but only to the *relative* wettability of two fluids. By making successive tests with many fluids, it is therefore possible to devise *ordinal* (or 'ranking') scales of wettabilities for fluids in relation to a given porous medium. This is in contrast to higher levels of measurement which are more commonly attained in physics, where the values obtained are isomorphic to the numerical structure which is known as arithmetic. In ordinal scales this is not true.

In any wettability tests of a porous medium there are three agents involved: two fluids and the porous medium. In order to set up an ordinal scale of wettability, two of these agents must be kept constant, the third being allowed to vary. The possibilities that suggest themselves are:

(a) One of the fluids and the porous medium is held constant, the other fluid being variable. An ordinal scale of wettabilities relative to the variable fluid can be set up.

(b) The two fluids are held constant, the porous medium being variable. It will be observed that some porous media will be wetted more eagerly by one of the fluids than by the other. This implies that an ordinal scale of wettabilities of various porous media relative to the two fluids can be set up.

As noted above, it is rather uncommon to deal with ordinal scales in physics. However, the behavioural sciences (such as psychology) have made much use of them. One of the features of ordinal scales is that they *may* be represented by a series of numbers, where increasing numbers correspond to higher orders of the property measured. Any ordinal scale, if expressed in terms of numbers, is invariant under a monotonic transformation of the numbers. If a statistical analysis of the numbers is to be undertaken, non-parametric methods (cf. Siegel 1956; Fraser 1957) have to be applied.

3.5.2 *Measurement of wettability*

As has been outlined above, there are two types of wettability scales that one ought to be able to set up: a wettability scale of a given porous medium relative to various fluids, and a wettability scale of various porous media relative to given fluids.

With regard to the wettability of a porous medium relative to various fluids one may note that, in order to arrive at a scale, one of the fluids must also be held constant. Different *pairs* of fluids cannot be compared directly. If the 'fluid' which is being held constant is the vacuum, one arrives at an order of wettability of a porous medium relative to various fluids.

A variety of experimental procedures have been described in the literature which in one way or another lead to an ordinal scale of wettabilities. These procedures either consist of the numerical measurement of certain quantities which can then be used as an indication of an ordinal wettability 'score' of the system under consideration, or compare two cases and yield the relative order of these two cases. A repetition of the latter type of test with many pairs of cases will then also produce an ordinal scale. In all these tests the problem of standardization will be of utmost importance so that one truly arrives in every case at a meaningful wettability 'score' that is usable for further discussions.

The tests proposed in the literature that came to the writer's attention are the following:

(a) *The measurement of displacement pressure*, proposed by Bartell and Osterhof (1927). The higher the displacement pressure to force one fluid against another into a porous medium, the smaller is the wettability by that fluid. This represents the 'capillary pressure type' of wettability measurement.

(b) *Imbibition tests*, which consist of immersing a fluid-filled porous medium in another fluid and measuring the rate at which the original fluid is displaced.

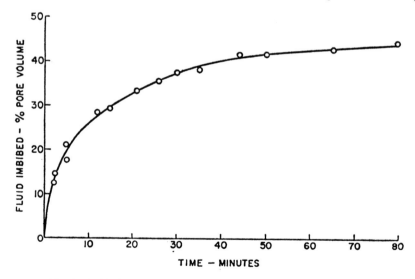

FIGURE 14 Typical imbibition curve, with water being imbibed against oil by sandstone (after Bobek, Mattax, and Denekas 1958).

Apparatus suitable for this purpose has been described, for instance, by Moore and Slobod (1956) and by Bobek, Mattax, and Denekas (1958). The greater the rate of imbibition, the higher is the wettability of the porous medium by the imbibed fluid. The test thus described appears to be very simple indeed. However, there are some difficulties. The rate of imbibition is by no means constant so that, experimentally, an imbibition *curve* is observed. A typical example of such a curve is shown in figure 14. Furthermore, this curve depends on many factors other than 'wettability,' such as the shape and size of the porous medium and its permeability (cf. sec. 4.2.2). Thus, problems of standardization are involved since 'wettability-values' obtained by imbibition tests can be compared only if the procedure has been standardized. No general efforts in this direction seem to have been reported in the literature.

(c) *Fractional wettability determinations*, which are based on the assumption (cf. Fatt and Klikoff 1959) that, microscopically, an elemental area of the internal surface of the porous medium is either wettable or non-wettable by one of the two fluids involved – with no in-between values being possible. The problem is thus one of determining what *fraction* of the internal surface is wettable by one fluid and what fraction by the other. The greater this fraction, the higher is the wettability relative to the fluid in question. The methods used to determine this fraction may be quite involved; Brown and Fatt (1956), for instance, describe a nuclear magnetic relaxation method, Holbrook and Bernard (1958) a dye adsorption method, and Donaldson et al. (1969) a method based on capillary pressure measurements leading to a numerical wettability scale.

The various methods discussed above are equivalent if the ordinal scales of wettabilities (either of a given porous medium relative to various fluids or of given fluids relative to various porous media) obtained thereby are monotonic transformations of each other. Comparisons of the various methods have been reported, for example, by Gatenby and Marsden (1957), but no systematic analysis of the existence or non-existence of an isomorphism between the various possible scales seems ever to have been attempted.

3.5.3 *Heat of wetting*

Owing to energy changes during the process of wetting of a porous solid, a certain amount of heat is liberated. This 'heat of wetting' can be used for internal area determinations as it should be proportional to the area which is being wetted by the fluid.

The heat of wetting due to the complete soaking of a porous medium by a wetting fluid can be estimated as follows (Wenzel 1938).

When the liquid spreads over an area S_1, a solid-air interface of that area is destroyed and thus an energy $S_1 \gamma_{SA}$ is gained, where γ_{SA} is the solid-air interfacial tension (see sec. 2.4.2). At the same time, a solid-liquid interface of the same area S_1 is formed, which means that an energy $S_1 \gamma_{SL}$ has to be expended. In addition, a liquid-air interface of area S_2 ($\neq S_1$) is formed, requiring the additional expenditure of an amount of energy equal to $S_2\gamma_{LA}$. Thus, the total amount of energy gained in the process (i.e., the heat of wetting) is

$$W = S_1\gamma_{SA} - S_1\gamma_{SL} - S_2\gamma_{LA}. \tag{3.5.3.1}$$

Introducing the value (2.4.2.2) for the contact angle θ, this becomes

$$W = \gamma_{LA}(S_1 \cos\theta - S_2). \tag{3.5.3.2}$$

In soaking a porous medium, the solid-liquid interface S_1 will be much larger than the liquid-air interface S_2, and therefore S_2 can be neglected. Furthermore, S_1 becomes equal to the total internal surface area S of the porous medium, and the expression for the heat of wetting thus becomes

$$W = \gamma_{LA}S \cos\theta. \tag{3.5.3.3}$$

This is the required connection between the heat of wetting W and the internal surface S of a porous medium.

In table v, we give some examples of heats of wetting that have been determined for various substances.

TABLE V

Heat of wetting against air for various solids in various liquids (in ergs cm^{-2}) (after Gregg 1951)

Liquid	Barium sulphate	Titanium dioxide	Silica	Graphite
Water	490	520	600	175
Ethanol	—	500	520	—
Benzene	140	150	150	—
Carbon tetrachloride	220	240	—	—

4
Darcy's law

THE FLOW OF HOMOGENEOUS FLUIDS IN POROUS MEDIA

The present chapter has as its subject one of the aspects of the flow of homogeneous fluids in porous media, viz., that for which the law of Darcy is valid. In the course of the chapter we shall first discuss Darcy's experiment as it was originally performed and then enumerate further experiments made to elucidate the laws of flow through porous media.

The experiments of Darcy and of later workers suggest a law of flow, which, however, is not uniquely defined. Therefore, there is considerable ambiguity in postulating a differential equation which would be equivalent to the results of the experiments. In fact, the differential equation which is now commonly called 'Darcy's law' is not an equivalent expression for Darcy's findings, although these do follow from it. However, they would equally well follow from other types of differential equations. This is especially true if generalizations of Darcy's law to anisotropic and compressible porous media are attempted. Some discussion will be devoted to this subject later in this review.

Another section of the chapter will be devoted to methods of measuring the coefficient occurring in Darcy's law, i.e., to methods of measuring the 'permeability.' The literature on permeability determination is extremely prolific as this quantity is important in many applications. The differential form of Darcy's law is also basic to the theory of the process of filtration. The chapter will therefore be concluded by a review of some aspects of filtration.

It is to be expected that Darcy's law will have limitations. Indeed, such limitations occur generally at high and low flow rates, as well as in relation to various other effects. The range of validity of Darcy's law and its limitations will be discussed in a later chapter of this book devoted to general flow equations.

4.2
EXPERIMENTAL INVESTIGATIONS

4.2.1 *Darcy's experiment*

The theory of laminar flow through homogeneous porous media is based on a classical experiment originally performed by Darcy (1856). A schematic drawing of this experiment is shown in figure 15. A homogeneous filter bed of height h is bounded by horizontal plane areas of equal size A. These areas are congruent so that corresponding points could be connected by vertical straight lines. The filter bed is percolated by an incompressible liquid. If open manometer tubes are attached at the upper and lower boundaries of the filter bed, the liquid rises to the heights h_2 and h_1 respectively above an arbitrary datum level. By varying the various quantities involved, one can deduce the following relationship:

$$Q = - KA(h_2 - h_1)/h, \qquad (4.2.1.1)$$

where Q is the total volume of fluid percolating in unit time and K is a constant depending on the properties of the fluid and of the porous medium. The relationship (4.2.1.1) is known as *Darcy's law*. The minus sign in the expression for Q indicates that the flow is in the opposite direction of increasing h.

Darcy's law can be restated in terms of the pressure p and the density ρ of the liquid. At the upper boundary of the bed (elevation above the datum level denoted by z_2), the pressure is $p_2 = \rho g(h_2 - z_2)$, and at the lower boundary (elevation above datum level denoted by z_1), the pressure is $p_1 = \rho g(h_1 - z_1)$. Inserting this statement into (4.2.1.1), one obtains (as $z_2 - z_1 = h$)

$$Q = - KA[(p_2 - p_1)/(\rho gh) + 1];$$

or, upon introduction of a new constant K', assuming ρ and g to be constants,

$$Q = - K'A(p_2 - p_1 + \rho gh)/h. \qquad (4.2.1.2)$$

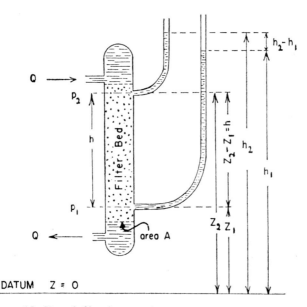

FIGURE 15   Darcy's filtration experiment.

The equations (4.2.1.1) and (4.2.1.2) are equivalent statements of Darcy's law.

The validity of Darcy's law has been tested on many occasions (e.g., Vibert 1939; Iwanami 1940; LeRosen 1942; Emmerich 1954). Thus, it has been shown that it is valid for a wide domain of flows. For liquids, it is valid for arbitrary small pressure differentials (Meinzer and Fishel 1934; Fishel 1935; Meinzer 1936; Schweigl and Fritsch 1942). It has also been used to measure flow rates by determining the pressure drop across a fixed porous plug (see Souers and Binder 1952). For liquids at high velocities and for gases at very low and at very high velocities, Darcy's law becomes invalid, as will be discussed later.

4.2.2 *The permeability concept*

Darcy's law in its original form is rather restricted in usefulness. The first task is to elucidate the physical significance of the constant K'. The constant is obviously indicative of the permeability of a certain medium to a particular fluid. It depends on the properties of both the medium and the fluid. Before about 1930 it was often called the 'permeability-constant'; its dimensions are $M^{-1} L^3 T^1$, with mass (M), length (L), and time (T) as fundamental dimensions.

A constant of the type K' is, however, not very satisfactory because one would like to separate the influence of the porous medium from that of the liquid. Nutting (1930) had already stated that one should have

$$K' = k/\mu, \tag{4.2.2.1}$$

where μ is the viscosity of the fluid and k is the '*specific* permeability' of the porous medium. But this relationship was not generally accepted until it was popularized by Wyckoff et al. (1933). The verification of it consists in the innumerable successful determinations of permeability that have been performed using it as a basis.

The dimension of (specific) permeability is a length squared, which suggests that the natural permeability unit in the c.g.s. system should be cm^2. Unfortunately, this unit has been adopted only by some physicists and chemists. In most branches of applied science, some other unit has been adopted specific to that branch. Thus, the oil industry, for example, uses the 'darcy,' with

$$1 \text{ darcy} = 9.87 \times 10^{-9} \text{ cm}^2. \tag{4.2.2.2}$$

In groundwater hydrology, it is customary to represent permeability in terms of the seepage velocity of the percolating water per unit drop of hydraulic head. According to country and inclination, this velocity may be expressed in terms of any combination of feet, centimetres, hours, minutes, or seconds. One has, for example,

$$1 \text{ cm/sec for water} = 1.02 \times 10^{-5} \text{ cm}^2. \tag{4.2.2.3}$$

Measured thus, the permeability is often called the seepage coefficient (hydraulic conductivity).

In order to compare permeabilities for various substances in various branches of applied science, it is best to transform everything into c.g.s. units.

Once the permeability concept has been defined, further experiments have to be directed towards determining the range of actual constancy of this permeability for a given porous medium. In other words, one has to determine the dependence of the permeability on external conditions of the porous medium.

The permeability would be expected to depend on external stresses that might be impressed upon the porous medium assuming that the latter is compressible. Indeed, this is a well-known fact in the theory and practice of filtration. Thus, Secchi (1936) has made experiments to determine the dependence of the permeability of a filter on the external pressure and has been able to show, not only that there is such a dependence, but also that it is subject to hysteresis. Another series of experiments was published by Ruth (1946). Tiller (1953a, b) showed that the dependence of the permeability on the fluid pressure p and the total pressure $\mathscr{T}$ (see sec. 1.7.1) can be represented by

$$k = K(\mathscr{T} - p)^{-m}. \tag{4.2.2.4}$$

This relationship was deduced from largely empirical investigations combined with notions of the Kozeny theory (see chapter 6). K and m are constants that have to be determined experimentally. The relationship is valid only if $\mathscr{T} - p$ is larger than some (experimentally determined) limiting value. Fatt and Davis (1952) have made similar investigations of the dependence of permeability on

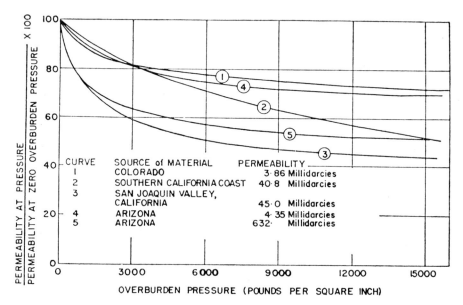

FIGURE 16   Curves showing the dependence of (specific) permeability on total external stress (i.e., overburden pressure) (after Fatt and Davis 1952).

stresses in connection with the study of petroleum well cores. Another set of
experiments has been performed by Grace (1953). In figure 16 we show some
curves demonstrating the dependence of permeability on the total external stress.

There is also the possibility that a directional variation of permeability occurs.
Thus, if a cube is cut out of a macroscopically homogeneous piece of rock, the
permeability may not be the same across all the faces. This effect has actually
been observed by Sullivan (1941), Pressler (1947), Johnson and Breston (1951),

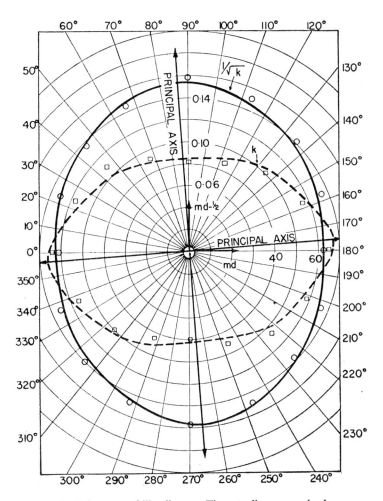

FIGURE 17   Polar permeability diagram. The actually measured values
(squares) fit on a complicated curve (dashed) which corresponds to an
ellipse (solid) if $k^{-\frac{1}{2}}$ is drawn instead of k. The circles correspond to the
values of $k^{-\frac{1}{2}}$ as measured; it is seen that the fit of these values with an
ellipse is excellent. This substantiates the tensor theory of permeability.
(After Scheidegger 1954)

and Griffiths (1950). In this connection, a very valuable set of experiments has also been performed by Johnson and Hughes (1948): the directional permeability was measured in intervals around 180° in sections of a number of natural rock pieces. The results are plotted in the form of polar permeability diagrams. Most polar diagrams have somewhat the shape of an ellipse; others, however, are a little like the figure 'eight' (see sec. 4.3.3). An example of a polar permeability diagram is shown in figure 17.

4.3
DIFFERENTIAL FORMS OF DARCY'S LAW

4.3.1 *Isotropic porous media*

Darcy's law, when the separation of the general constant into 'permeability' and 'viscosity' is taken into account, is expressible as follows:

$$q \equiv Q/A = - (k/\mu)(p_2 - p_1 + \rho gh)/h. \tag{4.3.1.1}$$

In this form, it applies to a horizontal bed of finite thickness h, being percolated by an incompressible liquid of (constant) density ρ. This form of the law has only a very restricted use.

The aim will, therefore, be to express (4.3.1.1) in differential form. A priori, there is no unique way of doing this. Naturally q will become a vector $\mathbf{q}$, which might be called the local 'filter-velocity' or 'seepage-velocity,' and the pressure difference in (4.3.1.1) must be somehow expressed by the pressure gradient.

A first possibility is suggested from (4.3.1.1) by letting h become infinitesimal. One then obtains:

$$\mathbf{q} = - (k/\mu)(\operatorname{grad} p - \rho \mathbf{g}), \tag{4.3.1.2}$$

where $\mathbf{g}$ is a vector in the direction of gravity (i.e., *down*ward) and of the magnitude of gravity. However, Darcy's experiment does not tell us what happens if the permeability and viscosity are variables. Thus, the coefficient might equally well have to be taken into the gradient:

$$\mathbf{q} = - \operatorname{grad}(kp/\mu) + k\rho \mathbf{g}/\mu. \tag{4.3.1.3}$$

The first possibility (equation 4.3.1.2) is equivalent to the introduction of a force potential ϕ (if it can be defined) inasmuch as the equation can be rewritten in an equivalent manner as follows (see Hubbert 1940):

$$\phi = gz + \int_{p_0}^{p} dp/\rho(p), \tag{4.3.1.4a}$$

$$\mathbf{q} = - (k\rho/\mu) \operatorname{grad} \phi, \tag{4.3.1.4b}$$

where z denotes the vertical coordinate. Correspondingly, in the second possibility, equation (4.3.1.3) is equivalent to the introduction of a velocity-potential ψ (if it can be defined) inasmuch as the equation can be rewritten in an equivalent manner as follows (see Gardner, Collier, and Farr 1934):

$$\psi = kp/\mu + \int_{z_0}^{z} k\rho g dz/\mu, \tag{4.3.1.5a}$$

$$\mathbf{q} = -\operatorname{grad} \psi. \tag{4.3.1.5b}$$

Both representations by potentials are valid only if the integrals are univalent.

It will never be possible to distinguish between the two differential representations of Darcy's law if experiments are performed which use only constant-viscosity fluids and porous media of homogeneous permeability. However, it is not even necessary to conduct a critical *real* experiment, as a simple 'thought-experiment' is able to provide the required answer. Let us envisage a linear porous medium where the permeability changes from one side to the other, but where the pressure is constant throughout (we neglect gravity). If equation (4.3.1.3) were valid, there would be flow in such a medium – a completely unthinkable occurrence. Hence (4.3.1.3) must be wrong and only (4.3.1.2) can be correct. It is thus a force-potential and not a velocity-potential which governs flow through porous media.

The differential form of Darcy's law, (4.3.1.2), is by itself not sufficient to determine the flow pattern in a porous medium for given boundary conditions as it contains three unknowns ($\mathbf{q}$, p, ρ). Two further equations are required for the complete specification of a problem. One is the connection between ρ and p of the fluid:

$$\rho = \rho(p), \tag{4.3.1.6}$$

and the other a continuity condition,

$$- P \, \partial\rho/\partial t = \operatorname{div}(\rho\mathbf{q}), \tag{4.3.1.7}$$

where as usual P is the porosity and t the time. With the help of these equations, one can eliminate all the unknowns except p, which leads to the equation

$$P\partial\rho/\partial t = \operatorname{div}[(\rho k/\mu) \, (\operatorname{grad} p - \rho\mathbf{g})], \tag{4.3.1.8}$$

where, again, $\mathbf{g}$ is a vector pointing *down*ward.

4.3.2 *Anisotropic porous media*

Early investigations on the subject of anisotropic media were made by Schaffernak (1933), Vreedenburgh and Stevens (1936), and Aravin (1937). The theory of flow through anisotropic media has been developed by Ferrandon (1948), Ghizetti (1949), and Litwiniszyn (1950), and a further review given by Irmay

(1951). Litwiniszyn arrived at his theory by analogy with the process of diffusion, whereas Ferrandon gave an actual theoretical derivation of the formulas. Both theories are essentially identical.

Ferrandon's theory (see Scheidegger 1954) assumes that the contribution to the quantity q_n of flow through a unit area in the direction $\mathbf{n}$ (components n_i), from elementary flow tubes parallel to the direction $\mathbf{m}$ (components m_i) whose combined cross-sectional area is equal to $cd\Omega$ ($d\Omega$ denoting the solid angle), is proportional to the gradient of a force-potential ϕ in the direction of $\mathbf{m}$ (cf. equation 4.3.1.4). Thus one has (the summation convention being applied)

$$dq_n = - kcn_im_i\rho(\partial\phi/\partial x_j)m_jd\Omega/\mu.$$

Here k and c are, of course, functions of m_i, such that one can set upon integration:

$$q_n = - n_i\rho(\partial\phi/\partial x_j) \int kcm_im_jd\Omega/\mu = - n_i(k_{ij}/\mu)\rho(\partial\phi/\partial x_j),$$

where $k_{ij} = k_{ji}$. This can be written in vectorial form as follows:

$$\mathbf{q} = - (\overline{k}/\mu)\rho \, \text{grad} \, \phi, \tag{4.3.2.1}$$

where $\overline{k}$ is a symmetric tensor with components k_{ij}. It can be properly referred to as the 'permeability tensor' of the porous medium.

The fact that the permeability in anisotropic porous media has the form of a symmetric tensor leads immediately to the following conclusions:

(i) In general the force potential gradient grad ϕ and filter velocity $\mathbf{q}$ are not parallel.

(ii) There are three orthogonal axes in space along which the force potential gradient and the velocity have the same direction. These axes are termed the 'principal axes' of the permeability tensor.

The task remains to relate the components k_{ij} of the permeability tensor to directional permeability measurements. Thus, let us investigate what happens in a narrow tube that was cut out in the direction $\mathbf{n}$ of an anisotropic porous medium. Obviously the filter velocity $\mathbf{q}$ must be parallel to $\mathbf{n}$; let it be denoted by q_n. The drop in ϕ along the tube, denoted by ϕ_n, is then given from equation (4.3.2.1) by

$$\phi_n = \mathbf{n} \, \text{grad} \, \phi = \mu\mathbf{n}\overline{k}^{-1}\mathbf{n}q_n/\rho.$$

If we define the 'directional permeability' k_n' by the expression

$$k_n' = \mu q_n/(\phi_n\rho), \tag{4.3.2.2}$$

we obtain at once

$$k_n' = 1/(\mathbf{n}\overline{k}^{-1}\mathbf{n}). \tag{4.3.2.3}$$

Let us now choose the principal axes of the permeability tensor as coordinates (the corresponding permeabilities being k_1, k_2, k_3) and denote the angles of **n** with those axes by α, β, γ. Then (4.3.2.3) yields

$$\frac{1}{k_n'} = \frac{\cos^2 \alpha}{k_1} + \frac{\cos^2 \beta}{k_2} + \frac{\cos^2 \gamma}{k_3}. \tag{4.3.2.4}$$

This is the central equation of an ellipsoid if

$$r = \sqrt{k_n'} \tag{4.3.2.5}$$

is plotted on the corresponding directions of **n**. One has thus the following theorem:

If the square root of the directional permeability k_n' is measured for all of the corresponding directions at a point of an anisotropic porous medium, then one obtains an ellipsoid. Its axes are in the direction of the principal axes of permeability, their length being equal to the square root of the principal permeabilities.

An *alternative way* to define directional permeability physically is by choosing a system in which the pressure drop is given by the boundary conditions, and by measuring that component of the velocity which is parallel to the pressure gradient. The directional permeability k_n'' is in this case defined by applying Darcy's law formally to the given pressure gradient and the velocity component parallel to it. Thus one has

$$k_n'' = \mathbf{n}\bar{k}\mathbf{n} \tag{4.3.2.6}$$

as a second definition of 'directional permeability.'

Let us again choose the principal axes of the permeability tensor as coordinate axes (corresponding permeabilities being k_1, k_2, k_3) and denote the angle of **n** with those axes by α, β, γ. Then (4.3.2.6) yields

$$k_n'' = k_1 \cos^2 \alpha + k_2 \cos^2 \beta + k_3 \cos^2 \gamma. \tag{4.3.2.7}$$

This is the central equation of an ellipsoid if

$$r = 1/\sqrt{k_n''} \tag{4.3.2.8}$$

is plotted on the corresponding directions of **n**. Thus one has the following alternative theorem:

If the inverse square root of the directional permeability k_n'' is measured for all of the corresponding directions at a point of an anisotropic porous medium, then one obtains an ellipsoid. Its axes are in the direction of the principal axes of permeability, their length being equal to the inverse square root of the principal permeabilities.

The fact that there are two different kinds of 'directional permeability,' denoted above by k_n' and k_n'', is somewhat disconcerting. Whether one should consider k_n' or k_n'' in any particular case depends, of course, on the type of measurement that has been made. Fortunately, it is possible to show that in most cases the difference between the two types of directional permeability is quite negligible (Scheidegger 1956).

In order to show this, let us consider the two-dimensional case. We then have

$$\frac{1}{k_n'} = \frac{\cos^2 \alpha}{k_1} + \frac{\sin^2 \alpha}{k_2}, \qquad\qquad (4.3.2.9)$$

$$k_n'' = k_1 \cos^2 \alpha + k_2 \sin^2 \alpha, \qquad\qquad (4.3.2.10)$$

and the ratio of the two types of directional permeability yields

$$k_n''/k_n' = \cos^4 \alpha + (k_1/k_2 + k_2/k_1) \cos^2 \alpha \sin^2 \alpha + \sin^4 \alpha$$

$$= 1 + \frac{(k_1 - k_2)^2}{k_1 k_2} \cos^2 \alpha \sin^2 \alpha. \qquad\qquad (4.3.2.11)$$

The right-hand side of this expression reaches a maximum value for $\alpha = 45°$; it is then equal to:

$$\left.\frac{k_n''}{k_n'}\right|_{max} = 1 + \frac{1}{4} \frac{(k_1 - k_2)^2}{k_1 k_2}. \qquad\qquad (4.3.2.12)$$

The maximum excess m over 1 of the ratio k_n''/k_n' (i.e., the maximum percentage of error introduced if k_n' is put equal to k_n'') is therefore given by

$$m = \frac{1}{4} \frac{(k_1 - k_2)^2}{k_1 k_2}. \qquad\qquad (4.3.2.13)$$

It is convenient to express this error m in terms of the excess n of the ratio k_1/k_2 over 1 ($k_1 > k_2$), which yields

$$m = \frac{1}{4} \frac{n^2}{1 + n}. \qquad\qquad (4.3.2.14)$$

This shows that, for excesses n smaller than 1, the corresponding error m is proportional to n^2 and therefore negligible; for excesses n larger than 1, the corresponding error m goes asymptotically to $n/4$. It is therefore seen that k_n' and k_n'' are, in approximation, freely interchangeable. The relative error introduced by using them thus increases with the excess of the ratio k_1/k_2 ($k_1 > k_2$) over 1, but never exceeds 1/4 of this excess.

Physical measurements of directional permeability have been reported by Johnson and Hughes (1948; see sec. 4.2.2). Unfortunately, these authors plotted k_n as a polar permeability diagram instead of $1/k_n^{\frac{1}{2}}$. They therefore obtained shapes resembling somewhat a circle or the figure 'eight.' Scheidegger (1954) recalculated the results of Johnson and Hughes and analysed their data in the light of Ferrandon's tensor theory. These investigations yielded a substantiation of the tensor theory of permeability. An example of the fit of measured values of $1/k_n^{\frac{1}{2}}$ to an ellipse is shown in figure 17.

4.3.3 *Compressible porous media*

After differential forms of Darcy's law have been derived for fixed porous media, they must be generalized for compressible porous media.

In order to treat the problem of flow through compressible porous media fully, one cannot separate the motion of the fluid from that of the bulk of the porous mass. One has thus a twofold problem: the porous medium is undergoing finite deformations (governed by corresponding equations of motion), and, at the same time, the fluid is being displaced. All these displacements are large so that the whole apparatus of finite strain theory has to be invoked, which makes the problem very complicated. Some work on this general problem has been done in connection with studies of consolidation of wet soil (Biot 1941, 1956; Florin 1948); the writer is not aware, however, of an occasion where the full finite strain mechanism has been introduced and consistently employed. Furthermore, in consolidation studies the flow of fluid contained in the pores is only of minor interest, the flow velocity is usually eliminated altogether, and one is left with equations which refer to the deformation of the porous mass as a whole. The standard way (Terzaghi 1951) to describe consolidation is to assume that the effective pressure (which, under constant overburden pressure, is a linear function of the pore pressure) affects the porosity *linearly*. The pore-water movement is then calculated according to Darcy's law which, under quite widely applicable conditions (cf. sec. 5.4.1), leads to a diffusivity equation. Because of the assumed *linear* connection between the pore pressure and porosity, one obtains a diffusivity equation for the description of the space-time dependence of the consolidation process under an imposed load. The above simple theory can be improved by introducing relationships between the permeability and pore pressure (cf. sec. 4.2), but it then becomes far more complicated.

In a study of flow through porous media it is, however, the flow of the fluids and not the consolidation of the medium which is of primary interest. In many practical cases, it is not necessary to treat the problem in its full generality. The displacements due to consolidation are usually small compared with the displacements due to the flow of fluid. Furthermore, it is often possible to predict at least the general geometry in which the consolidation will occur so that it is not necessary to solve a complete set of differential equations of motion to determine the flow. This leads to simplified descriptions of flow through compressible porous media which are fully adequate in most instances.

Attempts at such simplified descriptions of the flow through compressible porous media have been made by Froehlich (1937) and Jacob (1940). A more or less consistent theory of the subject was given later in a series of communications by Shechelkachev (1946a, b, c, 1959; cf. also Leïbenzon 1947, p. 84).

Shchelkachev's theory is based upon the assumption that (a) both fluid and medium are elastic bodies following Hooke's law, i.e. (see secs. 1.7 and 2.2.1),

$$d\rho/dp = \rho\beta_f, \tag{4.3.3.1}$$

$$dP/dp = \beta_m, \tag{4.3.3.2}$$

where ρ is the density of the fluid, P is the porosity of the solid, β_f and β_m are compressibility coefficients, assumed to be constant; and that (b) the permeability of the medium does not change during compression. Accordingly, the whole effect of the compressibility of the medium is due to an alteration of the continuity equation. This equation can be shown to become:

$$\text{div}(\rho\mathbf{q}) = -\left(P + \frac{\beta_m}{\beta_f}\right)\frac{\partial\rho}{\partial t}. \tag{4.3.3.3}$$

The proof of Shchelkachev's result runs as follows. Generally, one can write

$$\oint_S \rho\mathbf{q}ds = -\frac{d}{dt}\iint_V \rho P dV \tag{4.3.3.4}$$

for any volume V of surface S. The left-hand side (LHS) of this equation can be transformed by Gauss's theorem to yield

$$\text{LHS} = \iint_V \text{div}(\rho\mathbf{q})dV. \tag{4.3.3.5}$$

The right-hand side (RHS) can be transformed through the following steps:

$$
\begin{aligned}
\text{RHS} &= -\iint \frac{d}{dt}(\rho P)dV = -\iint\left(P\frac{d\rho}{dp}\frac{dp}{dt} + \rho\frac{dP}{dp}\frac{dp}{dt}\right)dV \\
&= -\iint\left(P\beta_f\rho\frac{dp}{dt} + \rho\beta_m\frac{dp}{dt}\right)dV \tag{4.3.3.6} \\
&= -\iint\left(P + \frac{\beta_m}{\beta_f}\right)\beta_f\rho\frac{dp}{dt}dV.
\end{aligned}
$$

However, by (4.3.3.1),

$$\beta_f\rho\, dp/dt = \partial\rho/\partial t \tag{4.3.3.7}$$

so that

$$\text{RHS} = -\iint\left(P + \frac{\beta_m}{\beta_f}\right)\frac{\partial\rho}{\partial t}dV. \tag{4.3.3.8}$$

Since RHS = LHS for all possible volumes V, the corresponding integrands must be equal, which leads to equation (4.3.3.3) as claimed. Shchelkachev's theory thus simply replaces P by $(P + \beta_m/\beta_f)$ in the normal continuity equation, and therefore also in the usual differential formulation (4.3.1.8) of Darcy's law. It will be shown later that assuming a finite compressibility of the medium is equivalent to assuming a 'super-compressibility' of the liquid.

Shchelkachev's theory is incomplete inasmuch as no dependence of the permeability on the prevailing fluid pressure has been assumed. We have mentioned above the investigations of Tiller (1953a, also 1953b, 1955), who postulated such a dependence (cf. equation 4.2.2.5). Tiller did not, however, deduce the continuity equation corresponding to the dependence of porosity on fluid pressure as assumed by him (which is different from 4.3.3.2), but proceeded directly to integrate the Darcy equation (4.3.1.8) with a variable permeability according to (4.2.2.5). He was only interested in the constant-rate case applicable to certain filtration experiments and hence assumed q to be a constant. In order to make a consistent hydrodynamic theory out of Tiller's results, one would have to deduce the corresponding consistent continuity condition. This condition would be different from Shchelkachev's owing to the difference in the assumption of the dependence of P as a function of p. These investigations have not yet been brought to a close.

If one has applications of the theory of compressible porous media to the flow of fluids in underground strata in mind, one can make certain types of simplifications in the general theory of consolidation which may be different from those considered above. This problem has been investigated by the writer (Scheidegger 1959). Accordingly, what is required for the practical study of the flow of fluids in compressible strata is a set of equations of motion for the fluid wherein due allowance is made for the concurrent deformation of the porous medium containing the fluid. There are two possibilities that come to mind: either the permeability and the stresses are locally isotropic, or they are not isotropic. One would think, offhand, that it might be possible to treat the porous medium as locally isotropic with regard to permeability and stresses, but investigations by Hubbert and Willis (1957) show that this cannot be done. The possibility that anisotropy occurs must therefore be investigated. In underground strata, the consolidation always occurs in the vertical direction.

The investigation of underground flow must be based upon an investigation of the stresses in and deformation of a porous medium as has been presented in sections 1.7.1 and 1.7.2. Using the notation employed there, one can then proceed to set up hydrodynamic equations of the fluid. The fluid will be connected with the porous medium during the latter's consolidation. What is of ultimate interest is only the motion of the fluid in terms of the parameters (cf. 1.7.2.4–5). As is well known, one needs three sets of conditions to make any flow problem determined: (i) Darcy's law, (ii) the constitutive equation of the fluid, and (iii) the continuity condition. The flow equations should be set up in such a fashion that the various quantities occurring therein can be conveniently measured.

Beginning with *Darcy's law*, we assume that it can be written as follows (cf. 4.3.2.1):

$$q_\alpha = -\frac{k_{\alpha\beta}}{\mu}\frac{\partial p}{\partial \xi_\beta},$$
(4.3.3.9)

where the summation convention is used. The quantity q_α is the filter velocity vector as referred to the filtration surface in the zero strain state, μ the viscosity, and p the pressure of the liquid. The quantity $k_{\alpha\beta}$ is the (variable) permeability tensor; there is some question as to whether or not it would have to be taken inside the gradient. This question also occurs in the discussion of incompressible porous media, and a dissertation of why the form given above may be preferable to other possibilities has been given in section 4.3.1.

The *constitutive equation* of the fluid is simply

$$\rho = \rho(p),$$
(4.3.3.10)

where ρ is the density and p the pressure. This is the same constitutive equation as that encountered in flow through incompressible porous media.

Finally, we turn our attention to the *continuity condition*. Expressing the condition that the mass flow out of a small volume $d^3\xi$ of porous medium (with vector-surface elements dA_α) must be equal to the change of mass inside yields

$$\oint \rho q_\alpha dA_\alpha = -\frac{\partial}{\partial t}\int_v (1 + \kappa)\rho P d^3\xi,$$
(4.3.3.11)

where P is the porosity at the point ξ and $1 + \kappa$ is the volume factor induced by the consolidation of the porous medium. In virtue of Gauss's theorem, the last equation can be written as follows:

$$\int \frac{\partial}{\partial \xi_\alpha} \rho q_\alpha d^3\xi = -\int \frac{\partial}{\partial t} [(1 + \kappa)\rho P]d^3\xi$$
(4.3.3.12)

or

$$-\partial(\rho q_\alpha)/\partial \xi_\alpha = \partial[(1 + \kappa)\rho P]/\partial t$$
(4.3.3.13)

This is the continuity condition.

Combining the three basic equations, one ends up with an equation for p only:

$$\frac{\partial}{\partial t}[(1 + \kappa)\rho P] = \frac{\partial}{\partial \xi_\alpha}\left(\frac{\rho}{\mu} k_{\alpha\beta} \frac{\partial p}{\partial \xi_\beta}\right).$$
(4.3.3.14)

This is the fundamental hydrodynamic equation.

Equation (4.3.3.14) can be further simplified if it can be assumed that the total change of bulk volume in any one specimen is entirely produced by a change in the pore volume, i.e., if it can be assumed that the volume of the solid matrix changes only negligibly. Then one can write (for the definition of $\mathcal{T}$, cf. equation (1.7.1.1)):

bulk volume $(p, \mathcal{T}) = $ bulk volume$_0$ $\{1 + \kappa(p, \mathcal{T})\}$,   (4.3.3.15)

pore volume$|_0 = P_0$ bulk volume$_0$,   (4.3.3.16)

$$P = \frac{\text{pore volume}(\mathcal{T}, p)}{\text{bulk volume}(\mathcal{T}, p)} = \frac{\text{pore volume}(\mathcal{T}, p)}{\text{bulk volume}_0\{1 + \kappa(p, \mathcal{T})\}},$$ (4.3.3.17)

but

pore volume $(\mathcal{T}, p) - $ pore volume (0)

$\quad = $ bulk volume $(\mathcal{T}, p) - $ bulk volume (0)

$\quad = $ bulk volume$_0$ $[(1 + \kappa(p, \mathcal{T})) - 1] = \kappa(p, \mathcal{T})$ bulk volume$_0$;   (4.3.3.18)

furthermore:

pore volume $(\mathcal{T}, p) = $ bulk volume$_0$ $(\kappa(p, \mathcal{T})) + $ pore volume$_0$,   (4.3.3.19)

pore volume $(\mathcal{T}, p) = $ bulk volume$_0$ $(P_0 + \kappa(p, \mathcal{T}))$;   (4.3.3.20)

hence:

$P = $ [bulk volume$_0$ $(P_0 + \kappa(p, \mathcal{T}))$]/[bulk volume$_0$ $(1 + \kappa(p, \mathcal{T}))$],   (4.3.3.21)

or

$P = [P_0 + \kappa(p, \mathcal{T})]/[1 + \kappa(p, \mathcal{T})].$ (4.3.3.22)

Therefore, in the simplified case at present under consideration, the fundamental hydrodynamic equation reduces to

$$\frac{\partial}{\partial t}[\rho(P_0 + \kappa(p, \mathcal{T}))] = \frac{\partial}{\partial \xi_\alpha}\left(\frac{\rho}{\mu} k_{\alpha\beta} \frac{\partial p}{\partial \xi_\alpha}\right).$$ (4.3.3.23)

Let us consider first the case of *locally isotropic stress and permeability*. Then, according to earlier remarks, the stress state is described by two parameters, viz., the fluid pressure p and the overburden pressure $\mathcal{T}$. The various parameters required in the flow equation (4.3.3.14) can be determined as follows:

(a) *Permeability*   Let a unit (at zero strain) volume be enclosed in a pressure cell. The parameters ξ_α then correspond to this state. Then the pressure is raised to $\mathcal{T}, p$ and one measures a certain flow. Let the volume reduction be by the relative amount $-\kappa(\mathcal{T}, p)$, i.e., the length reduction is $-\frac{1}{3}\kappa(\mathcal{T}, p)$ (neglecting squares of κ). What is measured is a superficial permeability k_{sup} according to the formula

$$\frac{Q}{A_0} = \frac{k_{\text{sup}}}{\mu} \frac{p_2 - p_1}{L_0},$$ (4.3.3.24)

where the index 0 indicates that the original (zero strain) area and length are used to calculate k_{sup}. The true permeability of the sample, assuming cubic volume changes, is given by

$$\frac{Q}{A_0(1 + 2\kappa/3)} = \frac{k_{true}}{\mu} \frac{p_2 - p_1}{L_0(1 + \kappa/3)};$$ (4.3.3.25)

hence

$$k_{true} = k_{sup} \frac{1 + \kappa/3}{1 + 2\kappa/3} = \frac{3 + \kappa}{3 + 2\kappa} k_{sup} = \left(1 - \frac{\kappa}{3}\right) k_{sup},$$ (4.3.3.26)

where the last equal sign has been obtained upon neglecting high-order terms.

However, the volume change in the field is not cubic, although it is cubic in the pressure cell. According to earlier remarks, the compaction occurs in the vertical (ζ) direction only. Assuming the true permeability under pressure to be the same in the field as in the pressure cell, we can express the condition for the 'superficial' flow q_{sup}, i.e., the flow in terms of zero strain coordinates of the field. We have

$$q_{sup\ x} = \frac{Q_{x(true)}}{L_0^2} = \frac{1}{L_0^2} \frac{p_2 - p_1}{L_0} \frac{k_{true}}{\mu} L_0^2(1 + \kappa),$$ (4.3.3.27)

and hence

$$k_{\xi\xi} = k_{\eta\eta} = (1 + \kappa)k_{true} = (1 + \kappa) \frac{1 + \kappa/3}{1 + 2\kappa/3} k_{sup}.$$ (4.3.3.28)

Neglecting squares of κ:

$$k_{\xi\xi} = \frac{3 + 4\kappa}{3 + 2\kappa} k_{sup} = \left(1 + \frac{2\kappa}{3}\right) k_{sup}.$$ (4.3.3.29)

Similarly

$$q_{sup\ z} = \frac{Q_{z(true)}}{L_0^2} = \frac{1}{L_0^2} \frac{p_2 - p_1}{L_0(1 + \kappa)} \frac{k_{true}}{\mu} L_0^2,$$ (4.3.3.30)

and hence

$$k_{\zeta\zeta} = \frac{1}{1 + \kappa} k_{true} = \frac{1}{1 + \kappa} \frac{1 + \kappa/3}{1 + 2\kappa/3} k_{sup}.$$ (4.3.3.31)

Neglecting squares of κ:

$$k_{\zeta\zeta} = \frac{1 + \kappa/3}{1 + 5\kappa/3} k_{sup} = \frac{3 + \kappa}{3 + 5\kappa} k_{sup} = \left(1 - \frac{4\kappa}{3}\right) k_{sup}.$$ (4.3.3.32)

The permeability tensor in the system ξ, η, ζ (ζ vertical) has therefore the following form:

$$k_{\alpha\beta} = \begin{pmatrix} 1 + \frac{2}{3}\kappa & 0 & 0 \\ 0 & 1 + \frac{2}{3}\kappa & 0 \\ 0 & 0 & 1 - \frac{4}{3}\kappa \end{pmatrix}.$$ (4.3.3.33)

It is therefore seen that, in order to determine $k_{\alpha\beta}$, one needs one permeability measurement (k_{sup}) and one measurement of the bulk compressibility for each possible pair of variables $p, \mathcal{T}$. During the depletion of the fluid content in underground strata it can be assumed that the overburden pressure stays constant; the measurements of k_{sup} and κ have therefore to be made with a given overburden pressure $\mathcal{T}$ for all values of p.

(b) *Porosity* Another parameter that enters into equation (4.3.3.14) is the combination $(1 + \kappa)\rho P$. It is customary to measure the differential quotient $(1/P_0)(d/dp)[(1 + \kappa)P]$ where the index 0 refers to the zero strain state. This differential quotient has been called the 'effective compressibility' (McLatchie, Hemstock, and Young 1958). It it is measured for one series of values $\mathcal{T}, p$ arriving at the field overburden pressure and fluid pressure from the zero strain state, and then for the given overburden pressure for all pertinent values of the fluid pressure p, and, furthermore, if the rheological equation of the fluid $\rho(p)$ is known, then the required values of $(1 + \kappa)\rho P$ can be obtained by integration. It is then possible to eliminate all unknowns except p from (4.3.3.14) and hence to solve any underground flow problem.

It is interesting to note the modifications that are required in the above scheme if one is to generalize it to take care of *anisotropic stresses and permeability*. The general equation of flow (4.3.3.14) is, of course, still applicable, but different means have to be found for measuring the required quantities.

In order to make the problem amenable to investigation at all, it is necessary to assume that the vertical direction is a principal direction for stress and for permeability. The other principal axes then lie somewhere in the horizontal plane.

In order to measure the quantities required in equation (4.3.3.14), a means must be found to restore in the laboratory the state in which the rock occurred in the field. In general, this requires a separation of the three principal stresses from one another. Thus, a piece of rock being taken from a well will, first of all, expand cubically (because it is taken out of confinement) while it is in transit to the laboratory. Then it must be enclosed in some device in which the total stresses $\mathcal{T}_{zz}$ and $\mathcal{T}_{xx}, \mathcal{T}_{yy}$ as well as the fluid pressure p can be brought up to the values which prevailed in the field. This reproduces the original state of the rock. For the following, it is useful to refer the parameters ξ_α to this state rather than to the zero strain state.

Afterward, during a test run, $\mathcal{T}_{zz}$ must be held constant, the lateral stresses $(\mathcal{T}_{xx}, \mathcal{T}_{yy})$ must be adjusted in such a manner that there is no lateral motion of the rock (which is the boundary condition prevailing in the field), and the vertical (k_z) and horizontal (k_x, k_y) permeabilities must be measured separately for all values of the fluid pressure p. This yields immediately the permeability tensor $k_{\alpha\beta}$,

$$k_{\alpha\beta} = \begin{pmatrix} k_x & 0 & 0 \\ 0 & k_y & 0 \\ 0 & 0 & k_z \end{pmatrix}. \qquad (4.3.3.34)$$

if ξ_α, the lengths, and the areas are referred to the original field state as defined above (rather than the zero strain state). The other quantity that is required in the flow equation is $(1 + \kappa)\rho P$; it can be measured just as in the isotropic case. Since the ξ_α now refer to the field state, the effective compressibility must be measured with this field state as basic state (P_0 referring to this state). It may be noted that is is now no longer necessary to obtain the bulk compressibility and the 'effective' (pore) compressibility separately, because an additional permeability measurement has been substituted for same.

Again, with the various quantities measured, all unknowns except p can be eliminated in the fundamental flow equation (4.3.3.14) and hence it is possible to solve any underground flow problem.

Finally, a simplified description of flow has also been achieved for some types of *plastic porous media*. Barenblatt and Krȳlov (1955) have shown that in plastic substances with hysteresis the equation for the pressure can sometimes be linearized. They showed that one then simply has *two* equations of the type (4.3.3.3), one for each branch of the hysteresis-cycle. Using the notation of section 1.7.1 and introducing the abbreviations

$$\theta = \tfrac{1}{3}\mathscr{T}_{jj}{}^F,$$ (4.3.3.35)

$$\beta_p = (\partial P/\partial p)_\theta,$$ (4.3.3.36)

$$\beta_\theta' = - (\partial P/\partial \theta)_{\text{loading}},$$ (4.3.3.37)

$$\beta_\theta'' = - (\partial P/\partial \theta)_{\text{unloading}}$$ (4.3.3.38)

the Barenblatt-Krȳlov equations are:
   (i) during a *decrease* of the fluid pressure

$$\partial p/\partial t = a_1{}^2 \operatorname{lap} p$$ (4.3.3.39)

with

$$a_1{}^2 = \frac{k(P_0)}{\mu[P_0\beta_f + \beta_p + \beta_\theta']};$$ (4.3.3.40)

   (ii) during an *increase* of the fluid pressure

$$\partial p/\partial t = a_2{}^2 \operatorname{lap} p$$ (4.3.3.41)

with

$$a_2{}^2 = \frac{k(P_0)}{\mu[P_0\beta_f + \beta_p + \beta_\theta'']}.$$ (4.3.3.42)

In these equations the contention of Gersevanov (cf. sec. 1.7.1) has been used with postulates that it is the trace of the fictitious stress tensor $\mathscr{T}_{ij}{}^F$ that causes consolidation (cf. equation 1.7.1.4). The Barenblatt-Krȳlov equations are applicable if the coefficients β_p, β_θ', β_θ'' are reasonably constant.

In concluding this section on compressible porous media, it should be stated once more that the complete theory of consolidation leads to a complicated problem. The basic formulation for the elastic case has been given by Biot (1941), simply by combining the stress-strain equations of section 1.7.2 (i.e., equations 1.7.2.3–4) with Darcy's law. However, for most practical cases, one of the simplified treatments outlined above will prove to be satisfactory.

4.4
THE MEASUREMENT OF PERMEABILITY

4.4.1 *Laboratory methods for isotropic permeability determinations*

Measurements of permeability can be performed using any of the forms of Darcy's law.

Thus, experiments are performed in which, in a certain system, a pressure drop and a flow rate are measured. The solution of Darcy's law corresponding to the geometry of the system and the fluid employed is calculated, and a comparison between the calculated and experimental results immediately yields the only unknown quantity k. Usually horizontal linear systems are used as they are most easy to calculate, but radial arrangements are also often used.

Physically, permeability measurements are very simple, involving chiefly questions of experimental technique. Methods have been discussed in general terms, for example by Kawakami (1933), Plummer et al. (1936), Manegold (1938), Koppuis and Holton (1938), and Eastman and Carlson (1940). A neat and simple apparatus has been described by Pollard and Reichertz (1952); it is shown in figure 18. The drawing is self-explanatory.

TABLE VI

Representative values of permeability for various substances

Substance	Permeability range (permeability in cm^2)	Literature reference
Berl saddles	1.3×10^{-3} to 3.9×10^{-3}	Carman (1938)
Wire crimps	3.8×10^{-5} to 1.0×10^{-4}	Carman (1938)
Black slate powder	4.9×10^{-10} to 1.2×10^{-9}	Carman (1938)
Silica powder	1.3×10^{-10} to 5.1×10^{-10}	Carman (1938)
Sand (loose beds)	2.0×10^{-7} to 1.8×10^{-6}	Carman (1938)
Soils	2.9×10^{-9} to 1.4×10^{-7}	Aronovici and Donnan (1946)
Sandstone ('oil sand')	5.0×10^{-12} to 3.0×10^{-8}	Muskat (1937)
Limestone, dolomite	2.0×10^{-11} to 4.5×10^{-10}	Locke and Bliss (1950)
Brick	4.8×10^{-11} to 2.2×10^{-9}	Stull and Johnson (1940)
Bituminous concrete	1.0×10^{-9} to 2.3×10^{-7}	McLaughlin and Goetz (1955)
Leather	9.5×10^{-10} to 1.2×10^{-9}	Mitton (1945)
Cork board	3.3×10^{-6} to 1.5×10^{-5}	Brown and Bolt (1942)
Hair felt	8.3×10^{-6} to 1.2×10^{-5}	Brown and Bolt (1942)
Fibreglass	2.4×10^{-7} to 5.1×10^{-7}	Wiggins et al. (1939)
Cigarette	1.1×10^{-5}	Brown and Bolt (1942)
Agar-agar	2.0×10^{-10} to 4.4×10^{-9}	Pallmann and Deuel (1945)

Special precautions have to be taken if the permeability is to be measured in highly compactible substances such as soils and clays, as the pore pressure may affect their consistency (cf. Wilkinson and Shipley 1969).

Permeabilities in various substances may have a wide range. In table VI, we give a compilation of some representative values. As was the case in table I referring to porosities, 'range' does not signify that the indicated values are the extreme limits of permeability which may be found in the substances; rather it signifies the range in which one is likely to find the permeability. As in table I, literature references are given indicating the provenance of the values shown.

4.4.2 *Anisotropic permeability measurements*

Special problems are encountered if it is suspected that the permeability in a porous medium is not isotropic. Then the permeability of a specimen has to be measured in various directions. Possible methods for measuring anisotropic permeability have already been touched upon in section 4.3.2, where it was shown that there are, in principle, two ways to do this. However, in practice, the results obtained by the two methods differ only slightly so that it is possible to use either of them indiscriminately.

Most equipment for measuring directional permeability described in the

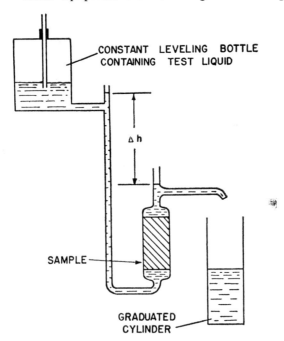

FIGURE 18 A permeability apparatus (after Pollard and Reichertz 1952).

literature uses a block of porous material, usually a cube, where the inlet and outlet for the percolating fluid are restricted to a small part of the surface (e.g., opposite faces on a cube). Equipment of this type has been described, for instance, by Maasland and Kirkham (1955) in connection with the analysis of soils, and by Hutta and Griffiths (1955) in connection with the analysis of oil well cores. If only the horizontal and vertical permeabilities of a cylindrical well core are desired, it is sufficient first to measure the vertical permeability across the faces and then to drill a hole along the axis and to use a radial flow system for the horizontal permeability determination.

A very interesting review of the anisotropy of permeability in underground strata has been given by Rühl and Schmid (1957). Accordingly, it turns out that in underground rock strata the horizontal permeability is, in the vast majority of cases, much greater than the vertical permeability. Correlations between the anisotropy of permeability with other properties of sedimentary rocks can be set up and may be useful in technological applications.

4.4.3 In situ permeability measurements

In many applications of hydrodynamics in porous media to the flow of fluids through underground strata, it is difficult to obtain a representative sample of the medium in question. It is therefore desirable to measure the permeability in situ. In principle, permeability measurements are performed in situ in the same fashion as in the laboratory: a flow system is created between an inlet and an outlet and, from the flow resistance observed, the value of the permeability is inferred. Any controlled tests that are carried out with the underground flow system under consideration are sufficient to determine the permeability. However, flow systems in the field are usually of necessity much more complicated than those chosen in a laboratory. The main task is, therefore, to integrate the differential flow equations for the flow systems under consideration. Methods for doing this will be discussed in chapter 5.

On this basis, many in situ methods of permeability measurement have been discussed in the literature. In connection with the surface properties of soil (air permeability), such methods have been described by Evans and Kirkham (1950) and Terletskaya (1954); in connection with groundwater studies, in situ tests have been proposed by Brown (1953), Kirkham (1955), Childs et al. (1957), Schoeller (1955, 1956), Shestakov (1955), Onodera (1958), and Jaeger (1959). It should be noted that, in the case of an artesian aquifer, the formation must be considered as elastic. Furthermore, the terminology in groundwater studies is somewhat different from that used above; since it can be developed only when the necessary solutions of Darcy's law have been deduced, we relegate the discussion of groundwater flow studies to section 5.5.3. Finally, the in situ determination of permeability has been of great importance in connection with the exploitation of oil fields. Accordingly, a prolific literature exists on the subject, the most important contributions probably being those of Miller, Dyes, and Hutchinson (1950), Horner (1951), and Hazelbrook, Rainbow, and Matthews (1958).

4.4.4 Averaging of permeability measurements

The measurements of permeability discussed above presuppose that the medium is homogeneous, if not isotropic. However, this is only rarely true. In most instances, a natural porous medium is very inhomogeneous, at least on a small, microscopic scale, although it may be possible to consider it as (statistically) homogeneous on a large, macroscopic scale.

What is wanted in practice is generally the 'average' or 'effective' permeability of the medium on a large scale. However, there are many instances in which measurements can only be performed on small-scale samples (such as oil well cores) of the large-scale system, so that widely differing 'local' permeabilities result. The question then arises of how to determine the 'effective' or 'average' permeability of a large-scale system from a series of 'local' permeabilities obtained by measurements on small samples.

It is immediately obvious that there are two extreme cases for the averaging of permeabilities. The *first extreme case* is obtained by considering a (horizontal) one-dimensional system of length L representing a porous tube sustaining a percolation flow of an incompressible fluid q per unit area. Over a piece of length L_1 we assume the permeability k_1, over a piece of length L_2 a permeability k_2, and so on. The pressure at the beginning of L_1 is p_0, at the junction of L_1 and L_2 it is p_1, and so on, Darcy's law then states that

$$q = \frac{k_1}{\mu} \frac{p_1 - p_0}{L_1},$$

$$q = \frac{k_2}{\mu} \frac{p_2 - p_1}{L_2}, \tag{4.4.4.1}$$

$$\cdots$$

$$q = \frac{k_n}{\mu} \frac{p_n - p_{n-1}}{L_n},$$

where it is to be noted that the value of q must be constant over the whole tube. Hence

$$\frac{L_1 \mu q}{k_1} = p_1 - p_0,$$

$$\frac{L_2 \mu q}{k_2} = p_2 - p_1, \tag{4.4.4.2}$$

$$\frac{L_n \mu q}{k_n} = p_n - p_{n-1}.$$

Adding all the equations yields

$$\mu q \left(\frac{L_1}{k_1} + \frac{L_2}{k_2} + \ldots + \frac{L_n}{k_n} \right) = p_n - p_0. \qquad (4.4.4.3)$$

This can be brought into a macroscopic form of Darcy's law using an 'effective' permeability k_{eff}:

$$q = \frac{k_{\text{eff}}}{\mu} \frac{p_n - p_0}{L} \qquad (4.4.4.4)$$

by setting

$$\frac{1}{k_{\text{eff}}} = \frac{L_1}{L} \frac{1}{k_1} + \frac{L_2}{L} \frac{1}{k_2} + \ldots + \frac{L_n}{L} \frac{1}{k_n}. \qquad (4.4.4.5)$$

Extending the argument, one can interpret L_i/L as the relative frequency f_i (as a fraction of 1) at which the permeability k_i was found in small-sample measurements, and state that

$$1/k_{\text{eff}} = \sum f_i/k_i. \qquad (4.4.4.6)$$

This composition law is the law of *harmonic mean* formation.

The *second extreme case* is obtained by considering a group of (horizontal) porous tubes in parallel, all of equal length L, but of various individual permeabilities k_i and cross-sectional areas A_i. The flow Q_i through the ith tube is then

$$Q_i = A_i q_i = A_i \frac{k_i}{\mu} \frac{p_{\text{end}} - p_{\text{begin}}}{L} \qquad (4.4.4.7)$$

and the flow Q through all the tubes is

$$Q = \sum Q_i = \sum A_i q_i = (\sum A_i k_i) \frac{p_{\text{end}} - p_{\text{begin}}}{\mu L}, \qquad (4.4.4.8)$$

which can be written formally as a Darcy type law by dividing by $A = \sum A_i$:

$$q = \frac{Q}{A} = \frac{1}{\mu} \left(\sum \frac{A_i}{A} k_i \right) \frac{p_{\text{end}} - p_{\text{begin}}}{L}. \qquad (4.4.4.9)$$

It is seen that the effective permeability of the system is

$$k_{\text{eff}} = \sum \frac{A_i}{A} k_i. \qquad (4.4.4.10)$$

Again extending the argument, one can interpret A_i/A as the relative frequency f_i (as a fraction of 1) at which the permeability k_i was found in small samples and state that

$$k_{\text{eff}} = \sum f_i k_i. \qquad (4.4.4.11)$$

This composition law is the law of *arithmetic mean* formation.

my case

The two extreme composition laws correspond to different spatial distributions of local sample-permeabilities. In practice, neither of these extreme cases will presumably prevail, and hence the effective permeability to be deduced from a series of local permeability measurements lies somewhere between their arithmetic and harmonic means. Empirically, the practice has evolved of employing the *geometric* mean for the averaging of permeability measurements:

$$\log k_{\text{eff}} = \sum f_i \log k_i. \tag{4.4.4.12}$$

The above argument is entirely heuristic. A careful analysis of the mathematical implications was undertaken by Matheron (1966, 1967, 1968), who showed that the effective permeability for a (stationary) flow system lies indeed somewhere between the arithmetic and harmonic means of the local permeabilities. The first case corresponds to the existence of a constant pressure gradient, and the second to the existence of a constant local flux everywhere in the porous medium. The commonly employed geometric mean gives correct results only in very rare cases. More specific statements cannot be made, since the statistics of the spatial distribution of the local permeabilities is evidently of paramount importance. From a mere 'list' of individual 'local' permeability measurements, the statistics of the spatial distribution can evidently not be inferred.

4.5
FILTRATION THEORY

4.5.1 *Build-up of a filter cake*

An important application of Darcy's law is to filtration. The process of filtration is conducted in such a manner that a slurry is filtered through a filter cake, which increases in size with the amount of material deposited. The rate of deposition of material is proportional to the throughput of filtrate.

Thus, for filtration at constant pressure, one has at time t a filter cake of thickness h. Darcy's law therefore yields for the throughput q (scalarly):

$$q = \frac{k}{\mu} \frac{p_2 - p_1}{h}.$$

However dh/dt is proportional to q, so that one has

$$\frac{dh}{dt} = cq = -\frac{k}{\mu} \frac{p_2 - p_1}{q^2} \frac{dq}{dt}. \tag{4.5.1.1}$$

This is a differential equation whose solution for q is

$$q = \sqrt{\frac{k(p_2 - p_1)}{2\mu(ct + c')}}. \tag{4.5.1.2}$$

The total volume V of filtrate filtered through a unit area from the beginning of the process of filtration is therefore

$$V(t) = \int_{t_0}^{t} q\,dt = (1/c)\sqrt{2(k/\mu)(p_2 - p_1)(ct + c')} + V_0. \tag{4.5.1.3}$$

This can be rewritten as follows, by including everything that is constant in the new constants G, K, and t_0:

$$(V + G)^2 = K(t + t_0). \tag{4.5.1.4}$$

This is the form in which the fundamental filtration equation was given by Ruth, Montillon, and Montonna (1933). These authors also checked their equation by a great number of experimental data.

The process of filtration has been investigated since by many authors (cf., e.g., Dickey and Bryden 1946); but the refinements achieved belong in a treatise on filtration rather than in one on flow through porous media. Some of these investigations, if they are concerned with the physics of the flow through porous media, will be discussed later in their proper context.

4.5.2 Colmatage

A generalization of the ordinary filtration theory can be obtained by considering the phenomenon of colmatage, which refers to the trapping of particles contained in a stream of fluid during its passage through a porous medium (Shekhtman 1961; Bodziony and Litwiniszyn 1962).

A theory of this process can be built upon the assumption that the concentration $K(t)$ of the pores of the porous medium blocked at time t by solid particles changes at a rate $\partial K(t)/\partial t$ which is proportional to the concentration of still unblocked pores. If A is the number of pores per unit volume that can be blocked at the initiation of the process ($t = 0$), the number of still unblocked pores at time t will evidently be $A - K(t)$. Thus, for a linear flow process (linear coordinate x),

$$\partial K(x, t)/\partial t = a\,C(x, t)\,[A - K(x, t)], \tag{4.5.2.1}$$

where a is a constant of proportionality (the colmatage coefficient) and $C(x, t)$ is the concentration (number per unit *bulk* volume of fluid plus porous medium) of particles in the fluid stream. The solution for constant C is

$$K(t) = A[1 - \exp(-aCt)].$$

However, the concentration C of particles in the liquid also changes during its passage through the porous medium. The mass balance equation immediately yields (see Trzaska 1966)

$$\frac{\partial C(x, t)}{\partial t} + \frac{\partial vC}{\partial x} = -\frac{\partial K(x, t)}{\partial t}, \tag{4.5.2.2}$$

where $v(x)$ is the pore velocity of the fluid.

The system of equations (4.5.2 1) and (4.5.2.2) describes the colmatage process. In applications it turns out that the coefficient a is not a constant, but a function of the pore velocity v. Solutions of this system of equations are available for various cases (cf. Trzaska 1966).

5
Solutions of Darcy's law

5.1
GENERAL REMARKS

Solutions of hydrodynamic problems concerning the flow of homogeneous fluids in porous media are obtained by solving equation (4.3.1.8) for particular boundary conditions. The differential equation in question is very closely related to the equations of diffusion and to that of heat conduction. Thus, solutions obtained with the intention of solving the heat-conduction equation can often be taken over directly for hydromechanics in porous media.

The treatment of the differential equation resulting from Darcy's law is, actually, a discipline of mathematics and has very little to do with the physics of the problem. General methods that are applicable may be found in appropriate mathematical texts such as that by Courant and Hilbert (1943). Much mathematical material which is directly applicable to the flow in porous media has also been accumulated by Carslaw and Jaeger (1959) in connection with their study of heat conduction. In the present chapter, we shall not give a comprehensive review of the methods that are applicable to the solution of the differential equation; that belongs in a study on the theory of functions and mathematical analysis. Rather, we shall give a review of physical conditions for which solutions have been achieved, with representative examples. Thus, solutions which have been obtained for problems other than the flow of fluids through porous media (such as heat-conduction problems) will not be found here although they might have direct analogues bearing upon the present study.

The physical conditions of flow for which solutions might be sought are (i) steady state flow, (ii) unsteady state flow, and (iii) gravity flow with a free surface. Of these, steady state flow solutions for incompressible fluids are most easily obtained; they are simply represented by solutions of the Laplace equation. Except for a few other special cases, however, Darcy's law leads to non-linear differential equations. The analytical methods of dealing with these are quite involved and lengthy so that efforts have been directed towards scaling phenomena and towards the experimental representation by analogous effects.

A comprehensive review of solutions of Darcy's law may be found in the book by Polubarinova-Kochina (1952), which has already been mentioned in the introduction. Some solutions have also been accumulated in the book by Muskat (1937) mentioned earlier. Furthermore, Matta (1957) analysed the general use of relaxation methods in connection with the solution of problems of flow through porous media, and Oroveanu (1963) studied flow through inhomogeneous porous media.

5.2
STEADY STATE FLOW

5.2.1 *Analytical solutions*

The steady state is characterized by the vanishing of the partial time derivatives of physical quantities such as the density, velocity, etc. Equation (4.3.1.8) thus reduces to:

$$\text{div}[(\rho k/\mu)\,(\text{grad}\,p - \rho g)] = 0 \qquad (5.2.1.1)$$

(note that g is a vector pointing *down*ward, but coordinates increase *up*ward). If furthermore, the fluid is incompressible and the porous medium homogeneous, one has

$$\text{lap}\,p = 0, \qquad (5.2.1.2)$$

where 'lap' denotes the Laplace operator. This is the well-known differential equation of Laplace applicable in many instances in physics, and the general methods used in solving it lead to valid solutions in the present case. Thus, if two-dimensional problems are considered, methods based upon the theory of functions of complex variables are applicable as well as the technique of using Green's function. A review of such methods is given in the books of Muskat (1937) and Polubarinova-Kochina (1952). Oroveanu (1963) discussed the general methods applicable to the treatment of systems with non-homogeneous permeability.

A great number of papers deal with the search for steady state solutions of Darcy's law relevant to particular applications. One of the most important applications is the study of the flow from strata into *wells*. Muskat (1937) and Polubarinova-Kochina (1952) discuss a series of such solutions in their books. Furthermore, Muskat (1943) studied the effect of casing perforations on well productivity, and Dodson and Cardwell (1945) investigated wells completed with slotted liners. Baker (1955) and Birks (1955) studied flow in fissured formations. Similarly, Luthin and Scott (1952) gave a numerical analysis of flow through aquifers towards wells in difficult conditions such as variable permeability. Van der Ploeg et al. (1971) investigated the steady flow into a well from an *elliptical* aquifer.

Related to the study of steady flow into wells is the problem of drainage of water-saturated strata by drainage tubes, etc.; Kirkham (1940, 1945, 1949, 1950a, b,

1954, 1958) has written a number of papers bearing upon that subject. Similar problems have also been investigated by Day and Luthin (1954).

Much work has also been done on the study of seepage underneath engineering structures, especially dams (with and without sheet-piling). Again, Muskat (1937) and Polubarinova-Kochina (1952) gave reviews of a wide variety of solutions that have been obtained.

Finally, the general methods for obtaining solutions referring to inhomogeneous porous media due to Oroveanu (see above) were applied by that author (Oroveanu 1961a, b, 1962) to fissured media. The steady flow through cavities was studied by Scheidegger (1953) and Gheorghitza (1969).

The details of these investigations are mainly concerned with particular engineering requirements and are of only limited interest in relation to fundamental physical concepts. The reader is therefore referred to the original papers and to the monograph of Polubarinova-Kochina (1952) for further study. As illustrative examples only a few solutions are presented here in some detail.

The first is the solution for two-dimensional radial flow of an incompressible fluid into a well which is completely penetrating the fluid-bearing medium. Assuming that the well is a cylinder of radius R_0, with pressure at the surface of the cylinder equal to p_0, and that the pressure at distance R_1 from the well is p_1, it is easy to verify that the required solution is

$$Q = \frac{2\pi k}{\mu \ln(R_1/R_0)} (p_1 - p_0), \tag{5.2.1.3}$$

where Q is the total discharge per unit time and unit penetration length of the well, k is the permeability of the medium, and μ is the viscosity of the fluid.

If the fluid in motion is compressible, then equation (5.2.1.1) does not reduce to the Laplace equation. However, it has been pointed out by Leĭbenzon (1947, p. 132) that for gases it does reduce to the Laplace equation if the following substitution is made:

$$\chi = \int_{p_0}^{p} \rho dp, \tag{5.2.1.4}$$

which leads to

$$\text{lap}\, \chi - \text{div}\,(\rho^2 \mathbf{g}) = 0.$$

For gases, the second term is much smaller than the first and can therefore be neglected. This also holds with horizontal flow of compressible liquids. One thus obtains

$$\text{lap}\, \chi = 0. \tag{5.2.1.5}$$

The study of the steady flow of compressible fluids in most instances is thus reduced to a discussion of the same differential equation as that encountered in the study of the steady flow of incompressible fluids. One has, for instance, in the linear case (coordinate x) for an ideal gas ($cp = \rho$)

$$\chi = \tfrac{1}{2}cp^2 = c' + bx.$$

With the boundary conditions $p(x = 0) = p_0$ and $p(x = L) = p_L$ one has

$$p^2 = p_0^2 + x(p_L^2 - p_0^2)/L.$$

If Darcy's law is now expressed at $x = 0$, one obtains

$$q_0 = -\,(k/\mu)\,(dp/dx)_0 = -\,(k/\mu)\,(p_L^2 - p_0^2)/(2p_0 L). \qquad (5.2.1.6)$$

This equation is commonly used in a permeability determination if a gas is used as percolating fluid. A similar formula can be deduced for a radial arrangement.

Finally, following Day and Luthin (1954), we consider the ideal case of a level, uniform porous medium of depth h underlain by a completely permeable substratum (see fig. 19). In practice the porous medium may be a sandy soil and the substratum a loosely packed gravel bed. Numerous long furrows extend parallel to one another across the surface of the medium and are separated by a distance a between centres. They are semicircular in cross-section and have a radius equal to r. The furrows are filled with fluid and are kept full by continuous additions (problem of irrigation of a field). It will be permissible to assume that a steady state is reached after fluid has flowed continuously into the substratum for a considerable time.

The pattern of flow, which is two-dimensional, can be deduced straight-forwardly. In a medium of homogeneous permeability, the formulation using the velocity potential may be used and one has

$$\psi = (k/\mu)\,(p + \rho gz), \qquad (5.2.1.7)$$

$$\mathbf{q} = -\operatorname{grad}\psi, \qquad (5.2.1.8)$$

$$\operatorname{lap}\psi = 0. \qquad (5.2.1.9)$$

The boundary conditions can be satisfied using the method of images (Kirkham 1949), and one obtains

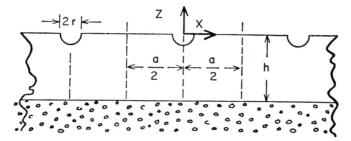

FIGURE 19   Geometrical arrangement corresponding to Day and Luthin's (1954) solution of a steady-state seepage problem (after Day and Luthin 1954).

$$\psi = A \sum_{m=-\infty}^{m=+\infty} \ln \left(\frac{\cosh[\pi(x-ma)/(2h)] - \cos[\pi z/(2h)]}{\cosh[\pi(x-ma)/(2h)] + \cos[\pi z/(2h)]} \right) + \psi_0, \tag{5.2.1.10}$$

where ψ_0 is the velocity potential at the top of the substratum, m is an integer having successive values running from $-\infty$ to $+\infty$, and A is a constant.

As is well known from the theory of the Laplace equation, one can introduce a stream function φ, connected with ψ by the Cauchy-Riemann differential equations

$$\partial\psi/\partial z = \partial\varphi/\partial x, \qquad \partial\psi/\partial x = -\partial\varphi/\partial z, \tag{5.2.1.11}$$

so that the lines $\varphi = $ constant represent streamlines. It is easy to verify that the expression for φ, corresponding to (5.2.1.11), is

$$\varphi = 2A \sum_{m=-\infty}^{m=+\infty} \tan^{-1} \left(\frac{\sinh[\pi(x-ma)/(2h)]}{\sin[\pi z/(2h)]} \right) + \varphi_0. \tag{5.2.1.12}$$

Thus, the flownet can be plotted. A solution conforming to some particular constants is shown in figure 20.

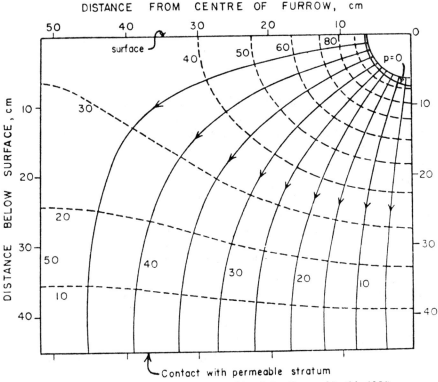

FIGURE 20   Streamlines in a particular seepage problem (after Day and Luthin 1954).

5.2.2 *Solutions by analogies*

The Laplace equation (5.2.1.2) occurs in many contexts in physics. Therefore solutions of that equation can be obtained by performing suitable experiments which are themselves governed by the Laplace equation. In this manner, it is often possible to set up an analogue to a certain problem of steady flow through porous media and thus to avoid the tedious job of solving the Laplace equation analytically. We shall discuss some of the possible analogies (Scheidegger 1953).

(*a*) *The flow of electricity* The steady flow of electricity in a conducting medium is governed by Laplace's equation

$$\text{lap } \Phi = 0, \tag{5.2.2.1}$$

$$\mathbf{i} = - \text{ grad } \Phi, \tag{5.2.2.2}$$

where $E = \Phi/\rho$ is the electric potential, ρ the conductivity, and $\mathbf{i}$ the current density. Furthermore, on the boundary of two media one of which is very much more conducting than the other, the potential will be constant. This gives a means of duplicating various boundary conditions of flow through porous media. Thus, in a model experiment one would use a trough of some electrolyte of conductivity ρ and insert electrodes, corresponding to the boundary conditions. In such an experiment ρ, E, $\mathbf{i}$ will be proportional to k/μ, p, q, respectively, if care is taken that corresponding units are used. In an actual experiment, one would also have to contend with polarization effects at the electrodes; it would therefore be advisable to use alternating current of moderately high frequency. Experiments of this type for the study of flow through porous media have been successfully completed by, for example, Ram, Vaidhianathan, and Taylor (1935), Lee (1948), Dolcetta (1948), Fil'chakov and Panchishin (1953), Crawford and Landrum (1955), Opsal (1955), and Kovacs (1956).

(*b*) *The flow of heat* The flow of heat in a heat-conducting medium is also governed by Laplace's equation; unfortunately it is not generally feasible to use this fact for constructing experimental analogies of the flow through porous matter, as it is quite difficult to measure heat flow accurately. However, many analytical solutions of Laplace's equation have been developed in connection with the study of heat flow (Carslaw and Jaeger 1959), and these can be at once applied to porous flow problems.

(*c*) *The distribution of stresses* It has been suggested that it might be possible to duplicate the flow lines by stress lines in stressed materials and to examine the latter by means of experimental stress analysis.

Indeed, a plane stress state could be used to represent two-dimensional flow. The trace of the stress tensor $\mathscr{T}$ fulfils Laplace's equation in two dimensions:

$$\text{lap Trace } \mathscr{T} = \text{lap } (S_1 + S_2) = 0. \tag{5.2.2.3}$$

Here S_1 and S_2 are the stress components in any two orthogonal directions in the point under consideration; S_3 is zero because of the assumption of a plane stress state.

Unfortunately, it is in general quite difficult to realize experimentally boundary conditions which would correspond to a flow-pattern. The proposition can be better realized if one restricts oneself to two dimensions; thus a soap film may be used to represent two-dimensional flow. The differential equation for a stretched membrane is of the form

$$\text{lap } z(x, y) = 0, \tag{5.2.2.4}$$

where z is the lateral ordinate as a function of x and y, two arbitrarily chosen Cartesian coordinates normal to z. Then the curves of equal height would correspond to equipotential lines in the flow. Such experiments have actually been conducted as part of the study of heat flow (Wilson and Miles 1950). In connection with flow through porous media, the analogy has been discussed by Hansen (1952).

(*d*) *The flow of viscous fluids*   Under certain circumstances, the Navier-Stokes equation (2.2.2.1), too, reduces to a Laplace equation. It is therefore possible to use the phenomenon of viscous flow for a representation of steady flow in porous media. An arrangement which has attained considerable popularity is the method of modelling two-dimensional flow in porous media by the flow of a viscous liquid contained between two parallel plates that are only a small distance apart. This type of analogue is usually called a *Hele-Shaw model* after Hele-Shaw (1897, 1898, 1899) who introduced it. This method has been applied by Günther (1940), Barron (1948), Todd (1955), Santing (1958), and many others.

(*e*) *Mechanical scaling*   It is possible, finally, to represent a large-scale flow phenomenon in porous media by a small-scale one. All that is necessary is to make the geometrical dimensions to scale. The pressures may be scaled in any desirable way since the Laplacian of ap is zero if that of p vanishes. A series of experiments using such scaling has been discussed in the book of Muskat (1937). The method has also been applied by Plummer and Woodward (1936), Kirkham (1939), Mosonyi and Kovacs (1956), and Nemeth (1956).

5.3
UNSTEADY STATE FLOW

5.3.1 *General remarks*

If one is to consider the general patterns of flow in porous media, one has to insert ρ as a function of p in the differential equation (4.3.1.8). In general, the equation then becomes non-linear.

Methods for obtaining solutions are often based on simplifications of the exact non-linear equation which lead to linear equations of the diffusivity type. We shall discuss first the theory of this linearization (sec. 5.3.2) and then review several typical solutions of the resulting equations (sec. 5.3.3). However, a linearization is not always possible, and in many cases, particularly if a gas is involved, the general non-linear equation has to be solved. This poses considerable analytical difficulties because, owing to the possibility that discontinuities (shock fronts) may evolve, even computers often do not lead to success. We shall discuss some cases in section 5.3.4. Finally, we discuss some 'experimental' solutions of the unsteady-state flow equation based on analogues.

5.3.2 *Theory of linear approximation of unsteady state flow*

It is possible to linearize the differential equation of unsteady state flow in certain approximations of two cases: (i) if the fluid is a liquid of constant compressibility and (ii) if the porous medium is elastic. These cases lead to identical differential equations, as pointed out by Shchelkachev (1946a, 1946b, 1946c), so that assuming a certain compressibility of the porous medium has the same effect as assuming a corresponding compressibility of the fluid.

The differential equation (4.3.1.8) becomes in the linear approximation, for a liquid with

$$\rho = \rho_0 \exp[\beta_f(p - p_0)], \tag{5.3.2.1}$$

according to (2.2.1.1), and upon neglect of the gravity term:

$$P\beta_f \, \partial\rho/\partial t = (k/\mu) \, \text{lap} \, \rho. \tag{5.3.2.2}$$

It should be noted, however, that the equation does *not* become linear if the gravity term is not neglected.

Similarly, it has been shown that assuming an elastic porous medium alters the continuity equation to read (4.3.3.3). If, therefore, this is inserted into Darcy's law and the same approximations made as above, one obtains

$$(P\beta_f + \beta_m) \, \partial\rho/\partial t = (k/\mu) \, \text{lap} \, \rho. \tag{5.3.2.3}$$

This can be shown as follows. Darcy's law simply states that

$$\mathbf{q} = - (k/\mu) \, \text{grad} \, p.$$

Inserting this into Shchelkachev's continuity equation (4.3.3.3) yields

$$\text{div}\left(\rho \frac{k}{\mu} \, \text{grad} \, p\right) = \left(P + \frac{\beta_m}{\beta_f}\right) \frac{\partial\rho}{\partial t}. \tag{5.3.2.4}$$

But we have, using equation (4.3.3.1),

$$\text{grad} \, p = \frac{dp}{d\rho} \, \text{grad} \, \rho = \frac{1}{\rho\beta_f} \, \text{grad} \, \rho.$$

Thus, we obtain from (5.3.2.4)

$$\text{div} \left(\rho \, \frac{k}{\mu} \, \frac{1}{\rho \beta_f} \, \text{grad} \, \rho \right) = \left(P + \frac{\beta_m}{\beta_f} \right) \frac{\partial \rho}{\partial t},$$

which, after multiplying through by β_f, yields (5.3.2.3), as claimed.

It is seen that (5.3.2.2) and (5.3.2.3) are identical if the terms $P\beta_f$ and $P\beta_f + \beta_m$ are identical. Thus, assuming an unrealistic 'super'-compressibility of the fluid may well account for a neglected compressibility of the porous medium.

It is often possible to neglect powers of β_f higher than the first in a Taylor expansion of (5.3.2.1):

$$\rho \simeq \rho_0 + \beta_f \rho_0 (p - p_0).$$

Using this approximation, it becomes obvious that equation (5.3.2.3) can be written as follows:

$$(P\beta_f + \beta_m) \, \partial p / \partial t = (k/\mu) \, \text{lap} \, p. \tag{5.3.2.5}$$

This shows that, under the assumptions outlined above, the same linear differential equation applies to the pressure as to the density.

As mentioned above, if gravity terms are kept in Darcy's law, the differential equations do not become linear in ρ. If the equations are to be linearized, some other approximations have to be made. Thus, it has been observed by Werner (1946a, b) that for small compressibilities the continuity equation for the fluid can, in analogy with (5.3.2.4), be written approximately as

$$\text{div} \, \mathbf{q} = - \bar{\rho}(P\beta_f + \beta_m) \, \partial \phi / \partial t, \tag{5.3.2.6}$$

with $\bar{\rho}$ denoting the average fluid density in the medium and ϕ the usual force potential (see 4.3.1.4a). Combined with Darcy's law (4.3.1.4b), one thus obtains (neglecting second-order terms in β):

$$\text{lap} \, \phi = (\mu/k) \, (P\beta_f + \beta_m) \, \partial \phi / \partial t, \tag{5.3.2.7}$$

which is linear.

It has been found that it is also possible under certain circumstances to linearize the equations for the unsteady flow of an ideal gas in a porous medium. The non-linear equation is (from (4.3.1.8), setting $\rho = \text{const.} \, p$), if gravity is neglected:

$$\text{lap} \, p^2 = \frac{2P\mu}{k} \frac{\partial p}{\partial t} . \tag{5.3.2.8}$$

This equation can be linearized under certain conditions to read

$$\text{lap} \, p^2 = \text{const.} \, \partial p^2 / \partial t \tag{5.3.2.9}$$

without committing a very great error. This has been demonstrated by Green and

Wilts (1952) for a linear semi-infinite porous medium which is originally filled with a gas at constant initial pressure p_0, the pressure at the finite end being lowered to p_1 at t_0 and maintained there. The analogous case of a radial arrangement has been discussed by Barenblatt (1956b). In both cases, the constant in (5.3.2.9) must be chosen judiciously.

Thus, in approximation, the same equation holds for the density ρ, the pressure p, and the force potential ϕ. It is always the heat-conductivity equation.

5.3.3 *Various specific solutions of the linearized equations*

Equations (5.3.2.2), (5.3.2.3), (5.3.2.7), and (5.3.2.9) are of the form of the heat-conductivity equation. Thus, the general methods applicable for solving problems of heat conduction are also applicable to transient flow in porous media. We shall not discuss here the analytical methods that may be used to treat (5.3.2.1), etc.; such methods may be found in treatises on mathematical methods (e.g., Courant and Hilbert 1943) and in Carslaw and Jaeger's (1959) book on heat conduction. It may just be mentioned, perhaps, that the Laplace-transformation has been found to be a useful tool for tackling such problems (Van Everdingen and Hurst 1949). This method was further generalized and it was shown that the type of boundary value problems that occur in production practice in the oil industry can be reduced to integral equations (Marsal 1970). Numerical methods for the solution of the heat-conductivity equation (and parabolic differential equations in general), based on an alternating-direction technique, were discussed by Douglas (1961).

Owing to limitations of space, it is impossible to give a detailed outline of all the methods that are applicable to the solution of the linearized non-steady-state equation. However, to demonstrate at least some representative cases, two solutions are given here in somewhat more detail. We first take the case of the linearized equation for a compressible porous medium in accordance with the theory of Shchelkachev (1946b). Considering the simplest, i.e., linear arrangement, we can write the fundamental equation as follows (5.3.2.5):

$$a^2\, \partial^2 p/\partial x^2 = \partial p/\partial t \tag{5.3.3.1}$$

with

$$a^2 = k/[\mu(P\beta_f + \beta_m)]. \tag{5.3.3.2}$$

Particular boundary conditions can be stated as follows:

$p = p_0$ for $x = 0$ at all times,

$p = p_1$ for $x = L$ at all times, $\qquad\qquad$ (5.3.3.3)

$p = p_0$ everywhere for $t = 0$.

The solution of this problem can be shown to be

$$p = p_0 - (p_0 - p_1)\frac{x}{L} + \frac{2}{\pi}(p_0 - p_1)\sum_{n=1}^{\infty}\frac{(-1)^{n-1}}{n}\exp\left(-\frac{n^2\pi^2 a^2}{L^2}t\right)\sin\left(\frac{n\pi}{L}x\right),$$

$$(5.3.3.4)$$

as one may verify easily by differentiation.

In order to give a series of 'standard' solutions, it is convenient to introduce the following dimensionless parameters:

$$\frac{x}{L} = \varepsilon, \qquad 2\frac{a^2}{L^2}t = F, \qquad \frac{p_0 - p}{\varepsilon(p_0 - p_1)} = W. \qquad (5.3.3.5)$$

In this notation, the solution of our problem becomes

$$W = 1 - \frac{2}{\varepsilon\pi}\sum_{n=1}^{\infty}\frac{(-1)^{n-1}}{n}\exp\left(-\frac{n^2\pi^2 F}{2}\right)\sin(n\pi\varepsilon). \qquad (5.3.3.6)$$

The solutions corresponding to this expression can then be plotted as a series of curves as shown in figure 21.

Of importance also are solutions of flow into wells. Thus we give second a classic solution which was obtained by Theis (1935), leading to an exponential integral function. This corresponds to a classic solution of the heat-flow equation.

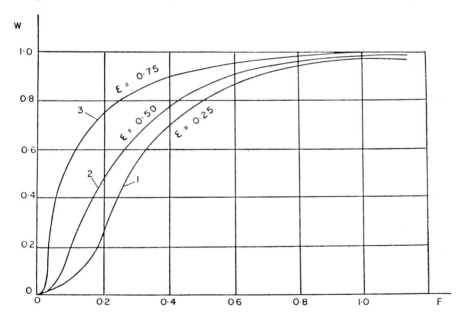

FIGURE 21   A series of solutions for linear non-steady-state flow in porous media (after Shchelkachev 1946b).

The fundamental equation (5.3.3.1) reads, in plane cylindrical coordinates,

$$\frac{\partial^2 p}{\partial t^2} + \frac{1}{r}\frac{\partial p}{\partial r} = \frac{1}{a^2}\frac{\partial p}{\partial t}. \tag{5.3.3.7}$$

For the boundary conditions

$$p = p_0 \quad \text{as } r \to \infty,$$

$$2\pi \lim(r\,\partial p/\partial r) = (\mu/k)\,Q \quad \text{as } r \to 0, \tag{5.3.3.8}$$

so that Q is the volume discharge rate of the well, and the initial condition

$$p(r, 0) = p_0 \quad \text{for } t \le 0, \tag{5.3.3.9}$$

the solution is ('Theis equation')

$$p = p_0 - \frac{\mu}{k}\frac{Q}{4\pi}\int_{r^2/(4a^2 t)}^{\infty}\frac{e^{-x}}{x}\,dx = p_0 + \frac{\mu}{k}\frac{Q}{4\pi}\left[\text{Ei}\left(-\frac{r^2}{4a^2 t}\right)\right]. \tag{5.3.3.10}$$

In addition, the possibility that progressive pressure waves might occur was studied by Werner and Norén (1951), Oroveanu and Pascal (1959), Geertsma and Smit (1961), and Pascal (1964). Non-steady flow in inhomogeneous porous media (containing fissures, holes, varying permeabilities) was investigated by Oroveanu (1961a, b, 1963).

5.3.4 General unsteady-state flow

If the approximations considered in the last paragraphs are not made, the unsteady-state equation becomes non-linear. Owing to analytical difficulties, only relatively few solutions of this case are available.

An interesting case occurs when one is considering non-homogeneous porous media. For a fractured porous medium, one treats the pore matrix and the fractures as two individual systems which are superposed. For each of these two systems one assumes Darcy's law (with corresponding individual permeabilities) to be valid. The connection between the two systems is then given by a source function. If the medium and the fluid are assumed to be slightly compressible, the system of equations becomes non-linear. The problem has been formulated and treated in this fashion by Barenblatt et al. (1960a, b).

One of the most important cases of general unsteady-state flow is the study of transient gas flow through porous media. The gravity term can then be neglected; however, the Darcy equation is nevertheless non-linear even if the simplest connection possible (i.e., $\rho = cp$ corresponding to an ideal gas) between the pressure p and density ρ of a gas is used.

Methods applicable in such cases (i.e., gas flow) have been discussed by Muskat and Botset (1931), Khristianovich (1941), Hetherington, MacRoberts, and Huntington (1942), Leĭbenzon (1945), Polubarinova-Kochina (1948),

MacRoberts (1949), Kalinin (1950), Piskunov (1951), Barenblatt (1952) Aronofsky and co-workers (Aronofsky and Jenkins 1952, 1954; Aronofsky and Ferris 1954; Aronofsky and Porter 1956), Green and Wilts (1952), Roberts (1952), Bruce, Peaceman, Rachford, and Rice (1953), Barenblatt (1953, 1954, 1956a), Douglas et al. (1955), Kidder (1957), and Fan and Yen (1968). Some of these solutions have been obtained using high-speed computing machines.

Again, owing to limitations of space, we must refer the reader to the original papers for the details of the above-mentioned investigations. As an illustration of some of the applicable methods, only one solution can be given here in some detail (after Aronofsky and Jenkins 1952). Consider a tube of porous medium with a constant cross-section and of infinite length. The initial and boundary conditions to be considered may be described as follows. Let p_0 represent the constant initial fluid pressure in the tube. The pressure at one end, $x = 0$, is suddenly increased (or lowered) to the pressure p_1. The problem is then to determine the flow rate along the tube at any instant of time. The fluid is supposed to be an ideal gas.

The problem is solved if it is possible to obtain, subject to the boundary conditions, a solution of the differential equation

$$\frac{\partial^2 (p/p_0)^2}{\partial x^2} = \frac{2P\mu}{p_0 k} \frac{\partial}{\partial t} \left(\frac{p}{p_0} \right)$$

(5.3.4.1)

which follows from (4.3.1.8) for an ideal gas in the one-dimensional case. Since this equation is non-linear, it cannot easily be treated analytically. Therefore, it appears that one should use a numerical approximation procedure. In order to use such a procedure, one replaces (5.3.4.1) by the following difference equation:

$$\left(\frac{p}{p_0} \right)_{x,\, t+\Delta t} = \frac{\Delta t}{(\Delta x)^2 a} \left[\left(\frac{p}{p_0} \right)^2_{x+\Delta x,\, t} + \left(\frac{p}{p_0} \right)^2_{x-\Delta x,\, t} - 2 \left(\frac{p}{p_0} \right)^2_{x,\, t} \right] + \left(\frac{p}{p_0} \right)_{x,\, t},$$

(5.3.4.2)

where $(p/p_0)_{x,\, t}$ is a dimensionless pressure at distance x and time t, $(p/p_0)_{x+\Delta x,\, t}$ is the same dimensionless pressure at distance $x + \Delta x$ and time t, $(p/p_0)_{x,\, t + \Delta t}$ is the dimensionless pressure at distance x and time $t + \Delta t$, and $a = 2P\mu/(p_0 k)$.

In order to ensure stability of equation (5.3.4.2) one has to restrict Δt such that

$$\Delta t \leq \tfrac{1}{4} a (\Delta x)^2.$$

(5.3.4.3)

Substituting the maximum Δt permitted by equation (5.3.4.3) in (5.3.4.2) yields the difference equation in the usable form

$$p_{x,\, t+\Delta t} = \tfrac{1}{4}(p^2_{x+\Delta x,\, t} + p^2_{x-\Delta x,\, t} - 2p^2_{x,\, t}) + p_{x,\, t},$$

(5.3.4.4)

where p_0 is unity. The difference equation can now be solved with ease for a given set of initial and boundary conditions, but the computations are tedious and lengthy, and call for the use of a high-speed (electronic) computing machine.

Applying the above procedure to the case under consideration, Aronofsky and Jenkins (1952) obtained the results presented in figure 22. The ordinate in this figure is a dimensionless pressure p',

$$p' = (p - p_0)/(p_1 - p_0), \tag{5.3.4.5}$$

which is the ratio of the pressure change at any point within the porous medium to the pressure change at the face, $x = 0$. The abscissa is the dimensionless parameter x',

$$x' = x \sqrt{P\mu/(4p_0kt)}, \tag{5.3.4.6}$$

which brings all the data into a single plot. It is necessary to use a family of curves to present the results as the ratio N,

$$N = p_1/p_0, \tag{5.3.4.7}$$

is allowed to take on different values.

It may be remarked that the solution represented in figure 22, although it was developed for an infinitely long tube, is also valid for a finite tube if one is interested only in events happening at one end of the tube before a substantial pressure change occurs at the opposite end of the tube. Many physical problems arise which meet this requirement.

5.3.5 Experimental solutions

The analytical difficulties met in dealing with the differential equations of section 5.3 are quite considerable even in the linearized case. This is especially true if the boundaries of the porous medium are irregular in shape: for instance, when it represents an oil field. Experimental methods for the solutions of such problems have therefore been developed.

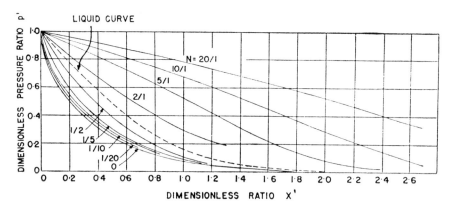

FIGURE 22   A series of solutions for non-steady-state gas flow (after Aronofsky and Jenkins 1952).

A tool often used for the study of the behaviour of compressible liquids in extended porous media is an electrical analogue (Bruce 1943; Morris 1951; Patterson, Montague, and Wiess 1951; Aleskerov and Makhmudov 1955). Such devices approximate the linearized differential equation. They are built in such a manner that a domain (or 'zone') of the porous medium is represented by a capacitor and a resistor in series. Such units are then connected in various ways so as to represent the extent of the porous medium. It can be shown that the behaviour of an array of such units is described by an equation approximately the same as (5.3.2.1). The proper sizes of capacitors and resistors are selected by trial and error to match a known past 'history' of the behaviour of the fluid in the porous medium (e.g., an oil field). Once a proper set of capacitors and resistors is found, the device is used to make predictions of the future behaviour. Unfortunately, a past history can be matched in many ways, which, in turn, may give somewhat different 'predictions,' so that there is the possibility of occasional prediction errors. In a similar fashion, Green and Wilts (1952) developed an electrical network analogue whereby the non-linear gas equations can be solved by successive approximation.

Finally, an experimental model for studying transient phenomena has been suggested by Landrum et al. (1959) in which the flow through porous media is simulated by the flow of heat.

5.4
GRAVITY FLOW WITH A FREE SURFACE

5.4.1 *Physical aspects of the phenomenon*

A peculiar special case of the determination of steady-state flow with gravity occurs when a fluid has a 'free' surface. In fact, this is a problem of multiple-phase flow; above the 'free surface' there will be another fluid. Nevertheless, if the other fluid is a gas, and the original fluid a liquid, one can disregard the motion of the gas relative to that of the liquid and assume that the gas pressure is constant in its entire domain. Thus one can express the assumption that there is a free interface between the two fluids by the condition that on this free interface the liquid pressure is uniform (equal to the atmospheric pressure) and that the displacement of the free surface is given by the normal component of the pore-velocity vector on any streamline intersecting it. Under steady-state conditions, this implies that any streamline having one point in common with the free surface lies *entirely* within it. In truth, no such free interfaces actually exist as there will always be a finite region within which the liquid saturation drops from 1 to zero. Nevertheless, we shall see later that sharp saturation discontinuities can exist such that the assumption of an actual 'free surface' may well be sensible.

A further peculiar phenomenon may occur if the notion of a free surface is

E

accepted. Should this free surface intersect an 'open' boundary of the porous medium, i.e., a boundary which is 'open' to the gas and upon which therefore the pressure is uniform and equal to that of the gas, then liquid will seep out from that boundary *below* the intersection with the free surface. Such a boundary is termed a 'surface of seepage.' The analytical condition for a surface of seepage is that the pressure in the fluid be constant and equal to that at the free surface. The physical picture of the surface of seepage has been analysed on many occasions, for instance by Hamel (1934), Emery and Foster (1948), Laurent (1949), and Childs (1956).

Thus, the analytical conditions of the flow with a free surface are fully determined. To summarize, one has to find a solution of the general flow equation with the boundary conditions such that there is a 'free surface' which moves with the streamlines and on which the fluid pressure is constant. On 'open boundaries' below the intersection with the free surface, the liquid pressure must be constant and equal to that on the free surface. On impermeable boundaries, of course, the condition is, as usual, that the normal component of the filter velocity vanish.

5.4.2 *Steady-state gravity flow*

Needless to say, it is extremely difficult to find analytical solutions conforming to the formulation of the free-surface problem. Considering first steady-state conditions, we note that it is now required that every streamline having one point in common with the free surface lie entirely within it. The flow equation becomes the Laplace equation (Hopf and Trefftz 1921). For this case, the books of Breitenöder (1942), Muskat (1937), Polubarinova-Kochina (1952), Harr (1962), Aravin and Numerov (1965), and Bear et al. (1968) reviewed some methods that have been applied successfully, notably one employing hodograph transformations which was developed chiefly by Hamel (1934). This method is based on the following considerations. In a medium of homogeneous permeability, the formulation using the velocity potential may be used:

$$\psi = (k/\mu)(p + \rho gz), \tag{5.4.2.1}$$

$$\mathbf{q} = -\operatorname{grad}\psi, \tag{5.4.2.2}$$

$$\operatorname{lap}\psi = 0. \tag{5.4.2.3}$$

In two dimensions (x horizontal, y vertical coordinates), this leads to the well-known possibility of representing everything in terms of complex numbers with $z = x + iy$. The hodograph transformation is characterized by setting

$$u = \partial\psi/\partial x, \tag{5.4.2.4}$$

$$v = \partial\psi/\partial y, \tag{5.4.2.5}$$

which again can be represented in a complex plane.

The hodograph method has the advantage that the free surfaces, whose shape is unknown in the original formulation of the problem, are determined in the hodograph plane: they are simply circles with known parameters. In addition, the surfaces of seepage are also determined prior to an actual analytical solution of the problem. Thus, the 'floating' boundary conditions of the original problem become fixed boundary conditions after a hodograph transformation has been made.

Owing to the difficulties of obtaining analytical solutions of the problem, graphical methods have been tried (Casagrande 1940; Nahrgang 1954; Cedergren 1967). For radial flow (vanishing azimuthal component of the velocity), the flow equation, using the notation of the velocity potential (cf. 5.4.2.1), reduces to

$$\frac{\partial^2 \psi}{\partial r^2} + \frac{1}{r}\frac{\partial \psi}{\partial r} + \frac{\partial^2 \psi}{\partial z^2} = 0. \tag{5.4.2.6}$$

It is well known from the theory of the Laplace equation that one can introduce a stream function φ, defined by the differential equations

$$\partial \varphi / \partial z = 2\pi r \partial \psi / \partial r, \tag{5.4.2.7}$$

$$- \partial \varphi / \partial r = 2\pi r \partial \psi / \partial z. \tag{5.4.2.8}$$

The stream function φ, in turn, satisfies the differential equation

$$\frac{\partial^2 \varphi}{\partial r^2} - \frac{1}{r}\frac{\partial \varphi}{\partial r} + \frac{\partial^2 \varphi}{\partial z^2} = 0, \tag{5.4.2.9}$$

as can easily be verified. The lines ψ = const. represent equipotential curves in the r–z plane, and the lines φ = const. represent streamlines. The streamlines and the equipotential lines form an orthogonal net of curves. Therefore, from equation (5.4.2.7–8) one can show that the differential of the potential can be expressed in terms of the differential of the stream function as follows:

$$d\varphi = 2\pi r \frac{dn}{ds} d\psi, \tag{5.4.2.10}$$

where dn is the differential along a potential line and ds is the differential along a streamline.

Using differences (denoted by Δ) instead of differentials, and postulating that

$$\Delta \varphi = \Delta \psi, \tag{5.4.2.11}$$

we have

$$\Delta s = 2\pi r \Delta n. \tag{5.4.2.12}$$

This initiates the possibility of obtaining a graphical solution which is particularly suited to the gravity flow problem. We have to construct a net of orthogonal curves satisfying (5.4.2.12) and the boundary conditions. This can be done by the method of trial and error (cf. fig. 23).

From the above graphical method it is only a short step to the solution of gravity flow problems by actual numerical calculations. Such calculations have been carried out, for instance, by Stallman (1956), Jeppson (1968a, b, c, d, 1969) and Neuman and Witherspoon (1970).

Again, because of the difficulties involved in obtaining rigorous solutions conforming to the assumptions basic to the present considerations, approximate procedures have been developed. The best known is that by Dupuit (1863) as modified by Forchheimer (cf., e.g., 1930). If it is assumed that for small inclinations of the free surface of a gravity flow system the streamlines may be taken as horizontal and, furthermore, that the corresponding velocities are proportional to the slope of the free surface, one readily arrives at the following differential equation for the height h of the free surface above the (horizontal) impermeable bed of the system (see Muskat 1937, p. 360):

$$\text{lap } h^2 = 0. \tag{5.4.2.13}$$

Unfortunately, the assumptions basic to this simple theory do not seem to be

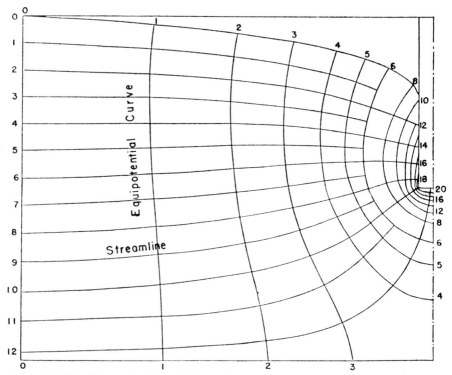

FIGURE 23   Flow lines and equipotential curves in a partially penetrating well (after Nahrgang 1954).

quite warranted so that the entire theory has been severely questioned (see Muskat 1937, p. 359).

It has been observed that the flux formulas obtained from the Dupuit-Forchheimer theory give much better results than the underlying assumptions might lead one to expect. Muskat (1937) showed that one can arrive at the Dupuit-Forchheimer *flux* formulas by another approximate theory which is free of the Dupuit assumptions. The Dupuit-Forchheimer theory has been discussed more recently by Leĭbenzon (1947), Reinius (1947), and Öllös (1958). In spite of its basic shortcomings, it is widely used for engineering calculations.

5.4.3 *Unsteady-state gravity flow*

The equations describing unsteady-state gravity flow are essentially non-linear. Even if the simplified (and therefore inadequate) assumptions of Dupuit and Forchheimer (see sec. 5.4.2) are used, one obtains non-linear equations. These assumptions of Dupuit and Forchheimer have been used by Boussinesq (1904) to construct a theory of the time variations of the free surface. The latter author thus arrived at the following equation (see, e.g., Leĭbenzon 1947):

$$P \partial h / \partial t = (k \rho g / \mu) \operatorname{div}(h \operatorname{grad} h), \tag{5.4.3.1}$$

where h is the height of the free surface above the base and the other symbols have their usual meaning. This equation can be simplified for certain cases so as to become linear (see sec. 5.5.2).

Problems involving non-steady-state gravity flow usually refer to groundwater. In this case, one has a standard fluid (density and viscosity constant), so that a simplified formulation of the equations of motion can be applied. Using a velocity potential ψ

$$u = \partial \psi / \partial x, \qquad v = \partial \psi / \partial y, \qquad w = \partial \psi / \partial z, \qquad \operatorname{lap} \psi = 0 \tag{5.4.3.2}$$

(where u, v, w are the projections of the seepage velocity onto the horizontal orthogonal directions x, y and the vertical direction z, respectively), one has (κ denoting the hydraulic conductivity; cf. 4.2.2.3)

$$\partial u / \partial x + \partial v / \partial y + \partial w / \partial z = 0, \tag{5.4.3.3}$$

$$\psi = - \kappa [(p/\rho g) + z] = - \kappa H, \tag{5.4.3.4}$$

where H is often designated as the 'head' in the groundwater system. The seepage velocities are connected with the pore velocities dx/dt, etc. by

$$u = P dx / dt, \qquad v = P dy / dt, \qquad w = P dz / dt. \tag{5.4.3.5}$$

If the equation of the free surface is written as

$$F(x, y, z, t) = 0, \tag{5.4.3.6}$$

then, after differentiation and insertion of (5.4.3.5), one obtains

$$P\frac{\partial F}{\partial t} + \frac{\partial F}{\partial x}u + \frac{\partial F}{\partial y}v + \frac{\partial F}{\partial z}w = 0. \tag{5.4.3.7}$$

If the equation of the free surface is written as

$$z = S(x, y, t), \tag{5.4.3.6a}$$

one has instead of (5.4.3.7)

$$P\frac{\partial S}{\partial t} + \frac{\partial S}{\partial x}u + \frac{\partial S}{\partial y}v - w = 0. \tag{5.4.3.7a}$$

At the free surface the pressure p must be constant. Thus we have, from equation (5.4.3.4),

$$\psi(x, y, z, t) + \kappa z = \text{const.} \tag{5.4.3.8}$$

The above is the general formulation of the problem. Solutions are difficult to obtain, and various approximate numerical and graphical methods have been tried. In the *method of successive changes of steady-state values*, the flow at each moment is assumed to be steady, but the boundary of the flow at each moment of time is taken to be changing. Therefore, the flow for an elementary lapse of time is equal to the change of area filled with fluid divided by this lapse of time. In the *method of finite differences*, the partial differential equations are made discrete and then solved by graphiconumerical analyses (Douglas and Rachford 1956; Bredehoeft and Pinder 1970; Neuman and Witherspoon 1971). In the *statistical method*, use is made of the fact that there is an analogy between groundwater-particle movement and the stochastic motion of a point. Thus, groundwater flow can be modelled by a Monte-Carlo procedure on a computer (Halek and Novak 1969). In the *wave-equation method* (Dracos 1962), it is noted that non-steady groundwater flow originates from the propagation of disturbances. This can be treated by wave-equation methods with large frictional effects.

5.4.4 Scaling and experimental investigations of gravity flow in porous media

Because of the difficulties in obtaining analytical solutions of the gravity flow equations, attempts have been made to make scale-model experiments. This can only be done if the appropriate scaling relations are known. For the deduction of the scaling relations, one needs the fundamental equations of motion of the process in question, but not their solutions, which are obtained in the *model*. The equations of motion contain *variables* and *parameters*, and one makes a transformation of the equations so that the new variables become dimensionless. Then all systems whose equations of motion can be transformed into identical dimensionless equations are dynamically similar. The scaling relations are obtained simply by setting the corresponding expressions equal to one another in equivalent systems under consideration, and by expressing the condition that the remaining dimensionless parameters in the dimensionless equations of motion are equal for equivalent systems. In this connection, the Pi-theorem of Buckingham (1914) states that a number of parameters equal to the number of independent fundamental dimensions can always be eliminated.

The above general principles were applied by Scheidegger (1963) to deduce the scaling relations for gravity flow with a free surface. One first has to write down the fundamental equations (5.4.3.2–7) in dimensionless form. It then turns out that the new dimensionless equations of motion contain no parameters. Thus, all free-surface groundwater flow processes with the same initial conditions are dynamically similar.

Denoting the ratio of a variable X in the prototype and the model by $R(X)$,

$$R(X) = X(\text{prototype})/X(\text{model}),\qquad\qquad(5.4.4.1)$$

one obtains the scaling relations:

$$R(H) = R(L) = R(p_0/\rho g),\qquad\qquad(5.4.4.2)$$

where H is the head, L any length, p_0 the free-surface pressure, ρ the density, and g the gravitational acceleration;

$$R(t) = R(PH/\kappa),\qquad\qquad(5.4.4.3)$$

where t is the time, P the porosity, and κ the hydraulic conductivity; and

$$R(v) = R(\kappa/P),\qquad\qquad(5.4.4.4)$$

where v is the velocity. It is evident that these scaling relations are not easy to satisfy.

Thus, experimental investigations other than scale models were also tried. Among these, efforts involving electrical analogues seemed the most hopeful, although the free surface cannot be duplicated in electrical models as there are no corresponding boundaries possible for electrical currents. Thus, the procedure is one of trial and error, of shaping the electrical model in such a manner that part of its boundaries corresponds to a free surface. Muskat (1937), in his book (p. 318), discussed the difficulties involved. Other investigations have been made by Wyckoff and Reed (1935), Fil'chakov (1949), and Bouwer (1964). The last-named author modified the usual analogue model and assumed that the electrical potential corresponds to the hydraulic head. Since the pressures at the free surface should be constant, the electrical potential measured at any point of the free surface should correspond to the elevation of that point. If the measured potentials exceed the elevations, the assumed water table is too low and vice versa. This yields the possibility of approaching the correct solution by successive approximations.

Instead of electrical models, the Hele-Shaw model has been applied by Todd (1954) and Zeller (1957) to gravity flow problems.

5.5
APPLICATIONS TO GROUNDWATER HYDROLOGY

5.5.1 *Notation*

On several occasions in the present monograph we have referred to applications of the physics of flow through porous media to groundwater hydrology. It was noted in this connection that the notation is somewhat different than in general flow theory, because one is dealing with a 'standard' fluid (water) that is assumed to be of constant viscosity and density, so that these quantities can be incorporated into others.

Let us summarize the notation (cf. Scheidegger 1962). Darcy's law (4.3.1.8) is generally written as

$$\mathbf{q} = -\kappa \operatorname{grad} H \tag{5.5.1.1}$$

with

$$\kappa = k\rho g/\mu, \tag{5.5.1.2}$$

$$H = p/(\rho g) + z, \tag{5.5.1.3}$$

where H is the hydraulic head (dimension a length), κ is the hydraulic conductivity (dimension a velocity), and z is a coordinate measured vertically upward. If a capillary pressure is present, one introduces the matric potential Ψ defined by

$$\Psi = p_c/\rho g, \tag{5.5.1.4}$$

which has again the dimension of a length.

The continuity equation cannot be formulated in general cases; one has to confine oneself to specific conditions.

An important quantity (Theis 1935) is the storage coefficient α, defined as the volume of water of an aquifer released from or taken into storage per unit aquifer surface area A per unit change in the hydraulic head H. Thus

$$\alpha = d(V_{\text{water}}/A)/dH. \tag{5.5.1.5}$$

One generally takes the conductivity κ and the storage coefficient α as the characteristic parameters of an aquifer. Unfortunately, the storage coefficient can mean very different things in different conditions as will be shown below.

5.5.2 *Unconfined aquifer*

We consider first an unconfined aquifer in which neither the water nor the rock is assumed to be compressible. The continuity equation then yields

$$\operatorname{div} \mathbf{q} = 0, \tag{5.5.2.1}$$

which gives, with Darcy's law,

$$\text{lap } H = 0. \tag{5.5.2.2}$$

The storage volume is directly proportional to the change in the height h of the free surface. Furthermore, h is approximately equal to H; then one has (P denoting the porosity)

$$V_{\text{water}}/A = (H - H_0)P \tag{5.5.2.3}$$

and hence from the definition of α (5.5.1.5)

$$\alpha = d(V_{\text{water}}/A)/dH = P. \tag{5.5.2.4}$$

It thus turns out that, in an unconfined aquifer, the storage coefficient is simply the porosity of the medium.

The time-dependence of the free surface, if the head is lowered in a well, has often been taken as given by a diffusivity equation. This follows directly from equation (5.4.3.1) if the variations of h are assumed to be small. Then one can assume the quantity h before the gradient in the bracket in (5.4.3.1) to be constant and equal to the thickness T of the aquifer. It can thus be taken out from under the 'div' sign. Using the notation (5.5.1.2) and (5.5.2.4), one obtains

$$\alpha \partial h/\partial t = T\kappa \text{ lap } h, \tag{5.5.2.5}$$

which is an equation that had been postulated originally by Theis (1935) from heuristic considerations. In approximation, h can also be replaced by H in this equation.

5.5.3 Artesian aquifer

Conditions are entirely different in an artesian aquifer, if the rocks are assumed to be slightly compressible. In general, this can no longer be represented easily in terms of the hydraulic head. A simplification can still be achieved in a *horizontal* artesian aquifer ($z = $ const.); equation (5.3.2.5) then yields

$$\beta_m \, \partial H/\partial t = (k/\mu) \text{ lap } H. \tag{5.5.3.1}$$

Multiplying this by $\rho g T$ (T representing the thickness) yields (using 5.5.1.2)

$$\rho g T \beta_m \, \partial H/\partial t = T\kappa \text{ lap } H. \tag{5.5.3.2}$$

It can be shown that

$$g\beta_m T = \alpha. \tag{5.5.3.3}$$

One has

$$V_{\text{water}}/A = P V_{\text{bulk}}/A. \tag{5.5.3.4}$$

Thus (cf. 4.3.3.2)

$$\frac{dV_{water}/A}{dp} = \frac{dP}{dp}\frac{V_{bulk}}{A} = \beta_m\frac{V_B}{A}.$$ (5.5.3.5)

In a horizontal aquifer $p = \rho g H$; hence, from the definition of α,

$$\alpha = \frac{dV_{water}/A}{dH} = \rho g\frac{V_{bulk}}{A}\beta_m,$$ (5.5.3.6)

and with $V_{bulk}/A = T$, we get (5.5.3.3) as claimed.

Thus we obtain again, from (5.5.3.2), the Theis (1935) equation

$$\alpha\,\partial H/\partial t = T\kappa\,\text{lap}\,H.$$ (5.5.3.7)

This makes it possible to find $\alpha/\kappa T$ from pumping tests in an aquifer in situ, if template solutions (for circular geometry one has Ei functions; cf. eq. 5.3.3.10) are compared with the measured values of the head as a function of time (Theis 1935; Jacob 1940; Youngs and Smiles 1963; Hantush 1964). The last-mentioned author has given a particularly useful compilation of type-functions.

There are only a few absolutely confined artesian aquifers, most of them being to some extent 'leaky'. Inflow of water occurs from above and from below. There exists a large amount of literature on leaky aquifers and on the interpretation of pumping tests made therein, as discussed in standard text books (cf. De Wiest 1965). A recent review of the applicability of current theories of flow in leaky aquifers has been given by Neuman and Witherspoon (1969), to which the reader is referred for further details. An automatic solution of the 'inverse' problem (direct calculation of the aquifer parameters from pumping tests) has been given by Emsellem and Marsily (1969, 1971).

5.5.4 Field applications

The chief application of the theory of gravity flow with a free surface in porous media is to the flow of water underground. In this connection, it is often difficult to trace the free surface and its behaviour, so that it requires considerable effort actually to test the theory.

A series of field measurements with the explicit aim of testing soil drainage theory has been reported by Kirkham and de Zeeuw (1952). The experiments were performed in recently reclaimed lands in the Netherlands and consisted of measuring rainfall, water table heights, soil permeability, and drainage outflows. Many startling discrepancies with what had been expected from theory were noted, but it is possible that these were due to the difference in the conditions in the field from those that had been assumed for the calculations.

In connection with field measurements of groundwater motion, it may be of interest to note that the motion may even be conditioned by general geophysical agents affecting the whole earth. The solid tides of the earth, by causing a compression in the rocks, may have a pronounced influence on the behaviour of wells. This has been noted by Kikkawa (1955) and, particularly, by Gulnick (1956) and Melchior (1956). The last two authors made a series of very interesting measurements on a well near Turnhout, Belgium, and found a pronounced correlation of its behaviour with the earth's tides. Similarly, the passage of earthquake waves also may cause an oscillation of the water level in wells. The theory of this effect ('phreatic seismograph') was given by Blanchard and Byerly (1935); some later observations of it in Hawaii were reported by Eaton and Takasaki (1959).

6
Physical aspects of permeability

6.1
EMPIRICAL CORRELATIONS

The concept of permeability as introduced in the earlier chapters of this book permits a phenomenological description of the flow through porous media in a certain velocity domain. However, an actual understanding of the phenomena can be obtained only if the concept of permeability can be reduced to more fundamental physical principles.

It seems intuitively clear that the property of 'permeability' should be linked with other properties of a porous medium, such as capillary pressure curves and internal surface area, since all such properties are the manifestation of the geometrical arrangement of the pores. However, to uncover the relationships will be possible only if one is able to understand exactly how all these properties are conditioned by the geometrical properties of the pore system.

A direct approach to finding relationships between the various properties of porous media is by an attempt to establish empirical correlations. A most obviously sought after correlation will be that between *porosity* and *permeability*. A simple consideration of theoretical possibilities of the structures of porous media, however, makes one realise that a general correlation between porosity and permeability cannot exist. It is obviously quite possible for two porous media of the same porosity to have entirely different permeabilities. Thus any correlation function between the two quantities cannot be unique. Therefore, most empirical correlations contain some other factors, usually vaguely identified with alleged geometrical quantities. They are, however, nothing but undetermined factors used in order to make the data fit the desired equations. There are even a series of claims for 'general' relationships, usually supposed to be true for 'average' porous media, whatever that means. In the early attempts (up to about 1933) of the oil industry to elucidate the behaviour of reservoirs, no distinction was even made between porosity and permeability, implying proportionality between the two properties.

Some of the claims for empirical relationships between porosity and permeability are listed by Jacob (1946), Franzini (1951), and Hudson and Roberts (1952). Thus, Mavis and Wilsey (1937) claim to have found that k is proportional to P^6 or P^5, Büche (1937) gives a similar relationship, and Rose (1945) sets k proportional to P^n (with n undetermined). Other equations have been proposed by Brevdy (1948) and Shuster (1952). Most of these proportionalities contain further factors when written as equations and stipulate that, in order to be applicable, the 'other factors' must be kept constant. Allegedly, these 'other factors' are representative of certain geometrical properties of the porous medium. The factor used mostly contains the 'grain diameter' δ in some form (usually to the power two), which, in turn, is thought to be equivalent to the 'pore diameter' (whatever this is). As neither of these quantities usually is properly defined, their introduction and proper 'adjustment' means nothing more or less than the introduction of an undetermined constant which is adjusted from case to case so as to make the 'correlations' properly valid. The introduction of these quantities would be significant only if there were an independent means of measuring them.

The influence of porosity upon permeability has been discussed in more general terms (i.e., without giving a specific equation, but by supplying curves or qualitative discussions) by Baver (1949) and Bulnes and Fitting (1945). A further discussion is due to Cloud (1941), who definitely finds that there is no sensible relation between porosity and permeability.

It is therefore obvious that no simple correlation between porosity and permeability can exist. The next more involved correlation that might be looked for is between *structure* and *permeability*. Again, of course, the meaning of 'pore structure' is somewhat arbitrary. One may understand that this means the 'pore size distribution,' as obtained from capillary pressure curves. If such correlations are found, one would have, in fact, obtained a correlation between the capillary pressure curve and permeability; whether this is also a correlation between the pore size distribution and permeability depends on whether one wants to consider capillary pressure as representative of pore size distribution, a problem which has been discussed in chapter 3. The correlations between capillary pressure and permeability are mainly based upon theoretical considerations which will be discussed later. The dependence of permeability on other 'structural' parameters of the pore space has been discussed, on an experimental basis, by Nelson and Baver (1940), O'Neal (1949), and Backer (1951). None of these investigations seem to show very promising results.

For unconsolidated porous media, correlations between *permeability* and *grain size distribution* have been attempted. Such correlations have been reported by Tickell (1935), Mavis and Wilsey (1937), Prockat (1940), Krumbein and Monk (1942), Sen-Gupta and Nyun (1943), and Pillsbury (1950). Most of these analyses were made on similar types of materials and therefore it was possible for the individual authors to claim such correlations. However, it seems hardly credible that general correlations should exist, i.e., that Raschig rings and pebbles that happen to be screened by the same set of sieves should have the same permeability. It has thus been pointed out by Makhl (1939) that a comparison of products on the basis of the coefficient of permeability is possible only if the tests are made under similar conditions and the products are of the same type. Cloud (1941) found that there is no correlation, and Griffiths (1952a) observed that, in oil sands, the grain size distribution is correlated with certain mineral compositions rather than with permeability. It so happens that specific minerals often form beds of certain characteristic permeabilities owing to their usual mode of breaking up and packing. In this manner, a correlation between grain size distribution and permeability might actually be obtained, but certainly not one that is expressible in mathematical terms; it is rather the case that certain minerals normally show both a characteristic grain size distribution and a characteristic permeability.

In view of the fact that a simple correlation between grain size distribution (as obtained from sieve analysis) and permeability does not seem to exist, an attempt has been made to introduce further parameters. Thus, the influence of a parameter indicative of the 'angularity' or 'roundness' of the grains on permeability has been investigated by Tickell et al. (1933) and Tickell and Hiatt (1938), and the influence of the 'packing' of the grains has been investigated by Martin et al. (1951).

Attempts have further been made to correlate the orientation of the grains with permeability. Such studies have been made by Griffiths (1950, 1952b), Griffiths and Rosenfeld (1953), Martin et al. (1951), and Heiss and Coull (1952). The outcome of these studies seems to be that orientation has a definite effect, although the results cannot be represented by a simple correlation. Nevertheless, this may explain why beds of the same materials, packed to the same porosity, occasionally have different permeabilities.

There have also been attempts to find experimental correlations between properties of porous media other than those mentioned above. Thus Bjerrum and Manegold (1927), Buchanan and Heymann (1948), Thornton (1949), Goring and Mason (1950), and Schopper (1966) discussed the question of the relationship between the electrical and mechanical flow paths in porous media and Kolb (1937) made an attempt to correlate water imperviousness with characteristics of adsorption of mortars and concrete.

6.2
CAPILLARIC MODELS

6.2.1 *The concept of models*

The empirical attempts to establish correlations between various dynamical properties of porous media all seem to be futile unless certain additional parameters are introduced. Theoretical considerations might, however, be able to attach physical significance to these parameters.

Such theoretical considerations will be based upon an analysis of the microscopic properties of flow. The flow through porous media, presumably, takes
place along flow channels with local (pore) velocity v. The pore velocity, on the
whole, must be larger than the filter velocity q, owing to the reduced space
available for the fluid to flow, as compared with the bulk volume of the
porous medium on the basis of which the filter velocity is calculated. A commonly
accepted hypothesis for the connection between pore velocity v and filter velocity
q is the following, known as the Dupuit-Forchheimer assumption:

$$v = q/P. \tag{6.2.1.1}$$

However, it should be noted that the 'pore velocity' has really no properly
defined meaning, because the actual velocity of the fluid must be expected to
fluctuate grossly within one flow channel and from one flow channel to another.
The Dupuit-Forchheimer assumption *defines* an 'average' pore velocity which
may, or more often may not, be identical with what a particular microscopic flow
theory would like to regard as the 'pore velocity'. It should therefore be noted
that the Dupuit-Forchheimer assumption cannot be regarded as basic, unless
it is shown that the 'pore velocity' under consideration *is* an actual statistical
average over all the local velocities, as implied by that very assumption.

The simplest way to try to establish correlations theoretically is to represent
the porous media by theoretical models which can be treated mathematically.
Only trial and error can show which models exhibit the characteristic phenomena
taking place in the porous medium and which do not. If a proper model is found,
it can be substituted for an actual porous medium and then one can predict by
calculation how the medium will behave under yet untried conditions. Relationships deduced from such models would be generally valid.

The simplest models that can be constructed are those consisting of capillaries.
Scheidegger (1953) gave a review of such models. These models aim at correlating
the permeability with either an 'average' pore size or with the pore size distribution (i.e., capillary pressure) curve.

6.2.2 Straight capillaric model

The simplest capillaric model of the linear case is one representing a porous
medium by a bundle of straight, parallel capillaries (see, for example, Scheidegger
1953) of uniform diameter δ. The model is illustrated in figure 24. The total
volume-flow Q through a capillary is then given by the well-known law of Hagen-
Poiseuille (see equation 2.2.2.2):

$$Q = - \frac{\pi \delta^4}{128 \mu} \frac{dp}{dx}, \tag{6.2.2.1}$$

where μ is as usual the viscosity and dp/dx is the pressure gradient along the capillary. If there are n such capillaries per unit area of cross-section of the model, the flow per unit area q (or the macroscopic or 'filter' velocity) will be

$$q = - \frac{n\pi\bar{\delta}^4}{128\mu} \frac{dp}{dx}.$$ (6.2.2.2)

As the flow can also be expressed by Darcy's law

$$q = - \frac{k}{\mu} \frac{dp}{dx},$$ (6.2.2.3)

it follows that

$$k = - n\pi\bar{\delta}^4/128.$$ (6.2.2.4)

The pore volume of the model (assuming unit cross-sectional area) is equal to $\frac{1}{4}n\pi x\bar{\delta}^2$, the length being denoted by x; thus the porosity is

$$P = \frac{1}{4}n\pi\bar{\delta}^2.$$ (6.2.2.5)

Eliminating n from equations (6.2.2.4) and (6.2.2.5) yields the equation

$$k = P\bar{\delta}^2/32.$$ (6.2.2.6)

If this equation is applied to an actual porous medium, $\bar{\delta}$ is a sort of 'average' pore diameter.

It is known that this equation does not correctly represent the connection between permeability and porosity in porous media as it is actually observed. Therefore, the factor 32 is commonly replaced by some arbitrary factor T^2, where T is called the 'tortuosity' or the like. Similarly, instead of $\bar{\delta}^2$ the 'average specific surface area' S can be introduced. The specific internal area of the model (i.e., the ratio of the area of the capillaries to the volume of the model) is given by

$$S = n\pi\bar{\delta}.$$ (6.2.2.7)

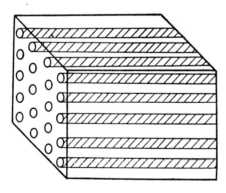

FIGURE 24   Straight capillaric model.

Eliminating n by means of equation (6.2.2.5), one obtains

$$S = 4P/\bar{\delta}. \tag{6.2.2.8}$$

Introducing this into the expression for the permeability, and taking all numerical factors together in the 'tortuosity' T, the equation (6.2.2.6) takes the form

$$k = P^3/(T^2 S^2). \tag{6.2.2.9}$$

A similar equation was proposed long ago by Krüger (1918) on experimental grounds; later a theoretical justification was given by Kozeny (1927a), using, however, an altogether different reasoning as will be shown later. Equations containing S, similar to (6.2.2.9), are now commonly called 'Kozeny equations.'

The present procedure of adding arbitrary factors does not help one to understand the phenomena in the porous medium, and it is therefore necessary to investigate what alterations would have to be made to the simple model of parallel capillaries to get a more adequate description of the flow phenomena.

6.2.3 Parallel type models

It is natural to stick first as closely as possible to the simple model of parallel capillaries treated in section 6.2.2.

First, one notices that this model gives a permeability in one direction only. All capillaries being parallel, there can be no flow orthogonal to the capillaries. The first modification of the relationship between porosity and permeability would therefore have to consist of putting one-third of the capillaries in each of the three spatial dimensions. The permeability is thus lowered by a factor 3 and equation (6.2.2.6) will read

$$k = P\bar{\delta}^2/96. \tag{6.2.3.1}$$

The inclusion of the factor 3, of course, invalidates the Dupuit-Forchheimer assumption that $v = q/P$, which is thus not valid in the present models.

Apparently, this is not much better than the original equation. One of the main difficulties is that the average diameter $\bar{\delta}$ of the pores of an actual porous medium does not have much meaning. It will be desirable to reduce $\bar{\delta}$ to some value calculated from the 'pore size distribution' $f(\delta)$ or from the 'differential pore size distribution' $\alpha(\delta)$ as defined in chapter 1. In fact, this means aiming at a connection between capillary pressure and permeability since the pore size distribution is usually measured (more or less accurately) by the capillary pressure method (see chapter 3). Thus, we shall construct a model where all the capillaries permitting flow in a given direction are parallel to that direction, but vary in pore diameter, leading from one face of the porous medium through to the other. We shall call such a model a 'parallel type model.' The pore size distribution of the model and an actual porous medium will be made identical. A model of this type was devised by Purcell (1949) in a somewhat different fashion. The following exposition is taken from an earlier review by the writer (Scheidegger 1953).

For a given pressure gradient equal to dp/dx, the mean velocity v of a flowing fluid through a capillary of radius δ is given by the Hagen-Poiseuille equation (6.2.2.1)

$$dp/dx = -32\,\mu v/\delta^2. \tag{6.2.3.2}$$

Now, the volume taken up by capillaries of pore diameter between δ and $\delta + d\delta$ parallel to the x-direction in a piece of the model of unit frontal area and thickness Δx is $\frac{1}{3}P\alpha(\delta)d\delta\Delta x$. (The factor $\frac{1}{3}$ expresses the fact that only one-third of the capillaries in the model are in one given direction, as explained above.) They have a frontal area of $\frac{1}{3}P\alpha(\delta)d\delta$. Thus the total quantity of fluid flowing through a unit area per unit time is

$$q = \tfrac{1}{3}P\int_0^\infty v\alpha(\delta)d\delta. \tag{6.2.3.3}$$

Inserting v from (6.2.3.2) yields

$$q = -\frac{dp}{dx}\frac{P}{96\mu}\int_0^\infty \delta^2\alpha(\delta)d\delta \tag{6.2.3.4}$$

and by comparison with Darcy's law

$$k = \frac{P}{96}\int_0^\infty \delta^2\alpha(\delta)d\delta. \tag{6.2.3.5}$$

Thus, we obtain almost the same expression as equation (6.2.3.1), except that the average pore radius $\bar\delta$ has now a more exactly defined meaning. It is given by the equation

$$\bar\delta^2 = \int_0^\infty \delta^2\alpha(\delta)d\delta. \tag{6.2.3.6}$$

No fundamental deviation from equations (2.2.3.1) and (6.2.2.6) is therefore obtained. Particularly, the specific surface area and a tortuosity factor can again be introduced, and thus one finally arrives again at a 'Kozeny' equation of the type of (6.2.2.9). It seems that parallel type models have not much advantage over the simple model of parallel capillaries; the essential consequences are identical.

6.2.4 *Serial type models*

A serious drawback of the parallel type models is that all the pores are supposed to go from one face of the porous medium right through to the other. This supposition is evidently quite remote from what happens in an actual porous medium.

The opposite extreme picture of the one-dimensional case would be obtained by assuming that all the pore space is serially lined up, so that each particle of fluid would have to enter at one pinhole at one side of a porous medium and travel through very tortuous channels through all the pores and then emerge at only one pinhole at the other face of the porous medium. Obviously, this picture is just as unreal as that treated in section 6.2.3 and a realistic model lies somewhere in between the extremes. We shall refer to such models as 'serial type models,' as capillaries of different pore diameter are put together in series one after another (Scheidegger 1953). This model is illustrated in figure 25.

We shall therefore assume a model of length x where there are n capillaries per unit area in each dimensional direction of a pore diameter $\bar{\delta}$ and length s. This model is assumed to represent a cylindrical porous medium of length x, porosity P, and 'average' pore diameter $\bar{\delta}$. In this instance we shall not specify more clearly what is meant by 'average' pore diameter, but think of it in rather indefinite terms as in section 6.2.2, where a simple model was discussed. Later on we shall try to find a more clearly defined expression in terms of the function $\alpha(\delta)$ of the porous medium.

A model of the above type is a 'serial type model' since the capillaries effecting the flow in any one direction may have contortions; in other words, s may be greater than x. Later on we shall vary the diameter of each channel along its length, but now we simply assume them all to be of the same 'average' diameter $\bar{\delta}$.

If we assume a pressure drop of $p_2 - p_1$ over the length of the above model (one-dimensional case), the mean flow velocity v through each capillary is given by the Hagen-Poiseuille equation (6.2.3.2), neglecting capillary pressure differentials:

$$\frac{p_2 - p_1}{x} = -32 \frac{\mu v}{\bar{\delta}^2} \frac{s}{x}. \tag{6.2.4.1}$$

Thus the total flow per unit area and unit time through the model, if there are n capillaries per unit area, is

$$q = n\frac{\pi}{4}\bar{\delta}^2 v = -\frac{n\pi}{128} \frac{x\,\bar{\delta}^4}{s\,\mu} \frac{p_2 - p_1}{x}. \tag{6.2.4.2}$$

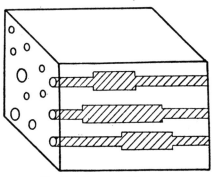

FIGURE 25   Serial type model.

The porosity of the model is (making allowance for the fact that *each* spatial dimension has n capillaries per unit area, by introducing the factor 3)

$$P = \tfrac{3}{4} n \pi \bar{\delta}^2 s / x. \tag{6.2.4.3}$$

Eliminating n from (6.2.4.2) and (6.2.4.3) yields

$$q = -\frac{1}{96} \left(\frac{x}{s}\right)^2 \frac{\bar{\delta}^2}{\mu} P \frac{p_2 - p_1}{x}. \tag{6.2.4.4}$$

Comparison of this with Darcy's law yields

$$k = \frac{1}{96} \frac{P\bar{\delta}^2}{T^2} \tag{6.2.4.5}$$

if we set

$$ds/dx = s/x = T. \tag{6.2.4.6}$$

Again, this is of the form of equation (6.2.2.6). The quantity T plays the role of the arbitrary factor which had to be introduced into equation (6.2.2.6) and which was termed there the 'tortuosity factor.' Actually $T = s/x$ would be an excellent measure of the tortuosity as it gives the ratio of the length of the flow channel for a fluid particle to the length of the porous medium, if the model satisfactorily represents a porous medium. Naturally, the pore channels then have to be intertwined so as to yield a 'spaghetti model' (Cornell and Katz 1953).

However, the average pore diameter $\bar{\delta}$ still has no clearly defined meaning. It is therefore necessary to modify the above model by varying the diameter of each capillary along its length according to the pore size distribution function $\alpha(\delta)$ of the medium which the model is supposed to represent. In the porous medium, a fraction $\alpha(\delta)d\delta$ of the pore space has a pore diameter between δ and $\delta + d\delta$. Thus, the length ds, over which each capillary of the model has to be made of diameter between δ and $\delta + d\delta$, is given by the condition that the ratio of the capillary volume along ds to the whole capillary volume must be equal to the ratio of the pore space between δ and $\delta + d\delta$ to the whole pore volume:

$$\frac{\tfrac{1}{4}\pi \delta^2 ds}{\displaystyle\int_0^s \tfrac{1}{4}\pi \delta^2 ds} = \alpha(\delta)d\delta. \tag{6.2.4.7}$$

However, the integral is obviously

$$\int_0^s \tfrac{1}{4}\pi \delta^2 ds = \frac{1}{n} \frac{Px}{n}, \tag{6.2.4.8}$$

where x is again the length over which the pressure drop exists and n is the number of capillaries per unit area. We obtain therefore

$$\tfrac{1}{4}\pi \delta^2 ds \equiv \tfrac{1}{4}\pi \delta^2 T dx = \alpha(\delta)d\delta \frac{1}{3} \frac{Px}{n}. \tag{6.2.4.9}$$

Furthermore, the continuity equation yields for the fluid velocities:

$$\tfrac{1}{4}\pi n\delta^2 v = q. \tag{6.2.4.10}$$

We assume now that the Hagen-Poiseuille equation (6.2.3.2) is valid for each infinitesimal length of the flow channel. This is, of course, an oversimplification of the facts since complicated effects will occur when a flow channel either contracts or expands. Nevertheless, this assumption is probably no worse than the assumption of a capillaric model in the first place. Thus, inserting (6.2.4.8–10) into the Hagen-Poiseuille equation, the following expression is obtained:

$$p_2 - p_1 = -\int \frac{32\mu v}{\delta^2}\, ds = -\int \frac{32\mu v}{\delta^2}\, T dx \tag{6.2.4.11}$$

$$= \frac{512\,\mu Px}{3\pi^2}\,\frac{1}{n^2}\,q\int_0^\infty \frac{\alpha(\delta)}{\delta^6}\, d\delta. \tag{6.2.4.12}$$

Comparison with Darcy's law yields

$$\frac{1}{k} = \frac{512\,P}{3\pi^2}\,\frac{1}{n^2}\int_0^\infty \frac{\alpha(\delta)}{\delta^6}\, d\delta. \tag{6.2.4.13}$$

This expression for the permeability seems to be quite different from anything obtained above. It appears as if the permeability would decrease for increasing porosity. However, this is not the case. It should be observed that n, the number of capillaries per unit area, has a peculiar influence: if the porosity is to be increased with n being held constant, then this can be done only by lengthening the flow channels, and therefore the permeability will be less.

It is therefore better to introduce the tortuosity T as defined in (6.2.4.6) instead of n. It follows from (6.2.4.8) that

$$\frac{\pi}{4} n\delta^2 = \frac{P}{3T}. \tag{6.2.4.14}$$

Multiplying this by $\alpha(\delta)d\delta$ and integrating gives

$$\frac{P}{3T} = \frac{\pi}{4} n\int \delta^2 \alpha(\delta)d\delta. \tag{6.2.4.15}$$

Thus, substituting n from this expression into equation (6.2.4.13) yields

$$\frac{1}{k} = \frac{96T^2}{P}\left(\int \delta^2 \alpha(\delta)d\delta\right)^2\int \frac{\alpha(\delta)}{\delta^6}\, d\delta. \tag{6.2.4.16}$$

This is identical with what was obtained above by assuming a more simplified model (equation (6.2.4.5)), if one sets

$$\frac{1}{\bar{\delta}^2} = \left(\int \delta^2 \alpha(\delta)d\delta\right)^2\int \frac{\alpha(\delta)}{\delta^6}\, d\delta. \tag{6.2.4.17}$$

The 'average pore radius' δ introduced above has thus again been given a more specific meaning as expressed by the last equation (6.2.4.17).

6.2.5 Branching type models

The various types of models discussed above yield, in certain instances, a satisfactory explanation of Darcy's law. However, they neglect the fact that the fluid flow path may branch and, later on, join together again. In order to take the phenomenon of branching into account, it is necessary to have a model in which one capillary may split into two or more. These models are called *branching type models*.

As the element of a capillaric model containing capillaries that split up into several, one may consider a pore doublet as introduced by Benner et al. (1943; cf. fig. 26), and later discussed further by Rose and Witherspoon (1956). Similarly, Fatt (1956) and Emersleben (1964) discussed *network models* of the flow of fluids through porous media in which a multitude of capillaries are arranged in the form of a regular network. The dynamic properties of the resulting structure are then deduced by calculation. This is a rather involved procedure which is best carried out by means of a high-speed electronic computer.

A more sophisticated type of network model was proposed by Schopper (1966), who introduced a statistical element into the configurations. This is done by introducing a network function describing the hydraulic resistance as a function of all the individual capillary resistances, whose properties are guessed at by making reasonable assumptions. The parallel type and serial type models occur then as limiting cases of a network model.

In a later paper, Rink and Schopper (1968) made actual mesh-type models of porous media whose properties were then followed up on a computer.

6.2.6 Comparison with tests

It remains to compare the performance of the models discussed above with actual tests on porous media. The first model of parallel capillaries has very little mean-

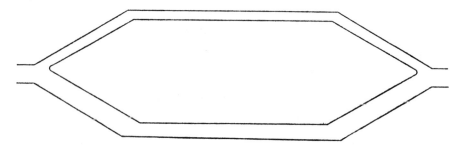

FIGURE 26   Pore doublet (after Benner et al. 1943).

ing, since the average pore diameter $\bar{\delta}$ is not properly defined. It is of course obvious that if $\bar{\delta}$ is *defined* by equation (6.2.2.6) any porous pedium will suit the model. But then $\bar{\delta}$ has very little connection with an average pore diameter as obtained from the pore size distribution and may at best be introduced as a 'Kozeny constant' or something similar.

A proper modification of the model of parallel capillaries which states plainly the connection of k and P with the pore size distribution is expressed by equation (6.2.3.5). In order to compare this equation with actual porous media, one needs the function $\alpha(\delta)$ for the latter. This function may be obtained from pore size distribution analyses, for example by the mercury injection method (see sec. 3.4.3). Then one can calculate the value of the integral I,

$$I = \int_0^\infty \alpha(\delta)\delta^2 d\delta, \tag{6.2.6.1}$$

and compare it with the expression $96k/P$ obtained from permeability and porosity data. At this point, one difficulty becomes apparent. This is that the integral is most affected by the value of $\alpha(\delta)$ for large pore sizes – and these cannot be measured by the mercury injection technique. If the integral is carried well into the region of large pores or vugs, its magnitude becomes out of all proportion and many powers of ten too large for what it should be, namely, equal to $96k/P$.

One can try to save the situation by saying that the vugs would have to be treated separately as enclosed cavities affecting the permeability, as discussed earlier (in sec. 5.2.1). Then it would have to be assumed that the integral must be cut off at some universal cut-off diameter; but an investigation shows that the cut-off diameter varies at least within a factor 100. This observation is substantiated by Henderson (1949), who shows that the permeability calculation from the pore size distribution based on the model of parallel capillaries does *not* work.

To come to the serial type models, it is at once apparent that the integrals in the final formula (6.2.4.16) are very sensitive to errors in $\alpha(\delta)$ for very low and very high values of the argument δ. These are just the values of δ for which the pore size distribution function is very difficult to measure. On the other hand, there is an additional parameter T (the tortuosity) so that there is no doubt that the model fits any porous medium, by a simple adjustment of T.

6.3
HYDRAULIC RADIUS THEORIES

6.3.1 *Principles of theory*

From the above description of the capillaric model theories it is apparent that more elaborate models must be used in order to obtain a satisfactory understanding of the hydromechanics of fluids in porous media. A set of theories is thus based upon the assumption that a porous medium is equivalent to a series of channels; the latter are, however, assumed to be somewhat more elaborate than in the capillaric models. These theories all make use of the fundamental observation that the permeability, in absolute units, has the dimension of an area or of a length squared. It may be argued, therefore, that a length should be characteristic for the permeability of a porous medium. Such a length may be called the 'hydraulic radius' of the porous medium, and is presumably linked with the hypothetical channels to which the porous medium is thought to be equivalent. A possible measure of a hydraulic radius would, for instance, be the ratio of the volume to the surface of the pore space.

Thus, the basic concept of the hydraulic radius theories is a consequence of dimensional considerations. One has to add, therefore, that the permeability, in addition to being proportional to some hypothetical square length, may also be dependent on any dimensionless quantities, notably on some function of the porosity (as the latter is dimensionless). Therefore the hydraulic radius theories assume the following basic expression for the permeability:

$$k = cm^2/F(P),$$
(6.3.1.1)

where m is the 'hydraulic radius,' $F(P)$ is the 'porosity factor,' and c is some dimensionless constant which could be incorporated into the porosity factor. The problem is, then, to find a physical significance for m and $F(P)$.

The assumptions basic to the hydraulic theories have been discussed by Carman (1941, 1948) and Klyachko (1948). According to them, the premises of the theories are:

(i) no pores are sealed off;
(ii) the pores are distributed at random;
(iii) the pores are reasonably uniform in size;
(iv) the porosity is not too high;
(v) diffusion (slip) phenomena are absent;
(vi) fluid motion occurs like motion through a batch of capillaries.

The simplest approach to the present problem is once more by construction of geometrical models. Leïbenzon (1947, pp. 31ff.) gave a good review of such attempts. Thus Slichter (1899) considered a bed of spherical particles as equivalent to a porous medium. On this basis he set m in (6.3.1.1) equal to the diameter δ of the spheres and deduced the following expression for the permeability:

$$k = \frac{n^2\delta^2}{96(1 - P)},$$
(6.3.1.2)

where n is the same quantity as defined on page 21 (cf. Leïbenzon 1947, p. 34). It is equal to 0.0931 for the densest packing of the spheres. Graton and Fraser (1935), Missbach (1938), and Dubinin (1941) extended the work on such models consisting of spheres. The modern version of the hydraulic radius theory, however, is based upon a completely different theoretical representation of porous media.

6.3.2 *The Kozeny theory*

Kozeny's work (1927a) is today a widely accepted explanation for the permeability as conditioned by the geometrical properties of a porous medium, although rather severe objections against it will be pointed out later. The Kozeny theory represents the porous medium by an assemblage of channels of various cross-sections, but of a definite length. The Navier-Stokes equations are solved simultaneously for all channels passing through a cross-section normal to the flow in the porous medium. Finally, the permeability is expressed in terms of the specific surface of the porous medium, which is, according to remarks in section 6.3.1, a measure of a properly defined (reciprocal) hydraulic radius. However, certain aspects are neglected, notably by *assuming that* in a cross-section 'normal' to the channel *there is no tangential component of the fluid velocity*. The Kozeny theory therefore neglects the influence of conical flow in the constrictions and expansions of flow channels, just as capillaric model theories do. The Kozeny equation was developed independently a few years after Kozeny in the United States by Fair and Hatch (1933).

In detail, the Kozeny theory proceeds as follows. Assume that a stream tube has a cross-section F and a pore cross-section f with

$$f = PF, \tag{6.3.2.1}$$

where P is, as usual, the porosity. Then one has for the volume of liquid passing during the time dt

$$Qdt = FdsP, \tag{6.3.2.2}$$

where Q is the volume of fluid passing during unit time and ds is the line element traversed in the time dt. Now, if v is the average pore velocity and q the filter velocity, one has

$$vf = qF = Q. \tag{6.3.2.3}$$

From (6.3.2.2) and (6.3.2.3) it follows that, if $v = ds/dt$,

$$q/v = f/F = P. \tag{6.3.2.4}$$

As outlined earlier, the last equation is known as the Dupuit-Forchheimer equation. In the Kozeny theory it comes out as a consequence of the Kozeny model of porous media and is therefore subject to the criticisms of that model as will be pointed out later. As an a priori assumption it was postulated by Dupuit (1863); see equation 6.2.1.1.

If one assumes a cross-section of the stream tube in the xy plane, the equation of motion is (cf. equation 2.2.2.1)

$$(\partial^2/\partial x^2 + \partial^2/\partial y^2) V_p = - \operatorname{grad} p/\mu, \tag{6.3.2.5}$$

where V_p is the velocity at the point x, y of the cross-section of the pores and μ the viscosity. Introducing new variables

$$\xi = x/\sqrt{f}, \qquad \eta = y/\sqrt{f},$$

one obtains the differential equation

$$(\partial^2/\partial \xi^2 + \partial^2/\partial \eta^2) V_p = - f \operatorname{grad} p/\mu, \tag{6.3.2.6}$$

which has the solution

$$V_p = f \frac{- \operatorname{grad} p}{\mu} \psi(x/\sqrt{f}, y/\sqrt{f}). \tag{6.3.2.7}$$

The function ψ is given by

$$\psi(x/\sqrt{f}, y/\sqrt{f}) = - \frac{x^2 + y^2}{4f} + \sum_{n=1}^{\infty} a_n \Phi_n(x + iy), \tag{6.3.2.8}$$

where Φ denotes a harmonic function. Assuming that the fluid sticks to the walls of the pores, one has the boundary condition

$$\psi(x/\sqrt{f}, y/\sqrt{f}) = 0, \tag{6.3.2.9}$$

or

$$\frac{x^2 + y^2}{4f} = \sum_{n=1}^{\infty} a_n \Phi_n(x + iy). \tag{6.3.2.10}$$

If equation (6.3.2.10) is written in polar coordinates r and ϕ, one obtains for the boundary $(r = \rho)$

$$\frac{\rho^2}{4f} = \sum_{n=1}^{\infty} a_n \Phi_n(\rho e^{i\phi}). \tag{6.3.2.11}$$

If the circumference of the cross-section of the pore is M, and the radius of a circle of the same circumference is ρ_m, with

$$\rho_m = M/(2\pi),$$

then it is possible to rewrite (6.3.2.8), observing the boundary condition, and one obtains:

$$\psi(r \cos \phi/\sqrt{f}, r \sin \phi/\sqrt{f})$$
$$= \frac{M^2}{16\pi^2 f} \left(\sum_{n=1}^{\infty} a_n \Phi_n(re^{i\phi}) \frac{4f}{\rho_m^2} - \frac{r^2}{\rho_m^2} \right) \equiv \frac{M^2}{16\pi^2 f} A. \tag{6.3.2.12}$$

Then one obtains for the average pore velocity v

$$v = \frac{\int V_p dF}{\int dF} = -\frac{f \operatorname{grad} p}{\mu F 16\pi^2} \frac{M^2 F}{f} A_m,$$ (6.3.2.13)

where

$$A_m = \int\int A r dr d\phi / \int\int r dr d\phi$$

is the mean value of A averaged over the area. If the expansion of the circumference is represented by $u_0 = M^2/f$, then it is clear that A and v become smaller as u_0 becomes larger. In the limiting case of maximum expansion of the circumference, $v = 0$, since all points of the cross-section will be on its edge. Equally, the average value A_m becomes smaller as the expansion of the circumference becomes larger; it must decrease faster than the expression $16\pi^2 c/u_0$, where c is a dimensionless number which depends only on the shape of the cross-section. The last condition is satisfied by the expression $16\pi^2 c/u_0^\zeta$ for all $\zeta > 1$.

Now, the pore cross-section of a virtual stream tube consists of a certain number n of single pore cross-sections, whose area f_1 we assume to be equal to all others. Then

$$f = nf_1 \quad \text{and} \quad M = nM_1,$$

where M_1 is the circumference of one single pore.

If we write equation (6.3.2.13) in the form

$$v = -(\operatorname{grad} p) F W/\mu,$$ (6.3.2.14)

then this expression must not change if the number of the pores is changed, i.e., if we change n. According to the above formulas one has

$$W = \frac{f}{F} c \frac{1}{u_0^{\zeta-1}} = \frac{c}{F} \frac{f}{M^{2\zeta-2}} = \frac{c}{F} \frac{f_1^\zeta}{M_1^{2\zeta-2}} \frac{1}{n^{\zeta-2}}.$$ (6.3.2.15)

It is obvious from this equation that W is invariant with respect to n, if the exponent ζ is made equal to 2.

Yet the filter velocity q, so far, is dependent on n. We have the equation

$$v = -\operatorname{grad} p \left(\frac{c}{\mu}\right)\left(\frac{f}{F}\right)\left(\frac{f}{M^2}\right) F,$$ (6.3.2.16)

and also the equation

$$q = -\frac{k}{\mu} \operatorname{grad} p = \frac{f}{F} v = -\frac{c}{\mu} \operatorname{grad} p \left(\frac{f}{M}\right)^2 \left(\frac{f}{F}\right).$$ (6.3.2.17)

Since f/M is independent of f/F and the latter is constant along a stream line, the expression f/M must also be constant along a stream line and hence in a stream tube. Thus, for a cylindrical stream tube of constant cross-section F, the quantities f and M must also be constant. However, the surface S_t in a stream tube of length L is given by

$$S_t = ML.$$

We can rewrite this as follows:

$$\frac{f}{M} = \left(\frac{f}{F}\right)\left(\frac{LF}{ML}\right) = \frac{P}{S},$$

where S is now the specific surface of the tube. Thus, we can write for (6.3.2.17)

$$q = -\frac{cP^3}{\mu S^2}\, \text{grad}\, p. \tag{6.3.2.18}$$

Comparing this with Darcy's law we obtain for the permeability

$$k = cP^3/S^2. \tag{6.3.2.19}$$

This relation, called the 'Kozeny equation,' shows that the filter velocity is inversely proportional to the square of the surface area per unit volume. The number c fluctuates, theoretically, only very little. We have for a circle $c = 0.50$; for a square $c = 0.5619$; for an equilateral triangle $c = 0.5974$; and for a strip $c = \frac{2}{3}$. The number c is called the 'Kozeny constant.'

One can now extend the Kozeny equation by introducing a 'tortuosity' T as an undetermined factor, in accordance with what was done for capillaric models. Thus, one can argue that equation (6.3.2.18) should refer to a 'reduced' pressure gradient

$$\text{grad}\, p_{\text{red}} = (1/T)\, \text{grad}\, p,$$

rather than to the 'apparent' one, indicating that the actual flow path is T times longer than the 'apparent' path straight across the porous medium; T is then called the 'tortuosity.' This changes the expression for the permeability (6.3.2.19) to read:

$$k = cP^3/(TS^2). \tag{6.3.2.20}$$

It may be noted that, owing to a difference in the models employed, the tortuosity T enters into the expression for the permeability in the Kozeny theory in a different manner than in the capillaric model theory (see equation 6.2.4.4).

The Kozeny theory is intended to have general applicability to all porous media, because the constants c and T, theoretically, involve only the detailed structure of the medium. However, it should be noted that the concept of tortuosity is really alien to the Kozeny theory. It does not come into play during the actual deduction of the Kozeny equation and can therefore be justified only a posteriori by the desire to have another arbitrary parameter.

6.3.3 Modifications of the Kozeny theory

The basic treatment of Kozeny has been modified on several occasions. In these modifications, the result that the permeability is essentially proportional to $1/S^2$, i.e., proportional to the square of a properly defined hydraulic radius, remains, but a great uncertainty over the correct porosity factor exists. Equations analogous to (6.3.2.20) are commonly called 'Kozeny' equations, whether or not the porosity dependence is that given by Kozeny. The various different porosity factors are usually obtained from more or less refined models.

Such modifications of the Kozeny equation have been proposed by Terzaghi (see Leïbenzon 1947, p. 38), Zunker (1932), and Bakhmeteff and Feodoroff (1937). A much used modification was postulated by Carman (1937, 1938a, 1939a), who set (Kozeny-Carman equation)

$$k = P^3/[5S_0^2(1 - P)^2], \qquad (6.3.3.1)$$

where S_0 is Carman's 'specific' surface exposed to the fluid, i.e., the surface exposed to the fluid per unit volume of *solid* (not porous) material. Equation (6.3.3.1) implies that the value of the Kozeny constant c is equal to $\frac{1}{5}$, as this gives, according to Carman, the best agreement with experiments. This is at variance with the calculations of Kozeny, according to which one ought to expect c to equal $\frac{1}{2}$. Here, therefore, is one of the facts which casts some doubts upon the Kozeny theory.

Another modification of the hydraulic radius theory appears to be due to Sullivan (1942), who introduced an orientation factor θ, which is defined as the average value of the square of the size of the angle between a normal to the walls forming the microscopic flow channel and the macroscopic direction of flow. The resulting formula is

$$k = c\theta P^3/[S_0^2(1 - P)^2], \qquad (6.3.3.2)$$

where c is a Kozeny constant that should be the same for all channels of the same geometric shape and that should not vary markedly from shape to shape.

Further discussions of the hydraulic radius theory may be found in papers by Wiggin, Campbell, and Maass (1939), Mitton (1945), Zhuravlevi and Sychev (1947), Baver (1949), Klyachko (1948), Arthur et al. (1950), Iberall (1950), and Loudon (1952). In these papers, the hydraulic radius theory has been carried to a high degree of refinement. Notably, Klyachko gives an analytical method of calculating the Kozeny constant. The theory has also been applied extensively to determine the internal structure of porous media, as will be shown later.

A modification of the hydraulic radius theory has been suggested by Mrosowski (1958). In this modification, a model is envisaged for which the main resistance to flow occurs at and around constrictions. The permeability is then a product of two factors; one corresponding to the resistance of a constriction and the other representing the statistics of the distribution of constrictions through the material. Basically the ensuing theory is still a hydraulic radius theory.

Similarly, Wyllie and Gardner (1958a, b) expanded the Kozeny-Carman theory to contain certain statistical features; they devised a capillaric model which they combined with a random interconnection of the pores. The capillaric model consists of a bundle of straight, capillary tubes the radii of which have a suitable range. This bundle is cut into thin slices of equal thickness which are then re-assembled in a random fashion. In this way, a new expression is obtained for the tortuosity. Wyllie and Gardner (1958b) claim that the general agreement with experimental evidence is fairly good although they have to state that 'it cannot be said that an unequivocal conclusion has been reached.'

6.3.4 *Experiments in connection with the hydraulic radius theory*

Many experiments have been based upon the Kozeny theory. Mostly, they are determinations of the constants occurring therein, particularly of the internal surface of porous media. The surface determination will be discussed later in detail, and we shall confine ourselves in this section to the discussion of other experiments.

An actual substantiation of the Kozeny equation is almost impossible because it contains three parameters (S, c, T), all of which would have to be determined by independent means in order to obtain a valid check. This, however, intro-duces many errors, especially in view of the fact that T is not even properly defined as it was introduced by the Kozeny equation itself.

Nevertheless, a number of experiments have been performed which claim to 'substantiate' the Kozeny theory, usually simply by measuring some of the constants occurring therein and showing that the values obtained are not unreasonable. Such experiments have been reported by Bartell and Osterhof (1928), Donat (1929), Kozeny (1932), Zunker (1932), Carman (1939a, b), Wiggin, Campbell, and Maass (1939), Fowler and Hertel (1940), Rigden (1943), Adams, Johnson, and Piret (1949), Blaine and Valis (1949), Kwong et al. (1949), Rose and Bruce (1949), Coulson (1949), Hoffing and Lockart (1951), and Sathapathy and Rao (1954).

An improvement over the above 'verifications' of the Kozeny equation has been attempted by Wyllie and co-workers (Wyllie 1951; Wyllie and Spangler 1952; Wyllie and Gregory 1955). Here an attempt was made to attach a more precise physical meaning to the 'tortuosity.' Wyllie postulated that the electrical and fluid flow paths should be identical if the electrical and pressure potential are analogous. Thus, the tortuosity could be measured by independent means (cf. sec. 1.6). However, inspection of Wyllie's work shows that there is some uncer-tainty regarding the proper expression of the tortuosity in terms of the electrical analogue, reflecting itself in the fact that the Kozeny constant is, in every case, adjusted a posteriori.

The Kozeny theory, therefore, contains some vague factors and it is the opinion of the writer that more than a qualitative description of the phenomena cannot be expected from it.

6.3.5 Criticism of the Kozeny theory

Some criticisms of the Kozeny theory from a theoretical standpoint have already been mentioned during the exposition of it above. In addition, Coulson (1949) made an extensive investigation of the data of previous authors, and on the basis of information obtained, from fluid flow through systems of packed spheres, prisms, cubes, cylinders, and plates, he concluded that more general formulations still must be developed to describe adequately flow phenomena in non-spherical systems. Similarly, Childs and Collis-George (1950) put forth severe criticisms of the Kozeny theory from other theoretical reasoning. They state that it seems to be generally admitted that hydraulic radius theories utterly fail to describe structured bodies such as, for example, 'stiff-fissured' clays. The structural fissures contribute negligibly both to porosity and to specific surface, and yet they dominate the permeability. Again, neither the porosity nor the internal surface is a directed quantity, and therefore the Kozeny formula cannot indicate anisotropic permeability, which nevertheless seems to be the rule rather than the exception in nature.

The whole question of the validity of the Kozeny equation was reviewed by Scheidegger (1962). Accordingly, many discrepancies with the Kozeny theory can be obtained from experiments, provided not too many undetermined 'fudge' factors are introduced into it. Thus, Kozeny himself (1927b), in an attempt to substantiate his formula experimentally, found discrepancies between calculated and measured surface areas of -69 to $+86$ per cent. It is a matter of taste whether one wants to consider these as merely experimental errors. More recently, Macey (1940) found enormous changes in permeability with porosity which are not explainable by the Kozeny equation. Similarly, Sullivan and Hertel (1940) and Sullivan (1941) showed that Kozeny's law breaks down, at least in the case of highly porous, fibrous media. Walas (1946) found that a modification in the Kozeny equation is necessary if it is to be applied to filtration, and Adamson (1950) observed that the Kozeny constant has to be adjusted a posteriori in every experiment, which is, of course, quite unsatisfactory. Wyllie and Rose (1950), in an attempt to justify the Kozeny equation, place much emphasis upon the 'tortuosity,' which, however, is a rather vague physical concept, and, if measured electrically, is in need of a proper justification for an application to fluid flow. Furthermore, Brooks and Purcell (1952), again claiming that their results substantiate the Kozeny equation, in fact find a large discrepancy when areas are measured by the gas-adsorption method and with the aid of the Kozeny equation. Finally, Kraus and co-workers (Kraus and Ross 1953; Kraus, Ross, and Girifalco 1953) attempted to correlate pressure drops and flow rates through beds of many materials with gross particle surface, as determined from geometric considerations, and with microstructure surface from nitrogen adsorption measurements. They found that both surfaces contribute to pressure loss, but by *different* mechanisms. Childs and Collis-George (1950) summarize the consequences of these experiments appropriately by stating that although most authors claim agreement with observed values of permeability, 'the fact seems to be that it has been impossible to secure a sufficiently wide range of variation of porosity and pore-size distribution to test the formulas severely, and the tests which have been reported above indicate a wide range of error. That the method of determining the specific area of powders by measurement of permeability has achieved considerable popularity [see sec. 6.4.1] may reflect only the essential similarity of most industrial problems involved.'

6.4
STRUCTURE DETERMINATIONS OF POROUS MEDIA

6.4.1 *Determination of surface area*

The Kozeny equation contains as a significant variable the specific surface area of a porous medium. Therefore it should permit one to calculate the specific surface area of a porous medium from permeability measurements, provided the other 'Kozeny constants' are known. For purposes of surface determination, it is generally assumed that $c = \frac{1}{5}$ (in equation 6.3.2.19) and the tortuosity factor is absent. This corresponds to the 'Kozeny-Carman equation' (6.3.3.1).

The determination of the surface area of porous media, mainly of industrial powders, on this basis has been chiefly advanced by Carman (1938b, 1939b, c) and by Lea and Nurse (1939). The method has since become very popular and many applications and refinements have been proposed. One of these refinements is to include a 'slip' term in the Kozeny equation as will be discussed in chapter 7. Reviews of these methods have been given by Blaine (1941), Zimens (1944), and Svensson (1949). Some of the values obtained from these investigations are shown in table II.

Very few of the surface determinations refer to any critical evaluations of the accuracy of the method. The Kozeny-(Carman) equation is usually assumed to be unquestionable and the results are presented as final values obtained by means of that equation.

6.4.2 *Determination of other geometrical quantities*

According to the simple geometrical models discussed earlier in this chapter, permeability measurements can be used, theoretically at least, to determine other geometrical quantities of porous media. This idea was taken up by Traxler and Baum (1936), who reported a determination of the 'average pore size' from permeability measurements. Similarly, Gooden and Smith (1940) reported calculations of the average particle diameter of powders from permeability, Kuhn (1946) claims to have determined the gel structure in a similar manner, and Wilson (1953) uses the Kozeny equation to determine a mechanism of agglomeration. Eyraud et al. (1963) used the movement of a gas through a porous medium to determine the distribution of pore radii. They believed that the pore size distribution was reflected in the variation of gas flow for an incremental change in the relative pressure.

These determinations are all based on rather specific assumptions about the porous medium, i.e., on some particular model. As the porous media under actual investigation were probably rather remote from such models, not too much confidence can be placed in the reported results.

6.5
DRAG THEORY OF PERMEABILITY

6.5.1 *Principles*

An approach to the physical explanation of permeability different from that of Kozeny was initiated by Emersleben (1924, 1925). This approach might be called the 'drag theory' of permeability. In it, the walls of the pores are treated as obstacles to an otherwise straight flow of the viscous fluid. The drag of the fluid on each portion of the walls is estimated from the Navier-Stokes equations, and the sum of all drags is thought to be equal to the resistance of the porous medium to flow (i.e., equal to μ/k, according to Darcy's law). It is to be expected that the drag theory will give good results for highly porous media, such as fibres, where the single particles can be actually regarded as solitary within the fluid. Newer versions of the drag theory have been given by Iberall (1950), Brinkman (1947, 1948, 1949), Mott (1951), and Happel (1959; see also Happel and Brenner 1965).

There are two types of specific drag theories, depending on the geometrical shapes of the 'obstacles' that are envisaged. The first approach is to take the obstacles as fibres, and the second to take them as spheres. Furthermore, an attempt can be made to employ the basic idea of the drag theory of permeability in general terms; i.e. to calculate the resistance of generally shaped obstacles in a steady stream of a viscous fluid. This leads to the general 'Stokes flow approach' in flow through porous media; it is called thus because Stokes was one of the first to calculate the resistance arising in steady-state flows of viscous fluids around variously shaped objects.

We shall discuss these various possibilities one by one below.

F

6.5.2 *Fibre theory*

We treat first the possibility of envisaging the 'dragging' obstacles as fibres. Iberall (1950) used such a model, taking a random distribution of circular cylindrical fibres of the same diameter and accounting for the permeability on the basis of the drag on individual elements. It is assumed that the flow resistivity of all random distributions of the fibres per unit volume will not differ, and that it is the same as that obtained with an equipartition of fibres in three perpendicular directions, one of which is along the direction of macroscopic flow. It is further assumed that the separation between fibres and the length of individual fibres are both large compared to the fibre diameter (high porosities), and that the disturbance due to adjacent fibres on the flow around any particular fibre is negligible.

If it is assumed that fluid inertial forces are negligible (low local Reynolds number), an equation can be derived by equating the pressure at two planes perpendicular to the direction of macroscopic flow to the viscous drag force on all elements between the planes. It is assumed that the pressure drop necessary to overcome the viscous drag is linearly additive for the various fibres, whether parallel or perpendicular to the flow.

The drag force per unit length of a single fibre surrounded by similar fibres all oriented along the direction of flow and with moderate separations has been given approximately by Emersleben (1925):

$$f = 4\pi\mu v, \tag{6.5.2.1}$$

where f is the drag force per unit length of fibre and v is the velocity of the fluid stream distant from the filament, i.e., the pore velocity.

If it is assumed that there are n filaments of unit length per unit volume, and that $n/3$ filaments are arrayed in each of the three perpendicular directions, the total drag force in a unit volume due to the $n/3$ filaments parallel to the flow can be equated to the pressure drop per unit length, so that

$$\Delta p / L = (4\pi n / 3)\mu v. \tag{6.5.2.2}$$

From calculations of the flow resistance for a cylinder perpendicular to a stream (see Lamb 1932), the drag force for such filaments is given by

$$f = \frac{4\pi}{2 - \ln(\mathrm{Re})} \mu v \tag{6.5.2.3}$$

in which $\ln(\mathrm{Re})$ is the natural logarithm of the local Reynolds number, defined as $\delta v \rho / \mu$, with δ the fibre diameter and ρ the fluid density.

The drag force on each of the two sets of $n/3$ filaments per unit volume, arrayed perpendicularly to the fluid flow, can be equated to the pressure drop per unit length, giving

$$\frac{\Delta p}{L} = \frac{4\pi n}{3[2 - \ln(\mathrm{Re})]} \mu v. \tag{6.5.2.4}$$

Linear superposition or simple addition of the three sets of pressure drops necessary to overcome the drag of the three sets of filaments results in a total pressure gradient of

$$\frac{\Delta p}{L} = \frac{4\pi n}{3} \frac{4 - \ln(\text{Re})}{2 - \ln(\text{Re})} \mu v. \tag{6.5.2.5}$$

The number of fibres n per unit volume, which is also equal to the total fibre length per unit volume, may be eliminated, as the apparent density of a fibrous pack in vacuum ρ_p is equal to the product of the fibre volume and the fibre density ρ_f or

$$\rho_p = \tfrac{1}{4}\pi \delta^2 n \rho_f. \tag{6.5.2.6}$$

Upon elimination of n, equation (6.5.2.5) reduces to

$$\frac{\Delta p}{L} = \frac{16\mu v}{3} \frac{\rho_p}{\rho_f \delta^2} \frac{4 - \ln(\text{Re})}{2 - \ln(\text{Re})}. \tag{6.5.2.7}$$

The velocity profile between fibres is assumed to be sufficiently flat for high porosities that the velocity may be taken as constant. The velocity v is therefore related to the volumetric rate of flow Q and the macroscopic cross-sectional area of a porous medium A by the (Dupuit-Forchheimer) relation

$$v = Q/(PA) = q/P, \tag{6.5.2.8}$$

$$q = Q/A. \tag{6.5.2.9}$$

It follows from the definition of the porosity P that

$$1 - P = \rho_p/\rho_f. \tag{6.5.2.10}$$

With the use of these two relations, equation (6.5.2.7) may be put in the form

$$\text{grad } p = q \frac{16\mu}{3} \frac{1 - P}{P\delta^2} \frac{4 - \ln[\delta q\rho/(\mu P)]}{2 - \ln[\delta q\rho/(\mu P)]}. \tag{6.5.2.11}$$

Although the derivation, as given, assumed an incompressible fluid, it can be readily shown that the derived equations are unchanged for a compressible fluid, flowing isothermally, if the volumetric flow q_m at the arithmetic mean pressure is used in equation (6.5.2.11). It is therefore applicable to both liquids and gases.

If equation (6.5.2.11) is compared with Darcy's law, one obtains formally for the permeability:

$$k = \frac{3}{16} \frac{P\delta^2}{1 - P} \frac{2 - \ln[\delta q\rho/(\mu P)]}{4 - \ln[\delta q\rho/(\mu P)]}. \tag{6.5.2.12}$$

This shows a very important point contiguous to the drag theory of permeability. It is noted that the permeability is not a constant but varies with the flow velocity. This slow variation of permeability with flow is quite characteristic of many instances in which the flow is nominally viscous. Although in general it is reasonable to assume that this effect is associated with fluid inertia, it is often difficult to account for it precisely. It is thus seen that the drag theory of permeability is not only a very different account of permeability, but even modifies the Darcy equation and thus leads to more general flow equations.

6.5.3 *Spherical obstacles*

As pointed out above, a different drag model can be obtained by taking the obstacles as spherical. We follow here the exposition of Brinkman (1947, 1948, 1949), who assumed that the particles in the fluid are spheres of radius R and that they are kept in position by external forces, as in a bed of closely packed particles which support each other by contact.

Following Brinkman (1949), in order to formulate an equation describing the flow of a fluid through such a swarm, one has to consider the forces acting on a volume element of fluid containing many particles.

On the one hand, there are normal and shearing stresses which act on the surface of the volume element. If no particles were present, these stresses would cause a force $\mathbf{F}_1 dV$ which is given by the Navier-Stokes equation (2.2.2.1):

$$\mathbf{F}_1 = - \operatorname{grad} p + \mu \operatorname{lap} \mathbf{v}, \tag{6.5.3.1}$$

where p is, as usual, the pressure, $\mathbf{v}$ is the flow velocity vector, and μ is the viscosity. The inertia terms have been neglected (i.e., the density ρ of the fluid has been assumed to be equal to zero), and the fluid has been assumed to be incompressible.

On the other hand, owing to the presence of particles, equation (6.5.3.1) has to be modified. In the first place, there may be particles occupying positions near the surface of the volume element dV and diminishing the area of contact between the fluid inside and outside the volume element. In the second place, if the particles are not closely packed, they may rotate and thus cause an increase in $\mathbf{F}_1$. At any rate, the velocity of flow will be a rapidly fluctuating function of the position in the fluid, having zero value at the surface of the particles and maxima in the spaces between them.

In anticipation of a consistent application of statistical mechanics to the hydrodynamics in porous media (see chapter 8), Brinkman (1949) defined a mean flow velocity by taking an average over a volume V_0 which is small compared to the total volume V but which contains many particles:

$$\mathbf{q} = \frac{1}{V_0} \int_{V_0} \mathbf{v} dV. \tag{6.5.3.2}$$

The particles which are kept in position by external forces exert a drag force on the velocity of flow of the fluid. It seems reasonable to assume that this force $F_2 dV$ is proportional to the mean velocity and to the viscosity of the fluid. Brinkman therefore put

$$F_2 = -(\mu/k)\mathbf{q}, \tag{6.5.3.3}$$

where k is a constant depending on the density of the particles and their radii, as will be discussed below.

Consequently, if only viscous forces are taken into account, F_1 and F_2 are the only forces acting on the volume element considered. For a stationary state of flow one therefore has

$$F_1 + F_2 = 0. \tag{6.5.3.4}$$

As an approximation Brinkman substituted the expressions given in equations (6.5.3.1) and (6.5.3.2) into equation (6.5.3.4). Expression (6.5.3.1) will only be a good approximation for a low particle density on account of the disturbing influence of the particles as discussed before. Thus, for low particle densities Brinkman put

$$- \operatorname{grad} p + \mu \operatorname{lap} \mathbf{q} - \frac{\mu}{k}\mathbf{q} = 0. \tag{6.5.3.5}$$

This equation may, of course, be extended in the usual way to include inertia terms. It is seen that the present theory again leads to a modification of Darcy's law through the inclusion of the term μ lap $\mathbf{q}$. For high particle densities the term μ lap $\mathbf{q}$ is negligible compared to $\mu\mathbf{q}/k$. This means that Darcy's law is the limiting form of equation (6.5.3.5) for low permeabilities.

Equation (6.5.3.5) is therefore a new description of the state of flow for both low and high particle densities. The permeability k in equation (6.5.3.5) has yet to be determined. Following Brinkman, an expression for this permeability may be found by calculating the drag force F_2 as defined by equation (6.5.3.3) for an assemblage of spheres of radius R. It may be assumed that this force is the resulting sum of the viscous forces exerted by the individual spheres on the fluid. It is found first by calculating the force exerted by the fluid on one of the spheres and then by summing up over all the spheres.

The fluid flow round one sphere is derived from equation (6.5.3.5). This means that the influence of the other particles is taken into account by the drag term in the equation. Brinkman emphasized that this treatment might be justified if the surrounding spheres are small compared to the central sphere, and may be applied by analogy to the case where all the spheres are of the same size. Thus, for an incompressible fluid, a solution of equation (6.5.3.5) has to be found which obeys the condition of incompressibility

$$\operatorname{div} \mathbf{q} = 0 \tag{6.5.3.6}$$

and certain boundary conditions at the surface of the particle. These boundary conditions are that the normal and the tangential velocity components at the surface of the particle should be zero. The condition for the normal component results from equation (6.5.3.6) in the usual way. The condition for the tangential component results from certain considerations on the continuity of the viscous stresses combined with an assumption about the variation of the permeability k near a boundary. Brinkman emphasized that the boundary condition for the tangential component can only be derived from some additional assumptions about the swarm of particles.

The actual condition chosen by Brinkman is consistent with the simplest assumptions. Calculation of the detailed flow pattern through the swarm of spheres might lead to a justification of equation (6.5.3.5) combined with the above boundary conditions. Brinkman stated, however, that the result might well be that the mean rate of flow near a boundary could not be represented by the theory discussed above.

With these boundary conditions a solution of equations (6.5.3.5) and (6.5.3.6) can be found which represents a uniform parallel state of flow infinitely far removed from the sphere (Brinkman 1947). By calculating the viscous forces, the force F on the sphere is eventually found to be given by

$$F = 6\pi\mu q_0 R\left(1 + \frac{R}{\sqrt{k}} + \frac{R^2}{3k}\right), \tag{6.5.3.7}$$

where q_0 is the mean rate of flow far from the sphere and R is its radius.

Equation (6.5.3.7) is a modification of Stokes' law as generalized for a dense swarm of spheres. For an isolated sphere (i.e., the permeability approaching infinity), the correction factor approaches the value 1.

An expression for the permeability can now be found by calculating the total drag force on a unit volume of the swarm containing n spheres. This force is equal to nF, where F is given by equation (6.5.3.7). But, on the other hand, by definition it is equal to $\mu q_0/k$. From this equality Brinkman finally found, for k,

$$k = \frac{R^2}{18}\left(3 + \frac{4}{1-P} - 3\sqrt{\frac{8}{1-P} - 3}\right), \tag{6.5.3.8}$$

$$\tfrac{4}{3}\pi n R^3 = 1 - P, \tag{6.5.3.9}$$

where P is, as usual, the porosity.

In conclusion, it may be mentioned that the Brinkman model (of an assembly of spheres) has been discussed in somewhat more detail by Happel and co-workers (Happel and Byrne 1954; Happel and Epstein 1954; Happel and Brenner 1965). The reader is referred to these papers for details.

6.5.4 Stokes flow approach

Finally, we discuss the general possibilities of calculating the resistance of variously shaped objects in a steady-state stream of a viscous fluid. For this purpose, one has to solve the Navier-Stokes equation (2.2.2.1) for the appropriate boundary conditions (sticking at the surface of the obstacles). One makes, in this connection, the assumption not only that the time term $\partial/\partial t$ is zero (befitting a steady state), but also that the inertia term $\mathbf{v}$ grad $\mathbf{v}$ in (2.2.2.1) can be neglected. Under such conditions, the Navier-Stokes equation for an incompressible fluid can be written (for notation see chapter 2)

$$\mathbf{F} - \frac{1}{\rho} \operatorname{grad} p - \frac{\mu}{\rho} \operatorname{curl} \operatorname{curl} \mathbf{v} = 0, \qquad (6.5.4.1)$$

$$\operatorname{div} \mathbf{v} = 0. \qquad (6.5.4.2)$$

Philip (1969) has given some solutions of the above equations, based on earlier work by Rayleigh (1893), for informative models of divergent-convergent flows in plane 'pores.' It seems to turn out that the notion that viscous flow in porous media takes place down a potential gradient on a microscopic scale must be abandoned: the velocity profile in a converging pore is quite different from that of Hagen-Poiseuille flow in a straight capillary.

Nevertheless, on a macroscopic scale, a linear equation (i.e., Darcy's law) is always the outcome if one starts on a microscopic scale with equations (6.5.4.1–2), i.e., with the Navier-Stokes equations in which the inertia and time terms have been neglected. Matheron (1966) and Whitaker (1969) have given a formal proof of this. In order to do this, appropriate assumptions about the statistical nature of the geometry of the porous medium have to be introduced, inasmuch as the 'macroscopic' viewpoint is obtained from the 'microscopic' one by statistical averaging. In effect, the permeability can then in principle be deduced from the function $f(s)$ introduced by Fara and Scheidegger (1961) (cf. sec. 1.1) to describe a porous medium, provided the medium is reasonably well behaved (it must be a realization of an ergodic and stationary random function so that the required averages exist).

However, if the inertia terms are *not* neglected in the Navier-Stokes equation, then, even in laminar flow through the pores, Darcy's law cannot be expected to hold: non-linear terms arise as will be discussed in chapter 7.

7
General flow equations

7.1
LIMITATIONS OF DARCY'S LAW

In previous chapters, Darcy's law has been accepted as fundamental for the deductions reached. However, it has been mentioned that Darcy's law may possibly be valid only in a certain 'seepage' velocity domain, outside of which more general flow equations must be used to describe the flow correctly.

Darcy's law represents a linear relationship between the filtration velocity **q** and the gradient of the pressure (head); moreover, the straight line representing this relationship passes through the origin of the coordinates. Any deviation from this type of relationship represents 'non-Darcian flows.' Kutilek (1969) made a summary of the various possibilities, which are represented here in figure 27.

The various possibilities shown in figure 27 are entirely heuristic. A more fundamental classification of the various flow regimes is obtained by seeking the physical causes underlying the deviations from linearity. A review of the literature reveals that such deviations may be expected owing to (i) high flow rates, (ii) molecular effects, (iii) ionic effects, and (iv) non-Newtonian behaviour of the percolating fluid itself. We shall discuss these various possibilities below.

7.2
HIGH FLOW VELOCITIES

7.2.1 *Experimental evidence*

It was observed very early that Darcy's law is valid only for very low percolation velocities. The domain of validity of Darcy's law has been called the 'seepage velocity domain.'

In order to characterize this seepage velocity domain, it is customary to introduce a 'Reynolds number' Re as follows:

$$\text{Re} = q\rho\delta/\mu, \qquad\qquad (7.2.1.1)$$

where q is the (scalar) filter velocity, ρ the density, μ the viscosity of the fluid, and δ a diameter associated with the porous medium, i.e., the average grain or pore diameter or some other length corresponding to the hydraulic radius theory. In this and the following formulas it is always understood that consistent units (e.g., c.g.s. units) are used.

Similarly, it has been customary to express the resistance to flow of the porous medium by the introduction of a 'friction factor' λ such that

$$\lambda = 2\delta \operatorname{grad} p/(q^2 \rho), \tag{7.2.1.2}$$

where, in addition to the previous symbols, grad p is the one-dimensional pressure gradient (a suitable modification could be made to account for gravity flow if so desired). In this representation, Darcy's law would be written as follows:

$$\lambda = C/\mathrm{Re}, \qquad C = \frac{-2\delta^2}{K} \tag{7.2.1.3}$$

where C is a constant. It is therefore seen that the flow is subject to Darcy's law if λ and Re (for one porous medium) are inversely proportional.

The representation of the flow by means of the Reynolds number and friction factor is, of course, dependent on the choice of the length δ. It is thus tied up with

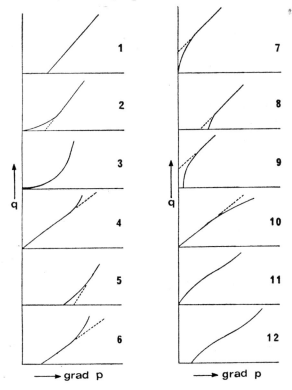

FIGURE 27   Twelve schematic flow curves for non-Darcian flow (after Kutílek 1969).

the notion of a hydraulic length as introduced by the hydraulic radius theory. We have listed above objections to the hydraulic radius theory; one of them was that the diameter δ is not really defined, either geometrically or by statistical reasoning. This same objection is true for the present representation. However, all experiments on the limitations of Darcy's law (at least at the high-velocity end) invariably seem to be tied up with the notion of the Reynolds number. Results obtained in this manner, therefore, must be regarded as subject to the same principal shortcomings as the hydraulic radius theory, including the Kozeny equation.

Nevertheless, many investigations have been directed towards finding the range of Reynolds numbers for which Darcy's law is valid. That the representation by Reynolds numbers was chosen originally is of course owing to the assumption of an analogy between the flow in tubes and the flow in a porous medium: i.e., the latter is thought to be equivalent to an assemblage of capillaries. Therefore one has been looking for a phenomenon in porous media similar to the onset of turbulence in tubes, which takes place at a definite Reynolds number. It was expected that above a certain number, which would be universal for all porous media, deviations from Darcy's law would occur.

Experiments for the determination of this 'critical' Reynolds number have been reviewed, for example, by Romita (1951), Hudson and Roberts (1952), and Leva and co-workers (1951). Original investigations have been made by a great number of people, notably Zunker (1920), Ehrenberger (1928), Fancher and Lewis (1933), Fancher, Lewis, and Barnes (1933), Hickox (1934), Schaffernak and Dachler (1934), Bakhmeteff and Feodoroff (1937, 1938), Carman (1937), Brieghel-Müller (1940), Chardabellas (1940), Gustafson (1940), Nissan (1942), Grimley (1945), Ruth (1946), Khanin (1948), Fox (1949), Cambefort (1951), Nielsen (1951), Plain and Morrison (1954), Stewart and Owens (1958), and Meadley (1962). The account of these researches in section 7.2.2 is essentially based upon the articles of Romita (1951) and Leva et al. (1951).

In these investigations, a great discrepancy regarding the 'universal' Reynolds number, above which Darcy's law would no longer be valid, is evident. The values range between 0.1 (Nielsen 1951) and 75 (Plain and Morrison 1954). The uncertainty by a factor 750 in the critical Reynolds number may reflect in part the actual indeterminacy of the diameter δ; however, it is probably essentially due to a fundamental failure of the hydraulic radius theory of porous media.

7.2.2 Heuristic correlations

The high-velocity flow phenomena occurring in porous media, described in section 7.2.1, can be put into mathematical terms in several ways. Without attempting to understand the physics of the effect, one can simply try to fit heuristically curves or equations to the experimental data so as to obtain a correlation between pressure drop and flow velocity.

In this instance, Forchheimer (1901) suggested that Darcy's law be modified for high velocities by including a second-order term in the velocity:

$$\Delta p/\Delta x = aq + bq^2 \qquad (7.2.2.1)$$

and later by adding a third-order term:

$$\Delta p/\Delta x = aq + bq^2 + cq^3. \qquad (7.2.2.2a)$$

Here, the pressure drop correlation is written for the linear case (linear dimension x); furthermore, p is as usual the pressure (gravity is neglected), q is the seepage velocity, and a, b, c are thought to be constants. The Forchheimer equations were postulated from semi-theoretical reasoning by analogy with the phenomena occurring in tubes. The third-order term was added to make the equation fit experimental data better.

The Forchheimer relation has been generalized further to contain a time-dependent term (Polubarinova-Kochina 1952):

$$\Delta p/\Delta x = aq + bq^2 + c\, \partial q/\partial t. \qquad (7.2.2.2b)$$

It turns out, however, that this term is small.

Another heuristic correlation was postulated by White (1935), who set

$$\Delta p/\Delta x = aq^{1.8}; \qquad (7.2.2.3)$$

this result was derived from an analysis of dry air flow through packed towers. The value of the exponent, however, seems not too well established, and thus Missbach (1937) set

$$\Delta p/\Delta x = aq^m, \qquad (7.2.2.4)$$

with m undetermined between 1 and 2. He further stated that as the diameter of the beds used for the packings became smaller, m approached 1 and the equation became Darcy's law. Similar experiments which bear out that there is a non-linear correlation between the pressure drop and percolation velocity have also been made by Wodnyanszky (1938), Spaugh (1948), Linn (1950), and Uguet (1951). The exact value of the exponent m, however, seems to vary from case to case so that no universal correlation could be achieved.

It is therefore obvious that a different type of correlation has to be established if there is to be any hope for it to be universal. An extremely popular correlation is that between the Reynolds number Re and friction factor λ (cf. equations 7.2.1.1 and 7.2.1.2 for their definition). Unfortunately, all such correlations are subject to limitations as the Reynolds number depends significantly upon the definition of a pore diameter, which cannot be achieved properly. Furthermore, if a true physical significance is to be attached later to correlations based on the Reynolds number, one has to introduce all the assumptions of the hydraulic radius theory.

A good review of the correlations that have been attempted between the friction factor and Reynolds number has been given by Romita (1951). Other general discussions of the subject have been given by Rose (1945a), Verschoor (1950), Leva (1949), Leva et al. (1951), and Hudson and Roberts (1952).

The number of papers proposing correlations between the friction factor and Reynolds number is very great. Fancher and Lewis (1933) represented the results of experiments as plots between λ and Re; their data were, however, still principally in the viscous range. Uchida and Fujita (1934; see also Fujita and Uchida 1934) passed gases through beds of broken limestone, lead-shot, and Raschig rings and expressed their results in the form of the equation

$$\operatorname{grad} p = \rho A \left(\frac{q^2}{2\delta}\right)^r \operatorname{Re}^s \left(\frac{\Delta}{\delta}\right)^t, \qquad (7.2.2.5)$$

where r and t are functions of the packing, and A and s are functions of both the packing and Reynolds number; Δ is the diameter of the vessel holding the packed bed. Experimental limits of $r, s, \ldots$ were also supplied.

Further investigations were carried out by Lindquist (1933). By a careful investigation of previous results, this author came to the conclusion that Darcy's law is valid for Re < 4; for 4 < Re < 180 he postulated the equation

$$\lambda \operatorname{Re} = a \operatorname{Re} + b \qquad (7.2.2.6)$$

with $a = 40$ and $b = 2500$. Givan (1934) made similar investigations and came up with an analogous relationship, but with the values $a = 34.2$ and $b = 2410$.

Similar investigations were undertaken by Fair and Hatch (1933), Mach (1935, 1939), Meyer and Work (1937), Johnson and Taliaferro (1938), Iwanami (1940), and Kling (1940). Kling obtained the result that Darcy's law holds up to Re = 10; for 10 < Re < 300 he postulated the correlation

$$\lambda = 94/(\operatorname{Re})^{0.16}.$$

Similar correlations have been postulated by Veronese (1941), namely:

$$\left. \begin{array}{l} \tfrac{1}{4}\operatorname{Re}\lambda = 1150 \text{ for Re} < 5, \\ \tfrac{1}{4}\operatorname{Re} 0.73\lambda = 720 \text{ for } 5 < \operatorname{Re} < 200, \\ \tfrac{1}{4}\lambda = 15.5 \text{ for Re} > 200. \end{array} \right\} \qquad (7.2.2.7)$$

The last expression is not valid for Re > 2000.

A series of further investigations was made by Holler (1943) and Gamson et al. (1943). The last authors give plots of the λ-Re correlation for air flow over wet and dry spheres and cylinders. They obtained separate curves for wet and dry packings in tubes. Furthermore, Allen (1944) obtained relationships for the flow of air, naphtha, and mineral oil through beds of granular adsorbents.

The above investigations are concerned with a correlation between λ and Re *only*. However, other variables also would be expected to have an influence. The porosity should be especially significant, for, if the idea of a Reynolds number is maintained, it should obviously be the pore velocity and not the filter velocity whose value is significant for the onset of 'turbulence.' According to the usual assumption (originally due to Dupuit), the pore velocity v is equal to q/P. Chilton and Colburn (1931a, b) disagreed with this, however, and postulatep that the porosity must be accounted for differently. Subsequently, Chalmers et al. (1932) introduced the porosity concept into the friction factor, which they defined as follows:

$$\lambda_c = \operatorname{grad} p \, \delta P^2 / (\rho q^2). \tag{7.2.2.8}$$

This expression for λ was used to obtain correlations.

A similar modification of the friction factor was introduced by Barth and Esser (1933), who wrote their friction factor λ_B as follows:

$$\lambda_B = \lambda P^3 / (1 - P). \tag{7.2.2.9}$$

With this friction factor, they postulated the following λ-Re correlation:

$$\lambda_B = 490/\mathrm{Re} + 100/(\mathrm{Re})^{\frac{1}{2}} + 5.85. \tag{7.2.2.10}$$

They claim that this correlation is valid for $5 < \mathrm{Re} < 5000$.

Bakhmeteff and Feodoroff (1937, 1938) defined λ by the earlier convention (7.2.1.2), but correlated the values of λ they obtained for gas flow through beds of lead-shot by bringing the porosity into their equations explicitly. Thus for laminar flow ($\mathrm{Re} < 5$) they wrote (see Leva et al. 1951)

$$\lambda = 710/(\mathrm{Re} P^{\frac{3}{2}}) \tag{7.2.2.11}$$

and for 'turbulent' flow ($\mathrm{Re} > 5$)

$$\lambda = 24.2/(\mathrm{Re}^{0.2} P^{\frac{3}{2}}). \tag{7.2.2.12}$$

For flow of gases and liquids through porous carbon, Hatfield (1939) claimed that the logarithm of λP^2 (where λ is defined as in 7.2.1.2) could be linearly correlated with the logarithm of the Reynolds number for the flow range $10^{-5} < \mathrm{Re} < 100$.

Further investigations to take the porosity effect into account were also made by Rose and Rizk (1949). Furthermore, Happel (1949) correlated $\lambda' = \lambda/(1 - P)^3$ with the Reynolds number and reported that for the 'turbulent' range

$$\lambda' = \mathrm{const.}/(\mathrm{Re})^{0.22}. \tag{7.2.2.13}$$

Finally, Romita (1951) also came up with a correlation taking the porosity effect into account, namely:

$$\lambda_r = 190/\mathrm{Re}_r + 20/\mathrm{Re}_r^{\frac{1}{2}} + 2.7 \tag{7.2.2.14}$$

with

$$\lambda_r = \tfrac{2}{3}\lambda P^{\frac{3}{2}}/(1 - P), \qquad \mathrm{Re}_r = \tfrac{1}{4}\mathrm{Re} P^{\frac{3}{2}}/(1 - P). \tag{7.2.2.15}$$

Romita claims that these relationships would be valid for $4 < \mathrm{Re} < 5000$.

There have also been attempts at establishing correlations between the flow variables involving the porosity and particle diameter δ which do not employ the representation by the friction factor and Reynolds number. Burke and Plummer (1928) suggested the flow equation

$$\operatorname{grad} p = K[(1 - P)/P^3]\rho q^2/\delta^2 ; \qquad (7.2.2.16)$$

and Zabezhinskiĭ (1939) postulated that

$$\operatorname{grad} p = Kq/\delta^n, \qquad (7.2.2.17)$$

where $n = 1$ for high fluid velocities and $n = 2$ for low ones. In both equations, K signifies some constant. Finally, Hiles and Mott (1945), investigating the flow of gases through beds of coke and other granular materials, arrive at the experimental correlation

$$\operatorname{grad} p = Kq^n, \qquad (7.2.2.18)$$

where n and K are experimental constants which depend on the size of the granules. The value of these constants has been determined for a series of conditions.

Just as porosity has been handled by various investigators in various ways, so the shape of the particles and other variables have been handled variously as factors influencing fluid flow (Leva et al. 1951). Wadell (1934) defined a shape factor for single particles as the ratio of the surface of a sphere having the same volume as the particle to the surface of the particle. Zeisberg (1919) published pressure-drop data for various types of packing. Chilton (1938, p. 2211) converted these data, as well as the data of White (1935), to values of friction factors for the various shapes for use in his previously published (Chilton and Colburn 1931b) equation. Blake (1922) correlated data on glass cylinders, Raschig rings, and crushed pumice by a linear plot on log-log coordinates of $(\operatorname{grad} p\, P^3/[\rho q S])$ versus $(\rho q/[\mu S])$, where S is the value of the surface area of packing per unit volume of packed tube (see Leva et al. 1951).

Similarly, Burke and Plummer (1928; see also Burke and Parry 1935), studying the flow through spherical lead-shot in tubes of various diameters, concluded that the pressure drop is a function of a modified Reynolds number. They obtained

$$\operatorname{grad} p = (K\rho q^2 S/P^3)(\mu S/\rho q)^{2-n}, \qquad (7.2.2.19)$$

where n is a function of the Reynolds number and K is again a constant.

Furnas (1931) reported on the effect of a large number of variables. However, he expressed his data in the form

$$\operatorname{grad} p = A/(\rho q)^B, \qquad (7.2.2.20)$$

where A and B are complex functions of particle size, bed porosity, and even of the fluid properties such as temperature, viscosity, density, and molecular weight. This, of course, is not very satisfactory.

Meyer and Work (1937) related the bed voidage for a given packing to some value P_n representing the loosest packing possible for the specified material and thus brought the notion of particle shape into their correlation.

Carman (1937) correlated the pressure drop data of other authors by means of the following dimensionally homogeneous formula (Leva et al. 1951):

$$\text{grad } p \,\frac{P^3}{\rho q^2 S_1} = C\left(\frac{\mu S}{\rho q}\right)^{0.1}, \qquad (7.2.2.21)$$

where $S_1 = S + 4/\Delta$ (Δ is the diameter of the tube containing the packing and S the specific surface area) and C is a constant depending on the particle shape. Oman and Watson (1944) correlated their pressure drop data in the 'turbulent' flow range of air flow through (dense and loose-packed) beds of a variety of substances by refining the correlation of Burke and Plummer (1928). They adjusted the exponent of porosity occurring in the friction factor, so as to obtain the best correlation.

Rose (1945a) analysed literature data for non-spherical particles and claimed that they correlate with those for spherical particles by use of appropriate shape factors. Brownell and Katz (1947) also correlated pressure drop data of other investigators as well as their own data on air flow through salt beds. They arrived at the following equation (see Leva et al. 1951):

$$\text{grad } p = fq^2\rho/(2\delta P^n). \qquad (7.2.2.22)$$

The factor f may be obtained from curves of Moody (1944) for flow through empty pipes as a function of a Reynolds number which is defined as

$$\text{Re}_M = \delta \rho q/\mu P^m. \qquad (7.2.2.23)$$

The exponents m and n are dependent on particle shape and bed porosity and are presented as experimentally derived curves. These investigations were later revised (Brownell et al. 1950) as it was found that the exponents m and n rose rapidly as the porosity approached unity. Therefore, in the revised correlation, the numerical values of the porosity functions are correlated directly, without assuming that the latter can be written as a simple power of P. The revised correlations use plots of these porosity functions against porosity with parameters of particle sphericity. Cornell and Katz (1951) used such relationships to predict the flow rates in industrial applications.

Finally, Leva and co-workers (Leva 1947; Leva and Grummer 1947; Weintraub and Leva 1948; Leva et al. 1951) also investigated the possibility of obtaining correlations. In the last paper mentioned, they arrived at the correlation

$$\text{grad } p = \text{const.} f\rho q^2\theta(1 - P)/(\delta P^3), \qquad (7.2.2.24)$$

where θ is a particle shape factor and f is called the 'modified friction factor.' For smooth particles such as glass and porcelain, it is given by the relation

$$f = 1.75(\delta \rho q/\mu)^{-0.1}. \qquad (7.2.2.25)$$

With rougher materials such as alundum or clay, it has the form

$$f = 2.625(\delta\rho q/\mu)^{-0.1}.$$ (7.2.2.26)

For still rougher packings, for example, aloxite or magnesium oxide granules, one finally has

$$f = 4.00(\delta\rho q/\mu)^{-0.1}.$$ (7.2.2.27)

Reviewing the correlations exhibited in the present section, one may note that there appear to be two critical Reynolds numbers at which the flow regime changes, although these cannot be universally defined. It stands to reason that the first change of flow occurs when the inertia effects in laminar flow become important, and the second, when true turbulence sets in. This is not generally properly emphasized (cf. sec. 7.2.3).

It is quite certain that the correlations reviewed above cannot *all* be universally valid, as practically any one of them contradicts all others. We shall see later that this variety is probably due to the fact that there is actually no physical basis for expecting that flows should be analogous if the Reynolds numbers are. The correlations, therefore, are at best valid each for an application to a set of very specialized porous media. In this instance, the correlations may be very useful for engineering applications to particular systems, but proper caution should be taken if they are to be applied in any other set-up than that for which they were originally obtained.

7.2.3 *Theoretical equations*

The heuristic correlations of the flow variables in the 'turbulent' flow region do not really shed any light upon the physics of the phenomenon. Not only are they based upon such vague concepts as the Reynolds number, but, moreover, the great variety of correlations that have been claimed to be valid also shows that the heuristic approach may at best be able to serve as a basis for a calculation of flow rates in substances similar to those for which the correlations have been obtained. They seem to be quite useless, however, as a basis for a further examination of the physical principles of the phenomenon.

One will therefore be prompted to attack the problem from a theoretical standpoint. For such an attack, the same methods are available as have been used for an investigation of the flow through porous media in the 'laminar' region, i.e., where Darcy's law is valid. It is thus possible to apply dimensional analysis, or to extend either the hydraulic radius theory (i.e., the theory of capillaric models as well as the Kozeny theory) or the drag theories of permeability to 'turbulent' flow.

We shall now review the various attempts that have been made to explain 'turbulent' flow equations.

(*a*) *Dimensional analysis*   Starting with dimensional analysis, we may note that, in fact, the representation of friction factors versus the Reynolds number, discussed above, is an outcome of dimensional considerations. For the friction factor and the Reynolds number are both dimensionless groups and, therefore, one must be a numerical function of the other. Derivations of the form of this numerical function have been attempted only experimentally as was discussed in section 7.2.2. Now, it must be expected that there are more variables influencing the flow through porous media than have been discussed heretofore. In this instance, Rose and co-workers (Rose 1945a, b, c, 1949, 1951; Rose and Rizk 1949) made a thorough study of the possible variables that might influence flow and the dimensionless combinations in which they might occur in a flow equation.

They assumed that the linear (filter) velocity q through a bed of granular material must depend somehow upon the density ρ and the viscosity μ of the fluid. It must also depend on the height h of the bed, the diameter δ of the particles constituting the bed, the diameter Δ of the container into which the bed is packed, the porosity P of the bed, and the gravity g. It may also be reasonably believed that the resistance to flow, and hence the difference of hydraulic head H across the bed, depends upon the height e of the surface roughness of the pores. In addition, the resistance to flow must vary with the shape of the particles of which the bed is composed and with their size distribution. However, a suitable definition will render these last two variables expressible by two dimensionless quantities Z and U respectively. Thus, the relationship governing 'turbulent' flow in porous media may be written symbolically as follows:

$$H = \Phi(q^{\alpha} h^{\beta} \delta^{\gamma} \rho^{\zeta} \Delta^{\xi} \mu^{\theta} g^{\tau} e^{\eta} P^{\lambda} Z^{\sigma} U^{\omega}), \tag{7.2.3.1}$$

where Φ denotes 'a function of.' If finally the dimensions of mass, length, and time of the variables are inserted into equation (7.2.3.1) and the exponents are equated, in compliance with the standard procedure of dimensional analysis, one obtains the following equation of flow for an incompressible fluid:

$$\frac{H}{\delta} = \Phi \left\{ \left(\frac{q \delta \rho}{\mu} \right)^{\zeta} \left(\frac{h}{\delta} \right)^{\beta} \left(\frac{\delta g}{q^2} \right)^{\tau} \left(\frac{\Delta}{\delta} \right)^{\xi} \left(\frac{e}{\delta} \right)^{\eta} P^{\lambda} Z^{\sigma} U^{\omega} \right\}. \tag{7.2.3.2}$$

Unfortunately, dimensional analysis is not able to yield more information than is expressed by equation (7.2.3.2). The values of the exponents $\alpha, \beta, \ldots$ and the form of the function have to be determined experimentally. Rose and co-workers have made many experiments with this intent, but these run along the same lines as those discussed in section 7.2.2. No additional theoretical insight into the physics of the phenomenon can be gained in this manner.

With some additional assumptions, Leĭbenzon (1945a, b) and Sokolovskiĭ (1949a, b) deduced theoretically the form of one of the unknown quantities. The latter author wrote the flow equation in the following dimensionless form:

$$\text{grad } H = - \Phi(q) \mathbf{q}/q, \tag{7.2.3.3a}$$

where Φ is again 'a function' which has somehow to be determined.

Taking the steady-state flow of incompressible fluids, one can formulate the continuity equation, which, it is seen, is satisfied if

$$\Phi(q) = \frac{q}{K} \frac{1}{\sqrt{1 - (q^2/m^2)}}.$$

(7.2.3.3b)

It is easy to see that for $m = \infty$ (7.2.3.3) is equivalent to Darcy's law. If m is finite, one obtains a deviation from Darcy's law and Sokolovskiĭ claims that the equation is good in the range $0 \leq q \leq m$. The constants m and K, of course, have still to be determined experimentally and may depend on other variables.

Finally Lapuk and Evdokimova (1951), in the case of two-dimensional seepage into a sink, established the following law by dimensional analysis:

$$q = Ck^{\frac{1}{2}(3n-1)} \mu^{(1-2n)} \rho^{(n-1)} (dp/dr)^n.$$

(7.2.3.4)

This relates the velocity of filtration q to the permeability k, absolute viscosity μ, density ρ, pressure p, and radial distance from the sink r. Furthermore, C is a dimensionless coefficient related to the Reynolds number and n is called the 'exponent' of the filtration law, and may also be dependent on the Reynolds number.

(b) *Capillaric models*   The next possibility of deriving flow laws and of understanding them physically lies in the employment of capillaric models, i.e., the employment of analogies with flow through tubes. Such models have been mentioned, for example, by Blake (1922), Lindquist (1933), Ward (1939), Brieghel-Müller (1940), Meinzer and Wenzel (1940), Ishikawa (1942), Takagi and Ishikawa (1942), and Wentworth (1946). Scheidegger (1953) gave a systematic discussion of such capillaric models. In this connection we must note, as has been shown in section 2.2.2(d), that, even in tubes, the dependence of the pressure drop on the flow velocity becomes non-linear as soon as the inertia terms in the Navier-Stokes equations become important. This occurs in curved tubes long before the onset of turbulence. Since the curvature of the channels is neglected in hydraulic radius theories; the possible non-linearity in *laminar* flow is completely overlooked. A very interesting result of the experimental evidence adduced is that the critical Reynolds number above which 'turbulence' is believed to occur is much lower for porous media than for straight tubes (where it is around 2000; cf. chapter 2). The conclusion can only be that the alleged 'turbulence' is not turbulence at all, but an expression of non-linear *laminar* flow; the breakdown of Darcy's law at high flow velocities is thus primarily due to the emergence of inertia effects in *laminar* flow.

In our treatise, however, we have retained the terms 'laminar' and 'turbulent' as they are used in the literature; to indicate that *physically* this does *not* correspond to laminar and turbulent flow, we have put the terms in quotes.

The law for 'turbulent' flow in a tube is (this holds also for non-linear *laminar* flow)

$$dp = c' (\rho v^2/\delta)\,dx,$$

(7.2.3.5)

where v is the pore velocity and c' is a certain constant. All the other symbols have the meaning defined in section 6.2 of this monograph. 'Turbulent' flow will occur when the Reynolds number Re,

$$\text{Re} = \delta v \rho / \mu, \tag{7.2.3.6}$$

exceeds a certain value, usually assumed to be in the neighbourhood of 2000 for true turbulence, but lower in curved tubes for the emergence of inertia effects in laminar flow, rendering this flow non-linear. For smaller Reynolds numbers, the flow will be linear, as expressed by the Hagen-Poiseuille equation:

$$dp = - c \, (\mu v / \delta^2) \, dx. \tag{7.2.3.7}$$

The application of the non-linear flow laws to porous media can now be sought in exactly the same fashion as the Hagen-Poiseuille equation was applied to models of porous media in section 6.2. Thus, starting with the first simple model discussed there, namely that of parallel capillaries, it is at once apparent that the flow in all capillaries will be either non-linear or linear, according to the external pressure applied. For the 'turbulent' region, a calculation can be made similar to that of section 6.2.2 for laminar flow. We have for the porosity, if there are n capillaries per unit area in each spatial direction,

$$P = \tfrac{3}{4} n \pi \bar{\delta}^2 \tag{7.2.3.8}$$

and for the macroscopic velocity,

$$q = v \, (\pi/4) \, \bar{\delta}^2 n. \tag{7.2.3.9}$$

It should be noted that, because of the particular form of this model, the Dupuit-Forchheimer assumption that $v = q/P$ is not valid. Thus, we obtain from equation (7.2.3.5):

$$dp/dx = 9 \rho c' q^2 / (\delta P^2). \tag{7.2.3.10}$$

This model is obviously inadequate as no transition zone between 'laminar' and 'turbulent' flow can occur in it. Attention will have to be confined to models featuring different pore sizes.

Using first the discussion of the parallel type model in section 6.2.3, we obtain at once from equation (6.2.3.3)

$$q = - (\text{grad } p)^{\frac{1}{2}} \, (9c'\rho)^{-\frac{1}{2}} P \int_0^{\delta_R} \delta^{\frac{1}{2}} \, \alpha(\delta) d\delta - \text{grad } p \, \frac{P}{3\mu c} \int_\delta^\infty \delta^2 \alpha(\delta) d\delta, \tag{7.2.3.11}$$

where δ_R is defined as that value of δ which corresponds to the critical Reynolds number.

It is obvious that forumla (7.2.3.11) is remote from anything experimentally observed, for example the Forchheimer equation (7.2.2.1).

An attempt to introduce 'turbulence' into the serial models discussed in section 6.2.4 is therefore indicated. The two laws of motion (7.2.3.5) and (7.2.3.7) entail

$$p_2 - p_1 = - \underbrace{\int \frac{c\mu v}{\delta^2} \, ds}_{\substack{\text{'laminar'}\\\text{region}}} + \underbrace{\int \frac{c'\rho v^2}{\delta} \, ds.}_{\substack{\text{'turbulent'}\\\text{region}}} \tag{7.2.3.12}$$

Inserting (6.2.4.8), (6.2.4.9), (6.2.4.10) yields (neglecting capillary pressure differentials):

$$\frac{p_2 - p_1}{x} = - q \frac{16c\,\mu P}{3\pi^2\,n^2} \int_{\delta_R}^{\infty} \frac{\alpha(\delta)d\delta}{\delta^6} + q^2 \frac{64c'\,\rho P}{3\pi^3\,n^3} \int_0^{\delta_R} \frac{\alpha(\delta)d\delta}{\delta^7}. \tag{7.2.3.13}$$

Again, n can be replaced by the tortuosity T by virtue of equation (6.2.4.15). We then obtain

$$\text{grad } p = - q \frac{3c\mu T^2}{P} \left(\int_{\delta_R}^{\infty} \frac{\alpha(\delta)d\delta}{\delta^6} \right) \left(\int \delta^2\alpha(\delta)d\delta \right)^2$$

$$+ q^2 \frac{9c'\rho T^3}{P^2} \left(\int_0^{\delta_R} \frac{\alpha(\delta)d\delta}{\delta^7} \right) \left(\int \delta^2\alpha(\delta)d\delta \right)^3. \tag{7.2.3.14}$$

This expression appears to be almost of the form

$$\text{grad } p = Aq + Bq^2, \tag{7.2.3.15}$$

which would correspond to the Forchheimer equation (7.2.2.1). However, one should keep in mind that δ_R is a function of q, according to (7.2.3.6). Only if it is assumed that the integrals do not depend much on the actual value of δ_R will equation (7.2.3.14) 'explain' the empirical formula of Forchheimer. For true turbulence, the critical Reynolds number for the pore velocity must be taken as equal to about 2000. This is the Reynolds number at which turbulence occurs in a *straight* tube and according to the present models, this should also be the Reynolds number at which turbulence would set in in the porous medium, if the non-linearity observed were due to *true turbulence*. In order to make a proper comparison, the pore velocity v should be expressed in terms of the seepage velocity q. Using the Dupuit-Forchheimer assumption, one would have $q = vP$. However, it should be remembered that the Dupuit-Forchheimer assumption is not valid in the present models. If the flow channels are, as postulated, indepen-dent of each other in the three spatial directions, only one-third of the porosity is available for flow in any one direction, and one has correspondingly $q = vP/3$. Thus, expressed in terms of seepage velocity, the critical Reynolds number will be $P/3$ times the original one. For a medium with porosity $P = 0.2$, turbulence should therefore set in at a Reynolds number (calculated with q) equal to about 130. This is about fifty times too high if it is compared with the actual limits of Darcy's law. The only conclusion possible is that the non-linearity observed is not primarily due to the onset of turbulence, but due to the emergence of inertia effects in laminar flow owing to the curvature of the flow channels. Only at very high Reynolds numbers will true turbulence occur, and this causes the second change in flow regime observed experimentally. Furthermore, no universal Reynolds number can be defined for either the first or the second change in flow regime, which indicates that there is something seriously wrong with the models. We have pointed out in chapter 2 that the critical Reynolds number for the emergence of inertia effects is very much affected by any curvature of the tube, and therefore it must be expected that there is no such thing as a 'universal' Reynolds number of porous media at which non-linearity would set in. Depending on the curvature of the channels, the critical Reynolds number will be different in some flow channels than in others, even if the cross-sections are identical and even if the flow channels are put together to form porous media of identical porosity and 'tortuosity.' But, with this, any basis of calculating δ_R (necessary in the above formulas) is logically non-existent. Moreover, any representation of porous media using a Reynolds number seems somewhat unsatisfactory. This verdict, of course, also applies to the heuristic correlations discussed in section 7.2.2. The discussion of capillaric models plainly bears out this fact, viz., that there is no proper physical basis for assuming that the flows (and therefore friction factors) should be identical for identical Reynolds numbers, unless the curvature of the flow channels is somehow brought into the picture.

(c) *Kozeny theory*   If the occurrence of 'turbulent' flow in capillaric models is extended to the complexity achieved in the Kozeny-Carman theory, one arrives at a corresponding theoretical description of non-linear flow in porous media. This idea has been followed up chiefly by Ergun and Orning (1949; see also Ergun 1952a, 1953). These authors started from an equation postulated by Reynolds (1900) for the flow in linear tubes where the resistance to the motion of the fluid offered by friction is represented as the sum of two terms, proportional respectively to the first power of the fluid velocity and to the product of the density ρ of the fluid and the second power of its velocity, i.e.,

$$|\text{grad } p| = av + b\rho v^2. \tag{7.2.3.16}$$

The factors a and b are functions of the system. Ergun and Orning assume now that equation (7.2.3.16) is also valid for flow through porous media, with the filter velocity q instead of v, which again implies that the medium is assumed to be equivalent to an assembly of capillaries. The authors then proceed to interpret the two terms as representing viscous and kinetic energy losses, respectively. They then reason that the 'viscous' energy losses correspond to Darcy's law, i.e., in the hydraulic radius theory of permeability to the Carman expression (cf. equation (6.3.3.1)):

$$a = 5\alpha \frac{(1 - P)^2}{P^3} \mu S_0^2, \tag{7.2.3.17}$$

The kinetic energy losses would, however, be identical with those sustained by 'turbulent' flow in a cylindrical tube:

$$b = \beta/(2\delta).\qquad(7.2.3.18)$$

For a porous medium composed of cylindrical tubes, we have, furthermore,

$$\delta = \frac{4P}{S_0(1-P)}$$

and with the Dupuit-Forchheimer assumption that $q = vP$, we finally obtain

$$|\text{grad}\,p| = 5\alpha\,\frac{(1-P)^2}{P^3}\,\mu S_0{}^2 q + \frac{\beta}{8}\,\frac{1-P}{P^3}\,\rho q^2 S_0.\qquad(7.2.3.19)$$

The factors α and β indicate what part of the pressure drop is due to viscous and what part to kinetic energy losses. In this instance, they presumably depend on the critical Reynolds number for the system, and as the latter cannot be determined in any universal fashion, the coefficients α and β likewise cannot be determined universally. This seems to subject the present theory to the same limitations as those based upon correlations of the Reynolds number.

An approach similar to Ergun's has been proposed by Cornell and Katz (1953). These authors also start from the assumption of a quadratic equation between grad p and q; they identify the linear term with the Kozeny-type expression, and the quadratic one with a 'turbulent' flow expression. The Kozeny-type equation they use is written in terms of a hydraulic radius instead of S_0, and a tortuosity as measured electrically is introduced. Thus, everything is reduced to measurable quantities – except for the same two factors α and β (Cornell and Katz call them k_1 and k_2) necessary in Ergun's equation.

Another attempt to deduce the 'turbulent' form of the Kozeny equation was made by Irmay (1958). This author retained *all* the inertia terms of the Navier-Stokes equations in the microscopic flow channels and then built up a hydraulic radius model in the manner of Kozeny. This procedure is identical with that leading to equation (7.2.3.19) and duly leads to the same equation, except, however that by retaining *all* the inertia terms, one has, microscopically, an additional term in (7.2.3.16), which then reads

$$|\text{grad}\,p| = av + b\rho v^2 + c\rho\,\partial v/\partial t.\qquad(7.2.3.16a)$$

For the Kozeny model, one has $c = 1$ so that, with the Dupuit-Forchheimer assumption, Irmay comes to postulate that a term $(\rho/P)\partial q/\partial t$ be added on the right-hand side of equation (7.2.3.19). This corresponds to the equation already postulated by Polubarinova-Kochina (cf. 7.2.2.2b); as noted earlier, the additional term is so small as to be negligible in most cases.

The merit of Irmay's paper lies in the fact that it definitely emphasizes that the kinetic energy losses are not necessarily due to turbulance, but may be due to the emergence of inertia effects in laminar flow. At very high Reynolds numbers, of course, turbulence will set in; the factors contained in equation (7.2.3.19) will then suddenly assume new values. The basic form of the equation, however, will not be changed because it so happens that the kinetic energy losses due to inertia terms are of the same type in laminar as in turbulent flow.

(d) *Drag theory* Of the theories different from the hydraulic radius theory, the drag theory has also been applied to the problem of high-velocity deviation from Darcy's law. Nemenyi (1934) argued qualitatively that the inertia terms which were neglected in Emersleben's theory must effect something like turbulence in porous media at high flow velocities. Similar investigations have been made by Biesel (1950) and Shoumatoff (1952).

(e) *Stokes flow approaches* Finally, the proof that the high-flow-velocity deviations from Darcy's law are indeed due to the inertia terms in the Navier-Stokes equations was provided by Stark (1969). Following the general methods of the Stokes flow approach (cf. sec. 6.5.4), Stark was able to solve the Navier-Stokes equations for a variety of idealized porous materials. He duly obtained (by a numerical procedure) results corresponding to the solutions of the Forchheimer relation (7.2.2.1).

7.2.4 *Solutions of 'turbulent' flow equations*

Compared with the number of solutions of the 'laminar' (i.e., linear) flow equations available, only a few solutions of the 'turbulent' (i.e., non-linear) flow equations have been investigated.

In order to obtain analytical solutions for 'turbulent' flow through porous media, those equations which one wants to consider as basic have first to be written in a suitable analytical, i.e., vectorial, form. Making the assumptions that the flow is linear (i.e., of Darcy-type) up to a 'critical' Reynolds number, and non-linear (i.e., of Forchheimer-type) above, we can encompass these assumptions in a set of equations as follows (Engelund 1953):

$$- \operatorname{grad} p = F(|q|)\mathbf{q},$$
$$F(|q|) = \begin{cases} \mu/k & \text{for } \mathrm{Re} < \mathrm{Re}_{\mathrm{crit}}, \\ a + b|q| & \text{for } \mathrm{Re} > \mathrm{Re}_{\mathrm{crit}}. \end{cases} \qquad (7.2.4.1)$$

A suitable modification of these equations can be made to take account of gravity flow simply by introducing the 'hydraulic head' H instead of p.

It is, of course, quite hopeless to try to obtain analytical solutions of the system (7.2.4.1) in the general case of three-dimensional, non-steady-state flow. A particular three-dimensional solution has been reported by Uchida (1952). Restricting ourselves to the steady-state case in *two* dimensions, we note that a method of obtaining solutions has been developed by Engelund (1953), as follows.

In two dimensions, the system (7.2.4.1) becomes (Cartesian coordinates x, y, z):

$$-\partial p/\partial x = F(q)q_x, \qquad -\partial p/\partial y = F(q)q_y. \tag{7.2.4.2}$$

The equation of continuity is, for the steady state,

$$\frac{\partial q_x}{\partial x} + \frac{\partial q_y}{\partial y} = 0. \tag{7.2.4.3}$$

It is automatically fulfilled if we introduce the stream-function ψ defined by

$$q_x = -\partial\psi/\partial y, \qquad q_y = \partial\psi/\partial x. \tag{7.2.4.4}$$

From equation (7.2.4.2) it is evident that grad p and $\mathbf{q}$ are oppositely directed vectors. Since grad p is perpendicular to the surfaces $p = \text{const.}$, the same must hold for the filter velocity vector $\mathbf{q}$, so that the filter stream lines cross the equipressure surfaces at right angles.

It is evident from equation (7.2.4.4) that the vector grad ψ is perpendicular to $\mathbf{q}$, from which it follows that $\mathbf{q}$ is tangent to the curves $\psi = \text{const.}$ From this one may further conclude that these curves may be interpreted as stream lines. Further it may be concluded from equation (7.2.4.4) that

$$|\text{grad }\psi| = |\mathbf{q}| = q. \tag{7.2.4.5}$$

Hence the stream lines $\psi = \text{const.}$ and the contours $p = \text{const.}$ form an orthogonal system, just as in linear flow.

Equations (7.2.4.2) are, however, not linear and they are therefore inexpedient for direct solution, but it is possible to introduce new variables transforming the flow equations into linear equations. Thus, it is convenient first to introduce coordinates s and n, denoting the length of arc along curves $\psi = \text{const.}$ and $p = \text{const.}$, respectively, and next to consider an *infinitesimal element* of flow confined by two neighbouring stream lines and two lines of constant pressure. The equation of continuity can then be written

$$-\frac{1}{\Delta n}\frac{\partial(\Delta n)}{\partial s} = \frac{1}{q}\frac{\partial q}{\partial s}. \tag{7.2.4.6}$$

Since curl grad $= 0$, the flow equations (7.2.4.2) can be rewritten as follows:

$$\text{curl}\,[F(|\mathbf{q}|)\mathbf{q}] = 0. \tag{7.2.4.7}$$

By means of Stokes' theorem this condition may be expressed by the vanishing of the circulation around any closed curve, i.e., around the element

$$-\frac{1}{\Delta s}\frac{\partial(\Delta s)}{\partial n} = \frac{1}{Fq}\frac{\partial}{\partial n}(Fq). \tag{7.2.4.8}$$

Engelund then proceeds to introduce, as more convenient independent variables, the filter velocity q and the angle φ between the velocity vector $\mathbf{q}$ and the

x-axis. We obtain for the angle difference between the two stream lines of the element

$$\frac{\partial \varphi}{\partial n} \Delta n = -\frac{1}{\Delta s} \frac{\partial (\Delta n)}{\partial s} \Delta s. \tag{7.2.4.9}$$

Furthermore, we have

$$\frac{\partial}{\partial s} \left(\frac{\pi}{2} - \varphi \right) \Delta s = -\frac{1}{\Delta n} \frac{\partial (\Delta s)}{\partial n} \Delta n. \tag{7.2.4.10}$$

These expressions reduce to

$$\frac{1}{\Delta n} \frac{\partial (\Delta n)}{\partial s} = -\frac{\partial \varphi}{\partial n} \quad \text{and} \quad \frac{1}{\Delta s} \frac{\partial (\Delta s)}{\partial n} = \frac{\partial \varphi}{\partial s}. \tag{7.2.4.11}$$

Substitution into equations (7.2.4.6) and (7.2.4.8) yields

$$\frac{\partial \varphi}{\partial n} = \frac{1}{q} \frac{\partial q}{\partial s}, \tag{7.2.4.12}$$

$$-\frac{\partial \varphi}{\partial s} = \frac{1}{qF} \frac{\partial}{\partial n} (qF) = \left(\frac{1}{q} + \frac{F'}{F} \right) \frac{\partial q}{\partial n}, \tag{7.2.4.13}$$

where F' denotes the derivative dF/dq.

The quantities p and ψ can be introduced into these equations by substitution of $\partial \varphi / \partial n = \partial \varphi / \partial \psi \cdot d\psi / dn = q \partial \varphi / \partial \psi$, etc., and thus become

$$\frac{\partial \varphi}{\partial \psi} = -\frac{F}{q} \frac{\partial q}{\partial p}, \tag{7.2.4.14}$$

$$\frac{\partial \varphi}{\partial p} = \frac{1}{F} \left(\frac{1}{q} + \frac{F'}{F} \right) \frac{\partial q}{\partial \psi}. \tag{7.2.4.15}$$

These equations can be solved for q and φ as functions of p and ψ. We may, however, equally well express the functions p and ψ in terms of q and φ, and substitute these into the flow equations, which then become

$$\frac{\partial \psi}{\partial \varphi} = -\frac{q}{F} \frac{\partial p}{\partial q}, \tag{7.2.4.16}$$

$$\frac{\partial \psi}{\partial q} = \frac{1}{F} \left(\frac{1}{q} + \frac{F'}{F} \right) \frac{\partial p}{\partial \varphi}. \tag{7.2.4.17}$$

Furthermore, ψ can be eliminated from these equations by appropriate differentiation, which leads to

$$\frac{\partial}{\partial q} \left(\frac{q}{F} \frac{\partial p}{\partial q} \right) + \frac{1}{F} \left(\frac{1}{q} + \frac{F'}{F} \right) \frac{\partial^2 p}{\partial \varphi^2} = 0, \tag{7.2.4.18}$$

and thus one arrives at a single partial differential equation for p which describes linear as well as non-linear steady-state flow in porous media. Engelund (1953), who developed this equation, showed how it can be solved for particular cases.

In the extreme case for high Reynolds numbers, the quadratic term in the equation (7.2.4.1) is preponderant (truly turbulent flow), and (7.2.4.18) can be reduced to

$$\frac{\partial^2 p}{\partial q^2} + \frac{2}{q^2}\frac{\partial^2 p}{\partial \varphi^2} = 0. \tag{7.2.4.19}$$

Again, Engelund lists a set of particular solutions. As an illustrative example, we reproduce here Engelund's solution for symmetrical radial flow to a single well.

On account of symmetry, p is independent of φ and (7.2.4.19) becomes

$$\partial^2 p / \partial q^2 = 0, \tag{7.2.4.20}$$

from which

$$p = c_1 q + c_2. \tag{7.2.4.21}$$

Thus, p is in this case a linear function of q. To see how p depends on the distance r from the axis of the well, one puts

$$dp/dr = bq^2 = c_1 dq/dr. \tag{7.2.4.22}$$

The solution of this equation is

$$q = -c_1/(br) \quad \text{or} \quad c_1 = -2\pi qrb/(2\pi) = Qb/(2\pi), \tag{7.2.4.23}$$

where Q denotes the discharge per unit length of the well. Thus, one has

$$p = \frac{bQ}{2\pi}q + c_2 = \frac{Q^2 b}{4\pi^2 r} + c_2. \tag{7.2.4.24}$$

Solutions of Engelund's more general flow equation (7.2.4.18) are much more tedious to obtain. Engelund lists a variety of methods and applies them to a series of special cases which are of interest in the theory of groundwater flow into wells, drainage tubes, etc. The reader is referred for the details to the cited paper of Engelund (1953).

7.3
MOLECULAR EFFECTS

7.3.1 General remarks

Deviations from Darcy's law have not only been observed at high flow rates, for which they would be expected by analogy with flow in tubes. A category of such deviations is particularly manifest in the flow of gases, where the deviations are presumably due to molecular effects.

Thus, Fancher, Lewis, and Barnes (1933, 1934) and Manegold (1937) observed that air permeabilities are higher than liquid permeabilities in the same porous medium, as calculated from Darcy's law. In fact, this is an indication that Darcy's law is not valid for gases. Similar results were obtained by Brown and Bolt (1942) and Ruth (1946). A theory of this phenomenon is given later (cf. sec. 7.3.2). Concerning the characterization of the point where Darcy's law no longer becomes valid for gas flow, Calhoun and Yuster (1946) summarized the facts by stating that Darcy's law breaks down if the pore diameters become comparable with, or less than, the molecular mean free paths of the flowing gas. One can again observe an analogy between this breakdown of Darcy's law with the breakdown of Hagen-Poiseuille's law in capillaries: if the radius of the capillary is made smaller and smaller, the originally viscous flow of the gas undergoes a transition to slip flow and thence to molecular streaming as observed by Knudsen (cf. chap. 2). Of course, such an analogy is possible only in the light of the hydraulic radius theory of permeability and is therefore subject to the corresponding limitations.

Deviations from Darcy's law that might be explained by an analogy with Knudsen's findings for capillaries are not the only ones, however. There are other deviations that might also be due to molecular effects.

Comparing air and liquid permeabilities for a series of porous media, Calhoun (1946) found that neither air nor liquid permeabilities were constant as calculated from Darcy's law. Air and different liquid permeabilities were not in agreement: the average for liquids was usually somewhat lower than for air. Both liquid and air permeabilities depended either upon the mean pressure or upon the pressure gradient. Similarly, Grunberg and Nissan (1943) claim to have found that there is no correlation between the permeability of porous media to gases and to liquids. Permeabilities to gases depend mainly on the mean linear speed of the gas, whereas permeabilities to liquids depend also on the pore diameter and the specific surface tension. On the other hand, Bulkley (1931) was unable to detect deviations from Darcy's law for various fluids for pore diameters as small as 5.6μ.

In making experiments with only one fluid, it was found that the gas permeability increases with gas pressure. This is an effect opposite to that expected from analogy with Knudsenian flow in capillaries. Strange 'anomalies' were also observed by Bodman (1937) when experimenting with the flow of water through saturated soils. One particular result was that the permeability may change with time. Similarly, Grisel (1936) observed a change of permeability with time when passing water through cement. In addition, Ferguson et al. (1956) discussed the possible influence of physicochemical surface effects on fluid flow.

Explanations that might account for the observed effects will be discussed systematically in section 7.3.3. Apart from deviations which have already been accounted for by the drag theory of permeability (cf. sec. 6.5 above), it will be seen that adsorption, capillary condensation, and molecular diffusion can actually produce such 'anomalies' (in the light of Darcy's law) as listed above. In connection with such theories, some further experiments will be mentioned which were performed to substantiate the explanations. The above list of observed anomalies is therefore not complete.

7.3.2 *Gas slippage and molecular streaming*

The existence of molecular phenomena that can effect a limitation of Darcy's law has already been discussed in section 7.1. From such experiments, Sameshima (1926) deduced an experimental flow equation for gases through porous plates. He found that

$$1/q = C\mu^n M^{(1-n)/2}, \tag{7.3.2.1}$$

where M is the molecular weight and C and n are constants independent of the gas but dependent on the porous medium.

Adzumi (1937b) seems to have been the first to give an explanation of gas flow equations of the form (7.3.2.1) in terms of molecular slip. He constructed a theoretical model of porous media. The porous medium is represented by a bundle of parallel capillaries, each of which is made up of a series of short capillaries of various diameters. Using a modification of Knudsen's law (Adzumi 1937a; see equation (2.3.4)) for the (slip) flow through *one* capillary, Adzumi arrives at an equation for the flow of a gas through the porous medium which may be written

$$q = \frac{\pi \Delta p E}{8 A \mu} + \varepsilon \frac{4}{3} \sqrt{2\pi} \sqrt{\frac{R \mathfrak{T}}{M}} \frac{F}{Ap} \Delta p, \tag{7.3.2.2}$$

where A is the cross-sectional area of the porous medium, ε is the Adzumi constant ($= 0.9$; cf. equation 2.3.4), $E = nR_0^4/h$, and $F = nR_0^3/h$, with n being the number of pores in A, R_0 the average radius of the pores, and h the thickness of the porous medium. R, M, $\mathfrak{T}$ are the gas constant, molecular weight, and absolute temperature, respectively.

The Adzumi equation explains the occurrence of a 'slip term' in the flow equation by means of a simple capillaric model of the porous medium. It is, of course, quite hopeless to expect that E and F can ever be calculated for an actual porous medium; they therefore will have to be measured.

The train of thought originated by Adzumi has been taken up by many authors with the intention of making refinements on Adzumi's model, mainly by starting out with the Kozeny equation for the non-molecular part of the flow instead of with a model. This has been done by Arnell (1946, 1947; Arnell and Henneberry 1948) and by Carman (1947, 1950; Carman and Arnell 1948).

Similar investigations have been undertaken by Hodgins, Flood, and Dacey (1946), Holmes (1946), Keyes (1946), Rigden (1946, 1947), Brown et al. (1946), Barrer (1948), and Wilson et al. (1951). The following exposition of the available modifications of the Adzumi theory is based upon a review article by Carman and Arnell (1948).

Thus, Rigden (1947) applied the same considerations as used to deduce the Kozeny equation to the Poiseuille equation corrected for 'slip' and arrived at an equation having the form

$$q = \operatorname{grad} p \, \frac{Pc}{\mu} \left[\left(\frac{P}{S_0(1-P)} \right)^2 + \frac{2Px\lambda}{S_0(1-P)} \right], \tag{7.3.2.3}$$

where c is a Kozeny constant approximately equal to $\frac{1}{5}$, x is a constant having the value 0.874, and λ is the mean free path of the gas molecules. However, the slip correction factor used by Rigden was in the form valid for circular capillaries, whereas, for other media, the correction should be larger. Thus, the use of a slightly modified equation worked out by Lea and Nurse (1947) is preferable:

$$q = \operatorname{grad} p \, \frac{Pc}{\mu} \left[\left(\frac{P}{S_0(1-P)} \right)^2 + \left(\frac{2}{f} - 1 \right) \sqrt{\frac{\pi R \mathfrak{T}}{2 M}} \frac{\theta \mu P}{S_0(1-P)p} \right], \tag{7.3.2.4}$$

where f is the fraction of molecules evaporated from the surface, or $(1-f)$ is the fraction of molecules undergoing elastic collisions with the surface, and p is the mean pressure of the gas. Furthermore, θ is a shape factor which generally lies between 2 and 3. Finally, Arnell (1946), by analogy with Knudsen's flow law through capillaries, arrived at the equation

$$q = \operatorname{grad} p \, \frac{Pc}{\mu} \left[\left(\frac{P}{S_0(1-P)} \right)^2 + \frac{8}{3} \sqrt{\frac{2R\mathfrak{T}}{\pi M}} \frac{\mu C\alpha}{cS_0(1-P)p} \right], \tag{7.3.2.5}$$

where C is a variable factor having a value of approximately 0.9 and α is given by the equation $\log_{10} \alpha = 1.41P - 1.40$. Carman's and Arnell's equations are identical if $f = 0.79$. Experimental checks of these equations seem to give plausible agreement, but the tests and the equations are open to the same criticisms as is the Kozeny equation.

A number of years after Adzumi, Klinkenberg (1941) also came up with an explanation of the observed gas flow 'anomalies,' apparently without knowing about Adzumi's work. Klinkenberg constructed a very simple capillaric model of porous media and applied Warburg's slip theory to each capillary. In this way, he arrived at a flow equation for gases in porous media which allows for slip effects. The Klinkenberg equation has been tested by Grunberg and Nissan (1943), Calhoun (1946), Calhoun and Yuster (1946), Heid et al. (1950), Hoogschagen (1953), and Collins and Crawford (1953). The Klinkenberg theory has been compared to that of Adzumi by Rose (1948), who duly showed that the two theories are identical: Adzumi studied the dependence of the filter velocity

q upon the average pressure p_m, whereas Klinkenberg introduced the 'superficial' gas permeability k_a, defined by

$$\mathbf{q} = - (k_a/\mu) \operatorname{grad} p, \tag{7.3.2.6}$$

and investigated its dependence on the *reciprocal* mean pressure. Klinkenberg found that this dependence should be linear if slip is the correct explanation of gas flow:

$$k_a = b + m/p_m. \tag{7.3.2.7}$$

This last equation is, in fact, an equivalent expression of Adzumi's statements. The quantities b and m are constants.

There have also been attempts at explaining anomalies in gas flow which fall somewhat out of line with those discussed above. Deryagin, Fridlyand, and Krȳlova (1948) used a modification of Knudsen's flow law for capillaries and applied it to capillaric models of porous media. Iberall (1950) extended the drag theory of permeability to include slip flow, and arrived at an equation for gas flow of the type

$$q = c(1 + bp_0/p_m) \operatorname{grad} p, \tag{7.3.2.8}$$

where p_0 is the reference pressure at which q is measured, p_m is the mean pressure, and b and c are constants. The last formula is the same as that given by Klinkenberg. Iberall claims fair agreement of the theoretically computed constants b, c with experiments.

Turning now to solutions of molecular flow laws for particular cases, we note that, in order to achieve them, we have to choose a particular law and write it in differential form. Using the representation of Klinkenberg (7.3.2.7), we obtain, in conjunction with (i) Darcy's law, (ii) the pressure-density relation for an ideal gas, and (iii) the continuity equation:

$$\operatorname{div}\left[\left(p + \frac{m}{b}\right) \operatorname{grad} p\right] = \frac{P\mu}{b} \frac{\partial p}{\partial t}. \tag{7.3.2.9}$$

Equation (7.3.2.9) is a non-linear differential equation and therefore difficult to treat analytically. Thus, Collins and Crawford (1953) developed a numerical method which makes the problem accessible for high-speed computing machines. Introducing the new variable W ('equivalent pressure'),

$$W = p + m/b, \tag{7.3.2.10}$$

and substituting into equation (7.3.2.9), we obtain

$$\operatorname{div}(W \operatorname{grad} W) = \frac{P\mu}{b} \frac{\partial W}{\partial t}. \tag{7.3.2.11}$$

The problem of finding particular flow patterns of molecular streaming in porous media is therefore reduced to finding solutions, suitable for prescribed boundary conditions, of equation (7.3.2.11), which is still a non-linear differential equation. The numerical method consists in replacing the differential equation (7.3.2.11) by a difference equation (applicable to the one-dimensional case):

$$\left(\frac{W}{W_m}\right)_{x,\,t+\Delta t} = \frac{1}{4}\left[\left(\frac{W}{W_m}\right)^2_{x+\Delta x,\,t} + \left(\frac{W}{W_m}\right)^2_{x-\Delta x,\,t} - 2\left(\frac{W}{W_m}\right)^2_{x,\,t}\right] + \left(\frac{W}{W_m}\right)_x,$$

(7.3.2.12)

where the relation between the time step and the coordinate step has been chosen as follows (W_m is a constant):

$$\frac{1}{2}\frac{\Delta t W_m b}{(\Delta x)^2 P\mu} = \frac{1}{4}.$$

(7.3.2.13)

The above numerical method has been employed by Aronofsky (1954), who lists several cases for which he has obtained solutions; as an example, we present one of them here in some detail.

Consider a finite tube of porous medium charged with an initial gas pressure p_m (equivalent pressure W_m). The pressure at one end of the tube ($x = 0$) is suddenly lowered to a constant value p_0 (equivalent pressure W_0); the other end of the tube is sealed so that no flow occurs across the plane $x = L$.

By iterating equation (7.3.2.12) numerically on a computing machine, it was found that the pressure declined at the sealed end of the tube ($x = L$) are as shown in figure 28. In this figure, the equivalent pressure ratio at the sealed end

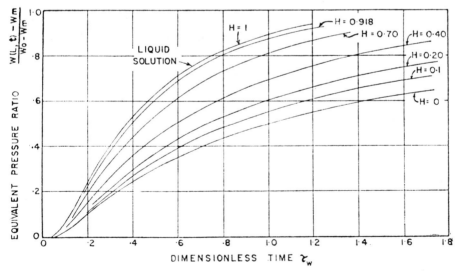

FIGURE 28   Pressure draw-down curves at the sealed end of a tube of porous medium charged with gas when the other end is suddenly opened to a constant pressure (after Aronofsky 1954).

$$\text{EPR} \equiv [W(L, t) - W_m]/(W_0 - W_m) \tag{7.3.2.14}$$

is plotted as a function of a dimensionless time parameter

$$\tau_w = bW_m t/(P\mu L^2). \tag{7.3.2.15}$$

A family of curves is shown for various values of $H = W_0/W_m \leq 1$. The spread of the curves can be taken as an indication of the non-linearity of the basic differential equation (7.3.2.11). The solution for molecular streaming approaches the laminar flow solution in the limit $H \to 1$. The laminar flow solution can be obtained by direct integration.

The various solutions for gas flow obtained by Aronofsky have been tested experimentally by Wallick and Aronofsky (1954), who found excellent agreement.

7.3.3 *Adsorption and capillary condensation*

The flow of a fluid through a porous medium, if adsorption occurs, has been discussed by Bull and Wronsky (1937) and later by Wicke (1938, 1939a, b; Wicke and Kallenbach 1941), who established the phenomenon as one of surface flow in the adsorbed layers. Other early studies are due to Šandera and Mirčev (1938), Barrer (1939, 1941), Penman (1940a, b), Glückauf (1944), and Cassie (1945). Wentworth (1944) investigated deviations from Darcy's law in assemblages of thin cracks (such as lava) and showed that

$$q \sim (\text{grad } p)^{1/N} \tag{7.3.3.1}$$

with $N < 1$. In this particular case, the rates were more than expected from Darcy's law as the molecules of the fluid within the sorbed layers are generally assumed to be more mobile. Further investigations on this subject have been made by many workers. In a series of papers, Carman and co-workers (Carman 1949, 1952; Carman and Malherbe 1950; Carman and Raal 1951) gave a comprehensive exposition of experiments and theories in connection with flow through adsorbents.

Thus, Carman and Malherbe (1950) approach surface flow as an example of diffusion along a concentration gradient. Following their exposition, we assume adsorption equilibrium at the end pressures, p_1 and p_2, of a cylindrical piece of porous medium with parallel end faces. The quantities adsorbed are then y_1 and y_2 millimoles per gram of adsorbent, where y_1 and y_2 are obtained from the adsorption isotherm for the temperature of the porous medium. As the weight of adsorbent per unit volume of porous medium is $\rho_p(1 - P)$ (ρ_p being the density of the solid material), the concentration gradient is equal to $\rho_p(1 - P)\Delta y/\Delta L$, where $\Delta y = y_1 - y_2$ and ΔL is the distance between the two faces of the porous medium. Then, if the flow rate by surface diffusion is Q millimoles per second,

$$Q\Delta L/A = K\rho_p(1 - P)\Delta y, \tag{7.3.3.2}$$

where K is a diffusion coefficient with dimensions cm^2/sec and A signifies the area of the faces of the porous medium. The concentration gradient is arrived at by assuming equilibrium with the gaseous phase at the two faces. When there is no surface flow, permeability is governed by equation (7.3.2.4). If now it is assumed that surface flow and gas flow take place independently and in parallel, the total flow is obtained by adding the surface term to (7.3.2.4). With properly chosen constants (K, K', K''), one obtains:

$$\frac{Q \Delta L}{A \Delta p} = K \rho_p (1 - P) \frac{\Delta y}{\Delta p} + \frac{K' P^2}{(1 - P) S_0 \sqrt{MR\mathfrak{T}}} + \frac{K'' p P^3}{\mu S_0{}^2 R \mathfrak{T} (1 - P)^2}. \tag{7.3.3.3}$$

The last two terms are the slip flow and viscous flow terms, respectively, for flow in the gas phase.

The assumption of independent surface, viscous, and slip flow cannot be accurate, since gas flow must be blocked to some extent by the adsorbed films. It turns out that, in the majority of cases, the flow in an adsorbed layer can indeed be expressed satisfactorily as diffusion along a concentration gradient, but if capillary condensation plays a major role, this is certainly not true, and a different mechanism must therefore take over.

The effect of capillary condensation can be estimated as follows (Carman 1952). Consider a cylindrical porous medium across which a condensable vapour is flowing under a pressure difference $\Delta p = p_2 - p_1$ across the medium, which is steadily maintained. As an approximation, one might assume that the condensate behaves as if it were bulk liquid in viscous flow across the porous medium. The driving force producing the flow, however, would then not be the maintained pressure difference in the gas, but a very much larger difference of capillary pressure resulting from this maintained pressure difference, presumably owing to action of surface tension at curved menisci. This pressure difference, in turn, can be estimated as follows.

Capillary condensation must be produced by a capillary potential, expressible as a negative pressure, or tension, $\bar{P}$, which must be applied to bulk liquid to reduce its vapour pressure from p_0 to that of the condensate, p. It is given by

$$\bar{P} = \frac{\rho_L R \mathfrak{T}}{M} \ln \frac{p_0}{p}, \tag{7.3.3.4}$$

where ρ_L is the liquid density at temperature $\mathfrak{T}$. It follows that

$$\Delta \bar{P} = \frac{\rho_L R \mathfrak{T}}{M} \ln \frac{p_1}{p_2}. \tag{7.3.3.5}$$

As long as Δp is less than half p_1 or p_2 (the end pressures), this is sufficiently approximated by

$$\Delta \bar{P} = \frac{\rho_L R \mathfrak{T}}{M} \frac{\Delta p}{p}, \tag{7.3.3.6}$$

where $p = \frac{1}{2}(p_1 + p_2)$.

G

A simpler derivation of equation (7.3.3.6) can be obtained by writing down the true thermodynamic relationship between $\bar{P}$ and p:

$$dp/d\bar{P} = \rho_g/\rho_L, \qquad (7.3.3.7)$$

where ρ_g is the gaseous density at pressure p. Equation (7.3.3.6) then follows by substituting Δp and $\Delta \bar{P}$ for differentials and assuming that the gas is ideal.

It follows that the value of $\Delta \bar{P}$ created by Δp across the sample is some hundreds of times larger than Δp.

Now, though $\bar{P}$ represents a tension which must be applied to bulk liquid to produce a vapour pressure, p, it does not follow that it actually exists in adsorbed films, since these are in a state different from bulk liquid. For a capillary condensate, however, it is usual to assume (in approximation) that it possesses the properties of bulk liquid, including the same surface tension, and that a reduction in the vapour pressure is wholly due to the tension, $\bar{P}$, which is produced by surface tension at curved menisci. It follows that a difference, $\Delta \bar{P}$, can be regarded as a *real* difference of pressure in the capillary condensate at the two ends of the sample. If we further assume that the viscosity is also that of bulk liquid, it is only a short step to envisage the flow of a capillary condensate as viscous flow through the pore space under the pressure difference, $\Delta \bar{P}$, as outlined above.

Thus, the flow of a capillary condensate can indeed be regarded as actual viscous flow under the pressure difference $\Delta \bar{P}$. Carman (1952) substantiated this by a series of experiments.

It is thus seen that flow in adsorbents can be visualized correctly, at least qualitatively, by assuming either capillary condensate flow or adsorbate flow along a concentration gradient. These two assumptions constitute two opposite and extreme cases and, in practice, one will find oneself somewhere in between the two. The theories for these extremes given by Carman are only approximate, but they do give at least a qualitative explanation for the observed phenomena. A gratifying aspect of these two theories is that they do not make use of any hydraulic radius models in connection with the respective surface flow terms. As was shown above, the derivation of the surface flow terms is based solely on thermodynamic reasoning and is therefore much more trustworthy than the derivation of the viscous and Knudsen terms also contained in the corresponding final equations.

The phenomenon of surface flow in porous media has received further study by Barrer and co-workers (Barrer and Grove 1951 a, b; Barrer and Barrie 1952). These investigations are only concerned with the sorbed (not the capillary-condensate) state of the fluid. By measuring the time-lag in setting up the steady-state flow, it was possible to measure the diffusion coefficient of gases and to obtain a mean pore radius and an internal surface of the porous medium.

Other investigations on the flow of fluids in adsorbents have been made by Flood et al. (1952a, b, c), who developed a simple flow equation for the correlation of their experimental data, and by Jones (1951, 1952). The latter author derived theoretical equations for the surface flow of an ideal two-dimensional gas through a porous material, treating the porous material as a bundle of capillaries. These equations explain earlier results of Tomlinson and Flood (1948) if the conditions are such that nearly a monolayer is sorbed.

In view of all these results, the necessity of adding a surface-flow term to the Darcy and Knudsen terms in flow through adsorbent porous media seems definitely established.

7.3.4 *Phenomenological diffusion*

When adsorption effects take place in a fluid streaming through a porous medium under a pressure difference, the appearance of the phenomenon is that of diffusion of the fluid through the solid; it is described by a diffusivity equation, with a variable diffusivity coefficient. The diffusion type of appearance of the phenomenon arises because the flow is determined by the adsorbate concentration gradient; the mass balance equation then leads immediately to a diffusivity equation. The specific form of the diffusivity coefficient depends on the model chosen. Various expressions have been proposed by Krischer (1963), Lykow (1958), and Philip and de Vries (1957).

Measurements of the diffusion coefficient as a function of adsorbate concentration have been reported by van der Kooi (1971) with an application to concrete roofs of buildings in mind. In fact, the moisture flow represents a form of immiscible multiple phase flow, and will be discussed in greater detail in that context (sec. 10.6.3).

7.4
IONIC EFFECTS

7.4.1 *Experimental evidence*

A group of anomalies relative to Darcy's law can be ascribed to the presence of ions in the percolating fluid. Such anomalies have been observed in soils, for example, by Nayar and Shukla (1943a, b, c, 1949), Shukla and Nayar (1943), Shukla (1944), and Sillén and co-workers (Sillén 1946, 1950a, b; Sillén and Ekedahl 1946; Ekedahl and Sillén 1947). In connection with the analysis of rocks this effect has been observed by, for example, Urbain (1941), Breston and Johnson (1945), Griffiths (1946), Miller (1946), Miller et al. (1946), and Heid et al. (1950). A particularly fine set of experiments has been reported by von Engelhardt and Tunn (1954). The change in permeability with the pH of the percolating solution is not a simple relationship: there are maxima and minima. It is generally thought that the permeability changes are brought about by an actual electrochemical reaction between the solution and the porous medium; in some instances (e.g., for rocks) clay swelling may also play a role.

There are indications, however, that ionic anomalies in the permeability of a porous medium are at least partly due to effects which are not at all chemical. Calhoun (1946) and Calhoun and Yuster (1946) analysed the ion exchange which is supposed to take place when salt solution is passed through porous media containing sodium ions, and found that there must be another effect besides mere ion exchange. Yuster (1946) made similar experiments with artifical quartz filters that contain no sodium or other ions, and was still able to observe an ionic effect.

In applications, it is particularly the clay content of the medium that seems to be affected by the presence of ions in the percolating fluid. This situation is of great importance with regard to the flow of fluids (brine) in underground strata. In a general manner, the relation between the clay content and capillary behaviour of sandstones has been described, for example, by Baptist and co-workers (Baptist and White 1957; Baptist and Sweeney 1955, 1957) and by Von Engelhardt and Tunn (1954).

Accordingly, the deviations from Darcy's law that do occur are of two types. First, any apparent permeability that one might define is no longer determined by the viscosity of the fluid and the geometry of the porous medium alone, but a third variable, indicative of the electrolytic behaviour of the fluid, enters. In the case of a clayey sandstone being percolated by brine, the permeability is generally found to increase with an increase in the strength of the salt solution. A typical graph demonstrating this effect is shown in figure 29. Second, Darcy's law itself is no longer exactly valid, inasmuch as the relationship between the pressure drop and the filtration velocity is no longer linear, even for a given porous medium and

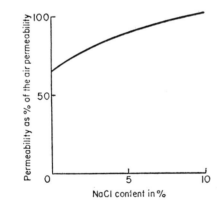

FIGURE 29   Change of permeability with electrolyte concentration (after Von Engelhardt and Tunn 1954).

fluid. The apparent permeability is found to increase with increasing flow velocity (cf. figure 30).

The effects discussed above have been observed mostly in sandstones containing clay which are being percolated by water. However, similar observations have also been made in other connections by Ruth (1946), Grace (1953), and Michaels and Lin (1954).

A possible explanation of these observations will be discussed below.

7.4.2 *Theory*

It stands to reason that the effects described above are due to the interaction of ions in the percolating fluid with the surface of the porous medium. In qualitative terms, such a mechanism has been postulated by Ruth (1946), Bridgwater (1950), and Duriez (1952); a more elaborate theory has been provided by Michaels and Lin (1955).

Accordingly, the ionic effects can be explained as follows: it is well known that at a solid-liquid interface an electrical double layer will be created with a potential difference, of value ζ, across it (the zeta potential). The movement of fluid past the solid boundary causes a downstream transport of the ionic charges in the

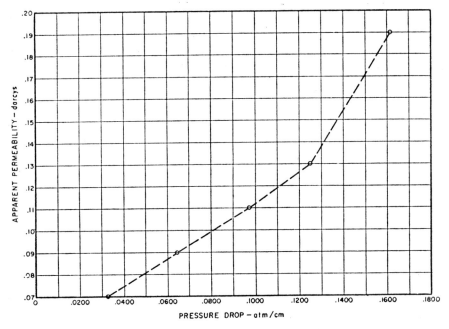

FIGURE 30 Change of apparent permeability with increasing flow velocity in a certain clayey sandstone which is being percolated by an electrolyte (drawn from data presented by Von Engelhardt and Tunn 1954).

upper part of the double layer. This displacement creates a streaming potential which causes a transport of ions ('electro-osmosis') in a direction opposite to forward flow.

The phenomenology of the process can be described by the thermodynamic theory of irreversible processes in its macroscopic form, according to which the over-all flow is expressed as a function of the over-all forces, using a set of conductance coefficients and Onsager's relations (Abd-el-Aziz and Taylor 1965; Spiegler 1958). However, a microscopic approach is more instructive.

Starting with the idea that the over-all flow q through a porous medium is a combination of the filtration velocity $q*$ that would be present if there were no electric effects, and an 'osmotic' flow q_{osm}, we can write

$$q = q* - q_{osm}.$$ (7.4.2.1)

Thus (Swartzendruber 1966, 1967) one can set, in one dimension,

$$q = \frac{k}{\mu}\frac{dp}{dx} - K'\frac{dE}{dx},$$ (7.4.2.2)

where K' is the linked transport coefficient for a flow caused by the streaming potential E. This equation can be rearranged as

$$q = \left(\frac{k}{\mu} - K'\frac{dE}{dp}\right)\frac{dp}{dx}.$$ (7.4.2.3)

Swartzendruber (1966) has shown that the ratio dE/dp may possibly not be constant, which leads to a non-Darcian type of flow.

The ionic streaming potential E that is set up in a porous medium when a liquid is mechanically forced through it is given by the equation (Kruyt 1952)

$$\frac{dE}{dp} = \frac{\varepsilon\zeta}{4\pi\mu\lambda},$$ (7.4.2.4)

where ζ is the zeta potential, ε the dielectric constant, and λ the combined specific conductance of the liquid and the surface of the porous medium. Then (7.4.2.3) becomes

$$q = \left(\frac{k}{\mu} - K'\frac{\varepsilon\zeta}{4\pi\mu\lambda}\right)\frac{dp}{dx}.$$ (7.4.2.5)

The constant K' can be determined by the use of a capillaric model (Michaels and Lin 1955). In a tube, one has, in analogy with (7.4.2.1), for the flow velocity v

$$v = v* - v_{osm}$$ (7.4.2.6)

with (Helmholtz's equation; see, e.g., Adam 1941, p. 353)

$$v_{osm} = \frac{\varepsilon\zeta}{4\pi\mu}\frac{dE}{dx},$$ (7.4.2.7)

where the symbols have the same meaning as above. Using a capillaric model for the porous medium of tortuosity T, this yields for each capillary

$$v_{osm} = \frac{\varepsilon\zeta}{4\pi\mu T}\frac{dE}{dx}.$$ (7.4.2.8)

Thus, for the filtration velocity q, using the Forchheimer relation

$$q = vP/T,$$

we have

$$q_{osm} = \frac{\varepsilon\zeta P}{4\pi\mu T^2}\frac{dE}{dx}$$ (7.4.2.9)

so that, by comparison with (7.4.2.2),

$$K' = \frac{\varepsilon\zeta P}{4\pi\mu T^2}.$$ (7.4.2.10)

Inserting this into (7.4.2.5) yields

$$q_{osm} = \frac{\varepsilon^2\zeta^2 P}{(4\pi)^2\mu^2\lambda T^2}\frac{dp}{dx}.$$ (7.4.2.11)

Writing (7.4.2.3) formally as Darcy's law, we set

$$q = (1/\mu)k_{app}\ \text{grad}\ p$$ (7.4.2.12)

with the apparent permeability k_{app}

$$k_{app} = k^* + k_{osm}.$$ (7.4.2.13)

Thus

$$k_{osm} = -\frac{\varepsilon^2\zeta^2 P}{(4\pi)^2\mu\lambda T^2}.$$ (7.4.2.14)

Here k^* denotes the permeability that would be present if there were no electric effects.

Equation (7.4.2.14) is the expression obtained by Michaels and Lin (1955). It explains why the apparent permeability k_{app} changes with the ion-concentration of the percolating fluid: simply because both the zeta potential and the specific conductance change. Thus, if the strength of the electrolyte is increased, the specific conductance is also increased and the term k_{osm} decreases in absolute value. This means that the apparent permeability increases if the strength of the electrolyte is increased, which is what has been observed experimentally (cf. fig. 29).

The above exposition explains the change of apparent permeability with electrolyte concentration only; it does not explain why Darcy's law breaks down even for one given electrolyte. However, a discussion of equation (7.4.2.14) permits one to obtain a qualitative understanding of this phenomenon. The theory of Michaels and Lin uses the zeta potential as it is present in a fluid at rest. If flow occurs, the equilibrium distribution of ions near the surface of the pore channels will be affected, inasmuch as part of the double layer will be destroyed. Thus, it is probable that, with increasing flow velocity, the observed zeta potential decreases, which effects an increase in the observed permeability. This is just what has been observed experimentally (cf. fig. 30).

7.4.3 *Flow through membranes*

The flow through membranes represents a special case of non-Darcian flow through a porous medium. In general, the fluid particles are charged, so that one has to deal with ionic flow. A membrane is, in fact, a very special 'porous medium,' and entire monographs have been devoted to the theory of flow through such media (cf. Schlögl 1964). The macroscopic flow is usually treated by means of thermodynamic considerations.

There are several types of models of membranes. For very fine pores, one generally assumes a fixed-charge model (Teorell 1935, 1937; Meyer and Sievers 1936) in which the charges on the porous matrix are assumed to be homogeneously distributed over the entire membrane; the mechanical resistance to flow is assumed to be very large.

For membranes with coarse pores, the theory becomes much more complicated, because the mechanical flow, in addition to the molecular flow, is now significant. For special shapes of pores, calculations have been made by Bjerrum and Manegold (1928).

A special type of membrane is an oil-film, i.e., a liquid phase which is immiscible with the flowing fluid (Beutner 1933).

For very thin membranes, continuous matter theory breaks down and individual steps have to be considered (Danielli 1937).

The theory of membranes is rather marginal to the theory of flow through porous media and therefore the reader is referred to the above-cited papers for further details.

7.5
THE FLOW OF NON-NEWTONIAN FLUIDS THROUGH POROUS MEDIA

7.5.1 *General remarks*

Finally, it is to be expected that the possible non-Newtonian behaviour of a fluid may also make itself felt during its flow through a porous medium. The relevant equations can best be deduced by considering capillaric models (Kutílek 1969).

One thus starts with the equation of flow through a single tube and then calculates the flow through a bundle of tubes. One can therefore follow exactly the same procedure as in section 6.2.2, changing, however, the flow equation (6.2.2.1) to suit the particular type of rheological equation of the non-Newtonian fluid that is under consideration. For such models, one has the general relations

$$P = \tfrac{1}{4}n\pi\delta^2, \tag{7.5.1.1}$$

$$S = n\pi\bar{\delta} = 4P/\bar{\delta}, \tag{7.5.1.2}$$

where n is the number of capillaries of diameter $\bar{\delta}$ per unit area. For the deduction of the porous media flow equations, specific rheological models of the fluid have then to be assumed.

7.5.2 Specific rheological models

If the shearing stress τ in the flow in each capillary depends on the flow velocity v at the distance r from the centre line of the capillary, i.e.,

$$\tau = -\mu \, dv/dr \tag{7.5.2.1}$$

(Newtonian behaviour), one duly ends up with Darcy's law and the Kozeny equation.

If (7.5.2.1) is slightly changed to read

$$\tau = -\mu \, (dv/dr)^a \tag{7.5.2.2}$$

corresponding to a pseudoplastic liquid, one finds (Kutílek 1969) that

$$q = K|\operatorname{grad} p|^{1/a}, \tag{7.5.2.3}$$

where K is a constant.

If Eyring's model is used (cf. 2.2.3.3),

$$\tau = \frac{1}{B}\sinh^{-1}\left(-\frac{1}{A}\frac{dv}{dr}\right), \tag{7.5.2.4}$$

one obtains (Kutílek 1969)

$$q = \frac{M}{|\operatorname{grad} p|} - \frac{N}{|\operatorname{grad} p|^2}\frac{1}{2}(e^{Q|\operatorname{grad} p|} - e^{-Q|\operatorname{grad} p|})$$
$$+ \frac{R}{|\operatorname{grad} p|^3}\left[\frac{1}{2}(e^{Q|\operatorname{grad} p|} + e^{-Q|\operatorname{grad} p|}) - 1\right] \tag{7.5.2.5}$$

with M, N, Q, R constants.

The rheological model of Ellis,

$$dv/dr = (\varphi_0 + \varphi_1\tau^{\alpha-1})\tau, \tag{7.5.2.6}$$

yields (Kutílek 1969)

$$q = C|\operatorname{grad} p| + D|\operatorname{grad} p|^{\alpha}. \tag{7.5.2.7}$$

The model of Reiner-Philipoff,

$$\frac{dv}{dr} = \left[\frac{1}{\mu_\infty + \dfrac{\mu_0 - \mu_\infty}{1 + (\tau/\tau_s)^2}} \right] \tau, \tag{7.5.2.8}$$

yields

$$q = \frac{M}{|\text{grad } p|} [F_1(|\text{grad } p|) + F_2(|\text{grad } p|) \ln F_2 - N] + K|\text{grad } p| \tag{7.5.2.9}$$

with

$$F_1 (|\text{grad } p|) = \frac{2P^2}{\rho^2} \ln \frac{\mu_\infty \rho^2 |\text{grad } p|^2}{\tau_s^2},$$

$$F_2 (|\text{grad } p|) = \frac{2P^2}{\rho^2} + \frac{2\mu_0 \tau_s}{\mu_\infty \rho^2 |\text{grad } p|^2},$$

and K, M constants.

Finally, Reiner's equation

$$\frac{dv}{dr} = \tau \left[\varphi_\infty - (\varphi_\infty - \varphi_0) \exp \frac{\tau^2 (d\varphi/d\tau)^2}{\varphi_\infty - \varphi_0} \right] \tag{7.5.2.10}$$

yields

$$q = M |\text{grad } p| + \frac{N}{|\text{grad } p|^2} e^{-Q|\text{grad } p|^2} - \frac{R}{|\text{grad } p|^4} \tag{7.5.2.11}$$

with M, N, Q, R constants.

The above equations can be compared with the empirical results of figure 27, and it becomes apparent that many of the types presented there can be produced by assuming appropriate rheological equations of the flowing fluid. However, with the increase in the number of constants in those equations, the procedure of comparison becomes more and more meaningless from a physical standpoint.

7.6
APPLICATIONS OF GENERAL FLOW EQUATIONS

7.6.1 *Permeability corrections*

The phenomena described by general flow equations may have the effect that the superficial permeability of a porous medium varies under different conditions. (By superficial permeability is understood the expression corresponding to k in Darcy's law if the law *were* valid.)

Thus, if 'permeability' measurements are made under conditions different from those for which they are to be applied, 'corrections' have to be made. This is particularly important if one is interested in the permeability of a porous medium to liquid, but the measurements are made employing a gas. A correction is best effected by using the representation of the slip effect given by Klinkenberg (1941; see sec. 7.3.2), where it was shown that the superficial gas permeability is a linear function of the inverse mean pressure of the gas. Since the cause of this dependence is supposed to be gas slippage, which in turn increases with increasing free path length, the effect should be nullified if the free path length is zero, i.e., if the mean pressure is infinite. Gas and liquid permeability should therefore be identical if the gas pressure is infinite. Using the equation of Klinkenberg, it is simple to extrapolate from two measurements at finite mean pressures to the infinite one. The routine application of this 'Klinkenberg correction' of gas permeabilities to obtain liquid ones has been described, for example, by Krutter and Day (1941) and Aronofsky et al. (1955).

7.6.2 Structure determinations of porous media

As with the Kozeny equation, those types of general flow equations which contain the specific surface S or S_0 can be used for structure determinations of porous media.

Of the 'turbulent' flow equations, Ergun (1952b) used his equation for the determination of particle density and geometric surface area of crushed porous solids.

Most applications, however, use the slip flow equations in one form or another; there are very many of them. Most of these applications are structure determinations of industrial powders. Owing to the fact that such powders are rather uniform to begin with, the success of the method does not necessarily imply the quantitative correctness of the formulas employed.

8
Statistical theory of flow through porous media

8.1
INTRODUCTION

8.1.1 *The problem*

We have described in the earlier chapters of this book some attempts to deduce Darcy's law from fundamental mechanical principles. The deduction of Darcy's law from fundamental mechanical principles is of importance for the understanding of flow through porous media. The basic mechanism of viscous flow is known: it is described by the Navier-Stokes equations. It should therefore be possible to achieve a correct description of the flow through porous media by integrating the Navier-Stokes equations for the geometrical boundary conditions represented by the pores.

Unfortunately, this approach has generally led nowhere. The boundary conditions, given by the configurations of the walls of the pores, are so complicated that any attempt at solving the Navier-Stokes equations for *these* boundary conditions is a priori doomed to failure. The obvious initial remedy of the situation has been to simplify the porous medium to such an extent that the Navier-Stokes equations *do* become solvable: this is the origin of and motivation for the many models of porous media built up from circular capillaries (see chap. 6). However, in this fashion, unrealistic oversimplifications are introduced and the results obtained are seldom significant. Evidently a different approach to the problem is called for.

8.1.2 *Statistical mechanics*

The difficulty of having to describe a large and complicated system whose differential equations of motion are known in principle but cannot be integrated on account of the complexity of the boundary conditions is not a new one in physics. An analogous situation was encountered in the last century when success was achieved in describing the kinetic theory of gases: the laws of collision and the equations of motion of the individual molecules are known, but the total effect cannot be built up from a consideration of the motions of the individual molecules. Thus, statistical mechanics was invented, in which the deterministic calculations of the path of each molecule were replaced by considerations *well, not exactly.*

of its probable behaviour. This 'probability' does not, in fact, exist, the motion of each particle being entirely deterministic. The 'probability' refers only to the incomplete knowledge of the investigator. It is simply a semantic trick to treat the motion of each particle as if it *were* stochastic because one does not know enough about it to treat it as deterministic – as it really is. Thus, one's incomplete knowledge of a complicated deterministic process is reflected by the fact that one makes a *stochastic model* of that process.

The general procedure is as follows. One considers a complicated system. This system is in a certain state, defined by the pertinent values of its characteristic variables. Instead of the state, one now considers a whole *ensemble* of states containing all those states that cannot be distinguished macroscopically because of one's ignorance about the microscopic details. Then the *expectation* value of an observable quantity is the *average* of that quantity over all states of the ensemble. For these expectation values, one tries to deduce predictions from what one knows about transitions between the states in the ensemble, based on the *microdynamic equations*. Often, in a time-stationary process, one can make the hypothesis that a particularly chosen system will, in time, pass through (or come arbitrarily close to) all the states that are possible at any one given time. Then ensemble averages and time averages of observables may be interchanged: this is the famous *ergodic hypothesis*.

Thus, in order to set up a stochastic model of a system, one must *first* choose an ensemble of states, *second* define microdynamic equations of motions on the microscale for transitions between the states, and *third* calculate the resulting comportment of observable quantities.

This is the general procedure. It has met with much success in the kinetic theory of gases, in quantum theory (where, incidentally, the 'probability' is real and not only semantic on account of the uncertainty relations), and in information theory, and it will now be applied to the flow of fluids through porous media. There are principally two types of stochastic models that can be envisaged: either the 'randomness' is directly ascribed to the fluid particles in an otherwise homogeneous medium (random-walk models), or one envisages deterministic flow along 'random channels' (random-media models). All specific stochastic models of flow through porous media proposed to date fall into one of these two classes. Differences arise from the construction of the basic ensembles of states and from the assumption of the basic equations of motion. Finally, a 'systemic' approach can also be made, in which no specific models are introduced, but only general relationships for expectation values of observables are employed.

8.2
RANDOM WALK MODELS

8.2.1 *Velocity dispersion model*

A simple velocity-dispersion model, based on an analogy with Brownian motion, was probably the first attempt (Scheidegger 1954) at a description of flow through porous media in terms of statistical mechanics. Such a model is obtained by considering a 'particle' of fluid, i.e., a volume of fluid which is small enough not to be separated into separate channels during its journey through a porous medium. The random walk is then produced by random 'knocks' (from the pore walls) on the particle; what happens in each time step is assumed to be independent of what happened at earlier times. The particle, during its walk, will thus have various velocities (velocity dispersion). One wishes to calculate its mean position after time t, assuming that the mean velocity would be that corresponding to Darcy's law. Because of the ergodic hypothesis (the velocity-distribution is time-stationary), this is equivalent to the mean position of a particle calculated over the ensemble of all possible random walks from a given point.

In the general terminology, one has thus the following formulation of the problem. The *ensemble* is that of all possible random walks in the time interval $(0, t)$ from a given point. The *microdynamic transition* from t to $t + dt$ is such that the mean velocity at any one time is constant as given by Darcy's law. The *expectation value* for the particle-position is then the average position over the ensemble.

In detail, the reasoning is as follows.

We split the time from 0 to t into N small intervals τ such that

$$N\tau = t. \tag{8.2.1.1}$$

In every time interval τ the particle will undergo a displacement $\boldsymbol{\xi}$ (with components ξ, η, ζ with respect to a Cartesian coordinate system). The probability that a given displacement $\boldsymbol{\xi}$ will take place in a time interval τ will be denoted by $\psi(\boldsymbol{\xi})$. By this definition, $\psi(\boldsymbol{\xi})$ is a time-averaged probability. Because of the ergodic hypothesis, this probability is assumed to be equal to the ensemble-averaged probability. Under the assumptions made, this probability taken over the ensemble will depend neither on the position of the particle in the porous medium nor on the instant. The probability $\psi(\boldsymbol{\xi})$ is assumed to be normalized so that

$$\int\psi(\xi, \eta, \zeta)d\xi d\eta d\zeta = 1. \tag{8.2.1.2}$$

The average (over the ensemble) displacement in the time interval τ, denoted by $\overline{\boldsymbol{\xi}}$, is given by

$$\overline{\boldsymbol{\xi}} = \int\boldsymbol{\xi}\psi(\boldsymbol{\xi})d^3\xi = (\bar{\xi}, \bar{\eta}, \bar{\zeta}). \tag{8.2.1.3}$$

It is very awkward to deal with probability distributions whose medians are not zero, so that it will be convenient to introduce a coordinate system in which the median of the probability distribution will be zero. We shall indicate this coordinate system by primes. Thus we set

$$\boldsymbol{\xi}' = \boldsymbol{\xi} - \overline{\boldsymbol{\xi}}. \tag{8.2.1.4}$$

In this primed system, the probability that the fluid particle undergoes a displacement ξ' will be $w(\xi')$. The average of ξ' is zero. The function $w(\xi')$ may depend on $\boldsymbol{\xi}$ as a parameter.

The fundamental question is now: What is the probability $w_N(x', y', z')$ over the ensemble that a fluid particle with initial coordinates $(0, 0, 0)$ in the primed system will have coordinates x', y', z' after the time $t = N\tau$?

Because of the assumptions made, viz., that the porous medium is homogeneous and isotropic, and the field of forces homogeneous, the original probability density $w(\xi', \eta', \zeta')$ will be split into a product of three identical probabilities:

$$w(\xi', \eta', \zeta') = w(\xi')w(\eta')w(\zeta'), \tag{8.2.1.5}$$

and, furthermore, these probabilities will be identical for every time step τ.

Thus, in terms of probability calculus, the problem is nothing but the composition of repeated trials. No matter what the original distribution w is, the central limit theorem states that the distribution after N trials is Gaussian, provided N is large. Thus, if we denote by σ the standard deviation of w, it follows that

$$w_N(\xi', \eta', \zeta') = (2\pi N\sigma^2)^{-\frac{3}{2}} \exp[-(x'^2 + y'^2 + z'^2)/(2N\sigma^2)]. \tag{8.2.1.6}$$

The quantities σ and τ are constants during the motion (although σ may depend on $\boldsymbol{\xi}$) such that one may set

$$\sigma^2/\tau = 2D, \tag{8.2.1.7}$$

where D is some 'factor of dispersion.' Using $N\tau = t$ yields

$$w(t, x', y', z') = (4\pi Dt)^{-\frac{3}{2}} \exp[-(x'^2 + y'^2 + z'^2)/(4Dt)] \tag{8.2.1.8a}$$

and, returning to unprimed coordinates,

$$\psi(t, x, y, z) = (4\pi Dt)^{-\frac{3}{2}} \exp\{-[(x - \bar{x})^2 + (y - \bar{y})^2 + (z - \bar{z})^2]/(4Dt)\}. \tag{8.2.1.8b}$$

This is the fundamental probability distribution describing the journey of a fluid parcel through the porous medium.

The next task is to find a connection between the average displacement $\bar{x}$ and the external field of forces. If the latter are assumed to be homogeneous and time-independent, then the laws of viscous flow state that the displacement ξ for every small time-step τ is given by

$$\bar{x}(t = N\tau) = N\boldsymbol{\xi}, \tag{8.2.1.9}$$

$$|\xi|/\tau = (B/\mu)|\text{grad } p| \cos \theta. \tag{8.2.1.10}$$

Here, grad p denotes the external force and $\cos \theta$ the angle between the vectors ξ and grad p; B is a certain factor indicative of the reciprocal resistance of the opening through which the fluid (of viscosity μ) moves during the time step τ.

Equation (8.2.1.10) will now be averaged over the ensemble. The average $\boldsymbol{\xi}$ will be in the direction of $-\operatorname{grad} p$ if the porous medium is assumed to be isotropic. Its magnitude will be equal to the average of the component of $\boldsymbol{\xi}$ in the direction of $-\operatorname{grad} p$, i.e., equal to the average of $|\boldsymbol{\xi}| \cos \theta$. Thus one has

$$\overline{\boldsymbol{\xi}}/\tau = - (\bar{B}/\mu) \operatorname{grad} p \overline{\cos^2 \theta}$$

or

$$\mathbf{v} = - (\bar{B}/\mu) \operatorname{grad} p \overline{\cos^2 \theta}, \tag{8.2.1.11}$$

where $\mathbf{v}$ is now the pore-velocity vector. Therefore, the probability distribution $\psi(\mathbf{x}, t)$ can now be expressed as follows:

$$\psi(\mathbf{x}, t) = (4\pi Dt)^{-\frac{3}{2}} \exp\left(- \frac{[x + t(\bar{B}/\mu) \operatorname{grad} p \overline{\cos^2 \theta}]^2}{4Dt}\right). \tag{8.2.1.12}$$

The average position of the median $\bar{x}$ of this distribution is

$$\bar{x}(t) = - \bar{B} \operatorname{grad} p \, t \, \overline{\cos^2 \theta}/\mu \tag{8.2.1.13}$$

and the average of the square-length $\overline{x^2}$ is (from the general theory of Gauss distributions)

$$\overline{x^2(t)} = 6Dt + \left(\frac{\bar{B}}{\mu} \operatorname{grad} p \overline{\cos^2 \theta} \, t\right)^2. \tag{8.2.1.14}$$

One can evaluate the factor D in terms of other quantities. Two procedures are possible: a dynamic and a geometric procedure.

(a) *Dynamic procedure* Newton's law of motion for a small particle of the fluid is

$$B(\mathbf{f} - \rho\ddot{\mathbf{x}}) = \mu\dot{\mathbf{x}}, \tag{8.2.1.15}$$

where $\mathbf{f}$ is the force per unit volume (apart from the viscous forces). If (8.2.1.15) is multiplied by $\mathbf{x}$,

$$B\mathbf{f}\mathbf{x} - B\rho\ddot{\mathbf{x}}\mathbf{x} = \mathbf{x}\dot{\mathbf{x}}\mu. \tag{8.2.1.16}$$

Since

$$\ddot{\mathbf{x}}\mathbf{x} = \frac{d}{dt}(\mathbf{x}\dot{\mathbf{x}}) - \dot{\mathbf{x}}^2 = \frac{1}{2}\frac{d^2}{dt^2}(\mathbf{x}^2) - \dot{\mathbf{x}}^2, \tag{8.2.1.17}$$

this yields:

$$B\mathbf{f}\mathbf{x} - \frac{1}{2}B\rho\frac{d^2}{dt^2}(\mathbf{x}^2) + B\rho\dot{\mathbf{x}}^2 = \frac{\mu}{2}\frac{d}{dt}(\mathbf{x}^2). \tag{8.2.1.18}$$

This equation can be averaged over the ensemble; the average of $\mathbf{f}\mathbf{x}$ is simply $\overline{\mathbf{f}\mathbf{x}}$; the average of $\mathbf{f}$ is then

$$\bar{\mathbf{f}} = - \operatorname{grad} p \overline{\cos^2 \theta} \tag{8.2.1.19}$$

since only the ordered component of **f** remains through the averaging process. Thus, averaging (8.2.1.18) yields:

$$\left(\frac{\bar{B}}{\mu} \operatorname{grad} p \, \overline{\cos^2 \theta}\right)^2 t - \frac{1}{2} \frac{\bar{B}\rho}{\mu} \frac{d^2}{dt^2} \overline{\mathbf{x}^2(t)} + \frac{\bar{B}\rho}{\mu} \overline{\left(\frac{B}{\mu} \operatorname{grad} p \cos \theta\right)^2}$$

$$= \frac{1}{2} \frac{d}{dt} \overline{\mathbf{x}^2(t)}. \tag{8.2.1.20}$$

This is a differential equation for $\overline{\mathbf{x}^2(t)}$ whose stationary solution is

$$\overline{\mathbf{x}^2(t)} = 2 \frac{\bar{B}\rho}{\mu} \left[\overline{\left(\frac{B}{\mu} \operatorname{grad} p \cos \theta\right)^2} - \left(\frac{\bar{B}}{\mu} \operatorname{grad} p \, \overline{\cos^2 \theta}\right)^2\right] t + \left(\frac{\bar{B}}{\mu} \operatorname{grad} p \, \overline{\cos^2 \theta}\right)^2 t^2.$$

$$\tag{8.2.1.21}$$

Comparing this with (8.2.1.14), one finally obtains:

$$D = \frac{\bar{B}\rho}{3\mu} \left[\overline{\left(\frac{B}{\mu} \operatorname{grad} p \cos \theta\right)^2} - \left(\frac{\bar{B}}{\mu} \operatorname{grad} p \, \overline{\cos^2 \theta}\right)^2\right]. \tag{8.2.1.22}$$

In this last equation, $\bar{B}$ and θ are given by the porous medium. Therefore, the expression for D may be written as follows:

$$D = (\rho a'/\mu^3) \, (\operatorname{grad} p)^2 = \operatorname{const.} |\mathbf{v}|^2, \tag{8.2.1.23a}$$

where now a' contains everything that depends on the porous medium. It is therefore a constant of the porous medium and may be termed the latter's dynamical dispersivity.

In a similar fashion, we can contract everything which in equation (8.2.1.13) refers to the porous medium only into one single constant k/P. We thus obtain:

$$\mathbf{q}/P = \mathbf{v} \equiv \bar{\mathbf{x}}(t)/t = -(k/\mu P) \operatorname{grad} p. \tag{8.2.1.24}$$

This corresponds to Darcy's law (which is thus shown to be valid for the median of the probability distribution); in the present theory it is therefore also appropriate to call k the 'permeability' of the porous medium.

The dynamic method is applicable if there is enough time in each flow channel for complete mixing by molecular sideways diffusion to take place. This is the condition which is necessary so that each small cross-section of a flow channel can be assumed to be a unit, so as to enable one to write down Newton's law of motion in the form of (8.2.1.15).

(b) *Geometric procedure* An alternative method of calculating the factor of dispersion is obtained by the remark that, if there is no interchange of particles pertaining to adjacent stream lines (incidentally, this corresponds to *true* laminar flow), then the geometrical distribution of ψ must be independent of $\bar{v}$. This means that ψ as a function of **x** and $\bar{\mathbf{x}}$ must be independent of the mean flow velocity (although, in a transient state, this distribution will be reached at an earlier time if the velocity is increased). The condition for this is that the exponent in (8.2.1.12) be independent of $\bar{v}$ if t is eliminated by means of (8.2.1.24). One obtains:

$$D = a'' [k/(\mu P)] |\text{grad } p| = a'' |v|, \tag{8.2.1.23b}$$

where a'' is another constant of the porous medium (preferably called its *geometrical dispersivity*). Equation (8.2.1.23b) is applicable if there is no appreciable molecular sideways diffusion from one stream line into another.

The above statistical theory refers to homogeneous forces only. It is, however, easy to effect a generalization to non-homogeneous forces by observing that the fundamental probability distribution (8.2.4.12) is a solution of the differential equation

$$\frac{\partial \psi}{\partial t} = \text{lap}(D\psi) + \text{div}\left(\bar{B}\psi \text{ grad } p \, \frac{\overline{\cos^2 \theta}}{\mu} \right). \tag{8.2.1.25}$$

Inserting the above-found values for $\bar{B}$, etc., we find that

$$\frac{\partial \psi}{\partial t} = \text{lap}(D\psi) + \text{div}\left(\psi \, \frac{k}{\mu P} \text{ grad } p \right), \tag{8.2.1.26}$$

which is also valid for non-homogeneous forces.

Finally, one has to formulate the continuity condition. At time $t = t_0$ there will be a certain distribution of the fluid in the porous medium. Let this distribution be denoted by $w(x_0, t_0)$, where the position coordinate is now denoted by x_0. Naturally, x_0 has the same range of values as x. At time t (arbitrary), the 'original' distribution of fluid $w(x_0, t_0)$ will have the probability-density $\psi(x, x_0; t, t_0)$ to be at the spot x at time t. Here, $\psi(x, x_0; t, t_0)$ is a 'key' solution of the differential equation (8.2.1.26). It satisfies the boundary condition and is a delta-function of the argument $x - x_0$ for $t = t_0$. It has, therefore, properties similar to a Green's function. Thus, the distribution of fluid at time t will be given by $w(x, t)$ as follows:

$$w(x, t) = \int \psi(x, x_0; t, t_0)w(x_0, t_0)d^3x_0. \tag{8.2.1.27}$$

The physical concept of continuity states that the distributions w must be proportional to the density p (or concentration of an injected spot of dye C) of the fluid times the local porosity, and thus equation (8.2.1.27) can be written as follows:

$$P(x)C(x, t) = \int \psi(x, x_0; t, t_0)C(x_0, t_0)P(x_0)d^3x_0. \tag{8.2.1.28}$$

This is the required continuity condition.

The continuity condition in the form of (8.2.1.28) is rather awkward to deal with. It would be preferable to have it in differential rather than integral form.

This aim can be achieved by taking the time-derivative with respect to t on both sides of equation (8.2.1.28):

$$P(\mathbf{x})\frac{\partial C(\mathbf{x}, t)}{\partial t} = \int \frac{\partial \psi(\mathbf{x}, \mathbf{x}_0; t, t_0)}{\partial t} C(\mathbf{x}_0, t_0)P(\mathbf{x}_0)d^3x_0. \tag{8.2.1.29}$$

However, the solution ψ is subject to the differential equation (8.2.1.26). One can thus insert $\partial \psi/\partial t$ from (8.2.1.26) into (8.2.1.29) to obtain:

$$P(\mathbf{x})\frac{\partial C(\mathbf{x}, t)}{\partial t} = \int \left[\text{lap}(D\psi) + \text{div}\left(\psi \frac{k}{\mu P} \text{grad}\, p \right) \right] C(\mathbf{x}_0, t_0)P(\mathbf{x}_0)d^3x_0.$$
$$\tag{8.2.1.30}$$

Upon letting $t = t_0$, this yields:

$$P\frac{\partial C}{\partial t} = \text{lap}(PCD) + \text{div}\left(C\frac{k}{\mu} \text{grad}\, p \right). \tag{8.2.1.31}$$

This is the macroscopic equation which describes 'dispersivity.' It is directly applicable to the spread of a pollutant (given by the concentration C) in a fluid streaming through a porous medium. If the factor D is assumed equal to zero, the fluid motion is identical with that described by Darcy's law. However, if this factor is not set equal to zero, then a macroscopic effect, dispersion, occurs. Individual particles of the fluid not only move along stream lines resulting from Darcy's law, but they are also dispersed sideways.

One may note that the diffusivity equation (8.2.1.26) is an outcome of the assumption that what happens in any one individual time-step is independent of what happens in any other time-step. This implies a random distribution of 'residence times.' If one makes such an assumption, then, in virtue of the Central Limit Theorem, one will always arrive at a Gaussian distribution for the probability ψ of a specific particle being at the position x at time t where $\bar{x}$ is the average position of the particle. One obtains:

$$\psi(x, t) = (4\pi Dt)^{-\frac{3}{2}} \exp[-(x - \bar{x})^2/(4Dt)]. \tag{8.2.1.32}$$

This automatically implies that the function expressed in 'mean' coordinates x' (i.e., coordinates in which there is no mean flow) is subject to a diffusivity equation:

$$\partial \psi/\partial t = D \text{ lap } \psi. \tag{8.2.1.33}$$

One can expect the use of other types of statistics to lead to different types of differential equations. At any rate, it should be noted that the diffusivity equation arrived at above has nothing to do with the specific details of the model that is considered; it comes out in any model that assumes 'random velocities.' As soon as a random distribution of velocities is assumed, explicitly or implicitly, the above diffusivity equation is the automatic and inevitable outcome. The diffusivity equation, in turn, implies that there exists an effect which was called *dispersion.* It is an effect which, in its phenomenological aspects, is very similar to diffusion, but it needs to be distinguished by having a separate name. Diffusion is usually associated with the intrinsic motion of the fluid molecules due to their thermal agitation, whereas dispersion is a mechanical effect due to the interconnections of the flow channels in a porous medium. Diffusion and dispersion, therefore, are due to entirely different physical causes. The fact that they show up phenomenologically in a similar fashion is more or less accidental.

The expressions found above for the constants a', a'', however, are not just a consequence of the type of statistics which was chosen, but are much more closely tied up with the specific assumptions upon which the present model is based.

The above model is a very primitive model and has been discussed at length only because it is the simplest velocity dispersion model and thereby conveniently illustrates the procedure. Refinements, mainly in the microdynamic equation of motion between states, soon make the mathematics and the presentation much more complicated. Naturally, 'refined' models also represent observed facts better. Thus, Todorović (1970) allowed for 'dead' spots in the flow. For sufficiently long times, he found that

$$D = \text{const.}\, v^n \quad \text{with } n > 1, \tag{8.2.1.34}$$

where v is, as usual, the pore velocity.

Summarizing, one can state that the velocity dispersion models discussed here all yield relations of the type

$$D = \text{const.}\, v^n. \tag{8.2.1.35}$$

In the dynamic model (cf. 8.2.1.23a) one has

$$n = 2, \tag{8.2.1.36}$$

in the geometrical model (cf. 8.2.1.23b) one has

$$n = 1, \tag{8.2.1.37}$$

and in the Todorović model one has

$$n > 1. \tag{8.2.1.38}$$

8.2.2 *Anisotropic dispersion*

The theory, thus far, treats the dispersion factor D as a scalar quantity. However, experimental evidence shows that, even in isotropic media, D cannot be treated as a scalar, inasmuch as the 'longitudinal' dispersion D_{long} (parallel to the flow direction) is different (viz. about eight times larger) from the 'lateral' dispersion D_{lat} (at right angles to the flow direction). One thus has to write (see Scheidegger 1961a) in the fundamental diffusivity equation describing dispersion (cf. 8.2.1.25)

$$\frac{\partial \psi}{\partial t} = \frac{\partial}{\partial x_i'} D_{ik} \frac{\partial \psi}{\partial x_k'}, \tag{8.2.2.1}$$

where the notation is as in section 8.2.1 and the tensorial summation convention is applied. The tensor D_{ik} must be a symmetric tensor:

$$D_{ik} = D_{ki},\tag{8.2.2.2}$$

as a consequence of Onsager's relations. The quantity

$$X_k = \partial\psi/\partial x_k'\tag{8.2.2.3}$$

represents a 'force' (the concentration gradient). The quantity

$$J_i = D_{ik}\,\partial\psi/\partial x_k'\tag{8.2.2.4}$$

represents a 'flux.' The flux-force equations are then of the form

$$J_i = L_{ik}\,X_k,\tag{8.2.2.5}$$

and from Onsager's principle it follows that

$$L_{ik} = L_{ki},\tag{8.2.2.6}$$

which leads to (8.2.2.2). The Onsager relations are somewhat naively applied in the above case, but the procedure is entirely analogous to that commonly used in the theory of heat flow through crystals (Nye 1957, p. 209).

The above theory can be carried further if it is assumed that one has only *geometrical* dispersivity (cf. eqs. 8.2.1.35 and 8.2.1.37). Then D_{ik} must be a linear function of v_j, which can be achieved by making the following working hypothesis (Scheidegger 1961a):

$$D_{ik} = a_{iklm}\,v_l v_m/|\mathbf{v}|.\tag{8.2.2.7}$$

This is, in fact, the simplest possibility. The tensor a_{iklm} is the dispersivity tensor, corresponding to the scalar dispersivity introduced in section 8.2.1. Since it is a tensor of rank 4, it would have 81 independent components. However, because of the symmetry in i and k (because D_{ik} is symmetric) and a further obvious symmetry in l and m, one is left with only 36 independent components. Thus, in a completely anisotropic porous medium, one has 36 dispersivity constants.

To deduce the general form of the dispersivity tensor for isotropic media, one requires numerical identity of the tensor elements for rotations by 90° about the 1- and 2-axes, and for an arbitrary rotation about the 3-axis. The result is then (Scheidegger 1961a)

$$a_{1111} = a_{2222} = a_{3333} = a_\mathrm{I},\tag{8.2.2.8}$$

$$a_{1122} = a_{1133} = a_{2233} = a_{2211} = a_{3311} = a_\mathrm{II},\tag{8.2.2.9}$$

$$\begin{aligned}
a_{1212} &= a_{1313} = a_{2323} = a_{2121}\\
&= a_{3131} = a_{3232} = a_{1221} = a_{1331}\\
&= a_{2332} = a_{2112} = a_{3113} = a_{3223}\\
&= \tfrac{1}{2}(a_\mathrm{I} - a_\mathrm{II}).
\end{aligned}\tag{8.2.2.10}$$

All other terms are zero. It is seen, thus, that the dispersion in an isotropic porous medium is anisotropic; it is described by the two dispersivity constants a_I and a_{II}. The physical meaning of these constants becomes obvious if one makes the following mental experiment. Let us consider an experiment where the velocity has the components $\mathbf{v} = (v, 0, 0)$. Then, the fundamental diffusivity equation (8.2.2.1) becomes

$$\frac{\partial \psi}{\partial t} = a_I v \frac{\partial^2 \psi}{\partial x_1'^2} + a_{II} v \frac{\partial^2 \psi}{\partial x_2'^2} + a_{II} v \frac{\partial^2 \psi}{\partial x_3'^2}. \tag{8.2.2.11}$$

Thus, the measured longitudinal D_{long} and transverse D_{lat} factors of dispersion turn out to be

$$D_{\text{long}} = a_I v, \tag{8.2.2.12a}$$

$$D_{\text{lat}} = a_{II} v. \tag{8.2.2.12b}$$

The above result was also deduced by Bachmat and Bear (1964) and by De Josselin de Jong (1969) using more specific models of dispersion.

8.2.3 *The effect of autocorrelation*

An interesting way of obtaining a statistical model different from that considered above is to change the basic statistics involved. Thus, one can, for instance, assume that there is a correlation between what happens to a fluid particle in successive time steps. This is termed 'autocorrelation.' Unfortunately it is difficult to estimate the effect of autocorrelation in the velocity-dispersion formulation discussed in section 8.2.1; a direct random-walk model is more suited to this purpose. If the random walk is likened to the walk of a drunkard, autocorrelation means that the drunkard has some memory. It can be shown (Scheidegger 1958) that, in this case, one does not end up with a diffusivity equation, but rather with a telegraph equation.

In detail, the random-walk model uses the approach to the problem of flow through porous media by the Lagrangian method. Lagrangian kinematics is characterized by the fact that the instantaneous position of any one particle $\mathbf{x}(t)$ as a function of time is considered as fundamental variable. We introduce statistics by considering the progress of a particle through a porous medium as a random-walk process.

In what follows, we shall assume that the flow is homogeneous with mean pore velocity $\mathbf{v}$; this is the case, for instance, in a linear motion in a long, packed pipe. Under these circumstances, it is useful to introduce a new coordinate system $\mathbf{x}'$ in which the mean flow can be assumed equal to zero; i.e., we have

$$\mathbf{x}' = \mathbf{x} - \bar{\mathbf{v}}t, \tag{8.2.3.1}$$

$$\overline{\mathbf{v}'} = 0. \tag{8.2.3.2}$$

Furthermore, all the discussion below will refer only to one coordinate x; the extension to the three-dimensional case is obvious. If anisotropy is involved, the coordinates must be chosen in the principal directions.

Now we assume that the flow of a fluid particle through the porous medium corresponds to a random-walk process. In order to investigate this, we split the time from 0 to t into n equal intervals τ such that

$$t = n\tau. \tag{8.2.3.3}$$

During each interval τ, the particle proceeds through the distance $\pm d'$, where the plus and minus signs are both equally probable and the prime denotes that we are dealing with deviations from the mean flow only. One can now ask for the probability $\psi(n, \nu)$ that the particle will reach the position $\nu d'$ after n steps. One has (Von Mises 1945, p. 145):

$$\psi(n, y) = \binom{n}{y}(1/2)^n \tag{8.2.3.4}$$

with

$$y = (n + \nu)/2, \tag{8.2.3.5}$$

where ν must be an integer of the series $-n, -n+2, \ldots, n-2, n$.

If the number of steps n is very large, this approximates a Gauss distribution:

$$\psi(n, \nu) \cong (\tfrac{1}{2}n\pi)^{-\frac{1}{2}} \exp(-\nu^2/2n), \tag{8.2.3.6}$$

where again $\nu = -n, -n+2, -n+4, \ldots, n-4, n-2, n$.

This means that the intervals for which ψ applies are all of length $2d'$. If one wants to express ψ as a density, one must therefore divide by $2d'$, which yields

$$\psi(t, x') = (\tfrac{1}{2}n\pi 4d'^2)^{-\frac{1}{2}} \exp(-\nu^2/2n), \tag{8.2.3.7}$$

where $x' = \nu d'$, $t = nT$. Hence

$$\psi(t, x') = [2(t/\tau)\pi d'^2]^{-\frac{1}{2}} \exp[-x'^2\tau/(2d'^2t)]. \tag{8.2.3.8}$$

It is customary to introduce a quantity D:

$$D = \tfrac{1}{2}d'^2/\tau; \tag{8.2.3.9}$$

then

$$\psi(t, x) = (4Dt\pi)^{-\frac{1}{2}} \exp[-x'^2/(4Dt)]. \tag{8.2.3.10}$$

The variance $\overline{x'^2}$ comes out as a linear function of t

$$\overline{x'^2} = 2Dt. \tag{8.2.3.11}$$

Furthermore, the Gauss function is an elemental solution of the diffusivity equation

$$\partial\psi/\partial t = D\,\partial^2\psi/\partial x'^2, \tag{8.2.3.12}$$

which describes the progress of the particles through the porous medium. It may be observed that this result is identical with that obtained in the earlier statistical model of flow through porous media.

The random-walk theory can be extended by assuming that there is a correlation between the directions that a particle possesses at time t and at time $t + \tau$. In order to take care of this case, one introduces the Lagrangian correlation coefficient $R(\tau)$ defined as follows (using again a coordinate system where $\overline{v'} = 0$):

$$R(\tau) = \overline{v'(t)v'(t + \tau)}/\overline{v'^2}. \tag{8.2.3.13}$$

The average square of the displacement is then

$$\overline{x'^2} = \overline{\left[\int_0^t v'(\tau)d\tau \right]^2} = \overline{\int_0^t \int_0^t v'(\tau_1)v'(\tau_2)d\tau_1 d\tau_2}$$

$$= \overline{v'^2} \int_0^t \int_0^t R(\tau_1 - \tau_2)d\tau_1 d\tau_2. \tag{8.2.3.14}$$

This can be further simplified to yield

$$\overline{x'^2} = 2\overline{v'^2} \int_0^t (t - \tau)R(\tau)d\tau,$$

which is a relation that was first found by Kampé de Fériet (1939) and is very useful in order to investigate limit cases. Introducing the autocorrelation time L_t,

$$L_t = \int_0^\infty R(\tau)d\tau, \tag{8.2.3.15}$$

one can, for instance, investigate the case

$$t \gg L_t. \tag{8.2.3.16}$$

In this case the average square displacement becomes

$$\overline{x'^2} = 2\overline{v'^2}L_t t - 2\overline{v'^2} \int_0^\infty \tau R(\tau)d\tau.$$

In this equation, the last term is a constant which can be neglected for large t; hence

$$\overline{x'^2} = 2\overline{v'^2}L_t t. \tag{8.2.3.17}$$

This is identical with the relation (8.2.3.10) with

$$D = \overline{v'^2}L_t, \tag{8.2.3.18}$$

which shows that the earlier simple model, in which the autocorrelation was neglected entirely, comes out from the more complete theory for large t. This, in fact, is how it should be since the autocorrelation must needs be insignificant if time intervals that are long compared with it are considered.

In the opposite limit case, viz., when t is very short, one has of course

$$\overline{x'^2}(t) = \overline{v'^2}t^2, \tag{8.2.3.19}$$

which simply expresses the fact that, for extremely short time intervals, the autocorrelation has the effect that there is no random process at all, but that every particle progresses for the short time interval considered with the velocity which it possesses. The time limit for which (8.2.3.19) applies can be obtained by investigating the time interval for which $R(t)$ does not significantly differ from 1. One sets

$$R(t) = 1 - \frac{\lambda^2}{2} \frac{1}{\overline{v'^2}} \overline{\left(\frac{dv'}{dt}\right)^2}, \tag{8.2.3.20}$$

which yields the condition

$$\frac{\lambda^2}{2} \frac{1}{\overline{v'^2}} \overline{\left(\frac{dv'}{dt}\right)^2} \ll 1, \tag{8.2.3.21}$$

or

$$\frac{1}{\lambda^2} \gg \frac{1}{2\overline{v'^2}} \overline{\left(\frac{dv'}{dt}\right)^2} = \frac{1}{2} \frac{d^2 R(0)}{dt^2}. \tag{8.2.3.22}$$

In the above analysis, the quantities D, v'^2, L_t, λ etc. have been taken as constants, which they are indeed in homogeneous flow. However, it must be expected that these quantities depend on the mean flow velocity and thus will change when the pressure drop is changed.

One would assume that the correlation function $R(t)$ depends on the pressure gradient, the geometry of the porous medium, etc. In discussing a statistical model, it would be indicated to introduce as few parameters as possible, and hence to assume the following representative correlation function:

$$R(t) = \exp(-|t|/A). \tag{8.2.3.23}$$

This is of the proper shape ($R = 1$ for $t = 0$, $R = 0$ for $t = \infty$) and introduces only one parameter A. One then obtains

$$L_t = A, \qquad \lambda = \sqrt{2A}. \tag{8.2.3.24}$$

In order to proceed from the correlation function to an actual distribution function corresponding to (8.2.3.10), one has to make the analogous transition from the random-walk problem to a continuous problem as has been done in connection with equation (8.2.3.10). This question has been analysed by Goldstein (1951) in relation to the kinematics of turbulent flow. We shall modify it here to make it applicable to the flow through porous media and also extend it so that the connection with the Lagrangian correlation function will be shown.

In order to trace the flow of a particle through a porous medium, we follow the treatment of the random walk of 'a drunkard with some memory.' In order to prevent the argument from becoming too involved, we consider the one-dimensional case and transform into a system (x') where the mean velocity $(\bar{v}')$ is zero. The extension of this to three dimensions is obvious.

We again split the time from 0 to t into small intervals so that

$$n\tau = t. \tag{8.2.3.25}$$

Since we are considering the case of a drunkard whose absolute velocity $|v'|$ stays constant, he will proceed by the distance $d = \pm v'\tau$ during each time step; the only thing that may change is the sign of the motion. After the elapse of the time $t = n\tau$ there are four different possibilities: if the drunkard moved to the right ($+$ direction), he may go on in this direction or reverse it; and similarly, if he moved to the left ($-$ direction), he may go on or turn back. The four probabilities are $p_{++}, p_{+-}, p_{-+}, p_{--}$. One has:

$$p_{++} + p_{+-} = 1, \qquad p_{-+} + p_{--} = 1. \tag{8.2.3.26}$$

Furthermore, if the case is isotropic, one has

$$p_{++} = p_{--}, \qquad p_{-+} = p_{+-}. \tag{8.2.3.27}$$

Because of the 'memory,' however, one has

$$p_{++} - p_{+-} = c \neq 0, \tag{8.2.3.28}$$

where c is a measure of the correlation. (It is assumed that there is no direct correlation between non-adjoining steps; i.e., the drunkard's memory does not last to the step before the last.)

One can calculate the Lagrangian correlation function for the above case. At time t, assume that the velocity is $+d/\tau$:

$$v'(t) = + d/\tau; \tag{8.2.3.29}$$

then at $t + \tau$

$$\overline{v'(t + \tau)} = p_{++} \frac{d}{\tau} - p_{+-} \frac{d}{\tau} = c \frac{d}{\tau}. \tag{8.2.3.30}$$

Similarly:

$$\overline{v'(t + 2\tau)} = c \frac{d}{\tau}(p_{++} - p_{+-}) \tag{8.2.3.31}$$

and hence

$$\overline{v'(t + n\tau)} = c^n d/\tau. \tag{8.2.3.32}$$

Therefore, the correlation function is

$$R(n\tau) = \overline{\{v'(t)v'(t + n\tau)\}}/\overline{v^2}(t) = c^n. \tag{8.2.3.33}$$

Denoting the probability that the drunkard has arrived at vd after n time steps by $\psi(n\tau, v)$, one has the following continuity conditions:

$$\psi([n + 1]\tau, v) = \alpha(n + 1, v) + \beta(n + 1, v),$$
$$\psi([n + 1]\tau, v) = \alpha(n + 2, v + 1) + \beta(n + 2, v - 1), \tag{8.2.3.34}$$

where $\alpha(n, v)$ denotes the probability that the drunkard arrived from the left at position vd at the end of time-step number n. Similarly, β denotes the probability that he arrived from the right.

Then, the correlation between subsequent time steps yields:

$$\alpha(n + 1, v) = p_{++}\alpha(n, v - 1) + p_{-+}\beta(n, v - 1), \tag{8.2.3.35}$$

$$\beta(n + 1, v) = p_{--}\beta(n, v + 1) + p_{+-}\alpha(n, v + 1). \tag{8.2.3.36}$$

From the set of equations (8.2.3.34)–(8.2.3.36), the α and β can be eliminated, and one has

$$\psi([n + 1]\tau, v) = p_{++}[\psi(n\tau, v - 1) + \psi(n\tau, v + 1)] = c\psi([n - 1]\tau, v). \tag{8.2.3.37}$$

Taking the limit with $(d/\tau)^2 \to \overline{v^2} = D/A$, $(1 - c)/\tau \to 1/A$, $d \to 0$, $\tau \to 0$, $n\tau \to t$, $n \to \infty$, one obtains the 'telegraph' equation:

$$\frac{\partial^2 \psi}{\partial t^2} + \frac{1}{A}\frac{\partial \psi}{\partial t} = \frac{D}{A}\frac{\partial^2 \psi}{\partial x'^2}. \tag{8.2.3.38}$$

Again, one can express the quantity A in terms of the Lagrangian correlation function. One has:

$$\frac{1}{A} = \lim \frac{1 - c}{\tau} = \lim \frac{1 - R(n\tau)^{1/n}}{\tau} = \lim \frac{1 - R(t)^{1/n}}{\tau}.$$

$$= \frac{1 - R(t)^{1/n}}{t/n} = t^{-1}\lim n[1 - R(t)^{1/n}]. \tag{8.2.3.39}$$

Using a well-known relation for the natural logarithm, this yields:

$$1/A = -t^{-1}\ln R(t).$$

Hence

$$R(t) = \exp(-t/A). \tag{8.2.3.40}$$

This shows that the telegraph equation proposed by Goldstein represents motion through a porous medium if the correlation function has the particular form given above.

The next step is to investigate the meaning of the constants D and A which have been introduced with the telegraph equation. We therefore specify the random-walk model of the porous medium somewhat more precisely. The porous medium shall consist of small capillaries of length d (projected onto the x-direction).

In the laminar flow domain, the displacement ξ_i during the time step τ is given by

$$\xi_i = \tau \frac{1}{\mu} b_{ik} \frac{\partial p}{\partial x_k}. \tag{8.2.3.41}$$

If we are interested in only one direction (say the x-direction), the displacement becomes (cf. 8.2.1.11)

$$\xi = \tau \frac{1}{\mu} b \cos^2 \theta \operatorname{grad} p, \tag{8.2.3.42}$$

where the quantity $b \cos^2 \theta$ depends on the flow channel under consideration and $\operatorname{grad} p$ is the one-dimensional pressure gradient. If we are to transform everything into a coordinate system where $\bar{v}' = 0$, we have

$$\xi' = \tau \frac{1}{\mu} (b \cos^2 \theta - \overline{b \cos^2 \theta}) \operatorname{grad} p \equiv \tau \frac{1}{\mu} \beta \operatorname{grad} p. \tag{8.2.3.43}$$

The quantity β depends only on the geometry of this flow channel:

$$\beta = (b \cos^2 \theta - \overline{b \cos^2 \theta}). \tag{8.2.3.44}$$

Since the random-walk model assumes that, during any one time-step τ, the motion is $\pm d$, this implies that

$$\frac{d}{\tau} = \pm \frac{1}{\mu} \beta \operatorname{grad} p. \tag{8.2.3.45}$$

The random-walk model is thus adequate if and only if the last relation is fulfilled.

We have now (from 8.2.3.9)

$$D = \frac{1}{2} \frac{d^2}{\tau} = \frac{1}{2\mu} \beta d \operatorname{grad} p, \tag{8.2.3.46}$$

which gives the connection between the macroscopic quantity D and the microscopic quantities d and β.

Similarly, one can calculate the Lagrangian correlation function for the model under consideration. One has obviously

$$R(t) = 1 \quad \text{for } t < \frac{d}{(1/\mu)\beta \operatorname{grad} p}, \tag{8.2.3.47}$$

$$R(t) = 0 \quad \text{for } t > \frac{d}{(1/\mu)\beta \operatorname{grad} p}. \tag{8.2.3.48}$$

Hence

$$L_t = \int_0^\infty R(t)\,dt = \frac{d}{(1/\mu)\beta\,\mathrm{grad}\,p}. \qquad (8.2.3.49)$$

This yields for A, in virtue of condition (8.2.3.23),

$$A = \frac{\mu d}{\beta\,\mathrm{grad}\,p}. \qquad (8.2.3.50)$$

Since the quantities β and d depend on the porous medium, the random-walk model makes the following measurable predictions:

$$D = \mathrm{const.}\,\frac{1}{\mu}\,\mathrm{grad}\,p, \qquad (8.2.3.51)$$

$$A = \mathrm{const.}\,\frac{\mu}{\mathrm{grad}\,p}. \qquad (8.2.3.52)$$

8.2.4 Analogy of laminar flow in porous media with turbulent flow in bulk fluids

The dispersion of the individual particles during laminar flow in porous media has also been interpreted in another manner (Yuhara 1954). The fluctuations in the velocities of the particles can be claimed to be analogous to the fluctuations of velocity during eddy motion in turbulent flow. Thus, the flow path of a particle in flow through a porous medium can be regarded as analogous to the trajectory of a particle in turbulent hydraulic motion. It might be expected, therefore, that methods yielding reasonable results for turbulent flow through pipes, etc., will also yield reasonable results if applied to the laminar flow through porous media.

Indeed, the flow path of a fluid particle in an ensemble of macroscopically identical porous media is a probability-phenomenon (over the ensemble) and therefore could be thought of as corresponding to the random flow path of a fluid particle during turbulent motion in a bulk mass of fluid. The appearance of a 'dispersivity' in the hydrodynamics in porous media may be thought of as analogous to the diffusivity of eddies, etc., in turbulent motion. It is by this analogy with turbulence that diffusivity equations have been suggested for the laminar flow through porous media without recourse to a proper statistical treatment.

It should be noted, however, that the analogy between laminar flow in porous media and turbulent motion in bulk masses of fluid is not very complete. This has not always been realized in applying it; moreover, it has even been claimed that the statistical hydrodynamics in porous media would suffer from a major flaw because it does not use the same velocity correlation tensors as are used in the statistical theory of turbulence.

Thus, let us introduce velocity correlation tensors in the statistical hydro-dynamics of laminar flow in porous media and investigate where this leads.

In analogy with the statistical theory of turbulence (cf. sec. 2.2.2), we shall therefore consider the local (pore) velocity $\mathbf{v}$ as fundamental dynamical variable. The commonly considered 'correlation tensor' is then defined as follows, the summation convention being applied (Scheidegger 1956) (cf. 2.2.2.4):

$$R_{ik}(\mathbf{r}) = \overline{[v_i(\mathbf{x}) - \bar{v}_i(\mathbf{x})][v_k(\mathbf{x} + \mathbf{r}) - \bar{v}_k(\mathbf{x} + \mathbf{r})]}. \tag{8.2.4.1}$$

Assuming narrow channels, the law of viscous flow (corresponding to the Hagen-Poiseuille equation) can be written as follows:

$$v_i = -\frac{b_{ik}}{\mu}\frac{\partial p}{\partial x_k}, \tag{8.2.4.2}$$

expressing the fact that, in narrow channels, the flow is proportional to the driving force. Assuming a homogeneous pressure gradient, the mean velocity can therefore be expressed by

$$\bar{v}_i = -\frac{1}{\mu}\frac{\partial p}{\partial x_k}\bar{b}_{ik} = -\frac{1}{\mu}k_{ik}\frac{\partial p}{\partial x_k}, \tag{8.2.4.3}$$

where

$$k_{ik} = \bar{b}_{ik} \tag{8.2.4.4}$$

turns out to be proportional to the permeability tensor (see sec. 4.3.3). Thus, the deviation of the velocity from its mean is

$$v_i - \bar{v}_i = -\frac{1}{\mu}\frac{\partial p}{\partial x_k}(b_{ik} - \bar{b}_{ik}). \tag{8.2.4.5}$$

Finally, forming the correlation tensor, we get

$$R_{ik}(\mathbf{r}) = \frac{1}{\mu^2}\frac{\partial p}{\partial x_l}\frac{\partial p}{\partial x_m}$$
$$\times \overline{[b_{il}(\mathbf{x})b_{km}(\mathbf{x} + \mathbf{r}) - b_{il}(\mathbf{x})\bar{b}_{km}(\mathbf{x} + \mathbf{r}) - b_{km}(\mathbf{x} + \mathbf{r})\bar{b}_{il}(\mathbf{x}) + \bar{b}_{il}(\mathbf{x})\bar{b}_{km}(\mathbf{x} + \mathbf{r})]}. \tag{6.2.4.6}$$

If we assume that the porous medium is isotropic, this formula can be simplified to yield

$$R_{ik}(\mathbf{r}) = \frac{1}{\mu^2}\frac{\partial p}{\partial x_l}\frac{\partial p}{\partial x_m}b_{iklm}, \tag{8.2.4.7}$$

where, for abbreviation, one has set

$$b_{iklm} = \overline{b_{il}(\mathbf{x})b_{km}(\mathbf{x} + \mathbf{r})} - k_{il}k_{km}. \tag{8.2.4.8}$$

As is easily seen, the tensor b_{iklm} is a tensor which depends on the porous medium only. It is obtained by forming the required averages. Equations (8.2.4.7–8) show that the velocity correlation tensor in laminar flow through porous media depends only in a trivial way on the dynamical variables (viz., on the pressure). The main dependence of this tensor is on the porous medium. It thus turns out that the velocity correlation tensor is unsuitable as a dynamic variable, and that the analogy between turbulent flow and flow through porous media is not very complete, in spite of the fact that the contrary has been claimed.

8.2.5 General flow regime

(a) *High flow region* Finally, it remains to apply concepts of random-walk models of porous media to the non-linear flow region (Scheidegger 1955). Following the same procedure as for linear flow (see sec. 8.2.1) one can introduce a probability-density $\psi(\mathbf{x}, t)$ for a particle to be at the spot $\mathbf{x}$ at the time t. As before, this probability density refers to a fictitious ensemble of similar porous media rather than to an actual stochastic flow process. If correlations between adjacent time steps are neglected (as in sec. 8.2.1), the same fundamental distribution for ψ is obtained as in equation (8.2.1.8):

$$\psi(\mathbf{x}, t) = (4\pi Dt)^{-\frac{3}{2}} \exp[-(\mathbf{x} - \bar{\mathbf{x}})^2/(4Dt)]. \tag{8.2.5.1}$$

The connection between the average displacement $\bar{\mathbf{x}}$ and the field of forces (denoted by grad p) is now given, if the simplest case of purely 'turbulent' flow is considered, by

$$(1/\tau) |\boldsymbol{\xi}| = -c\sqrt{|\mathrm{grad}\, p|}\,\sqrt{\cos\theta}/\sqrt{\rho}, \tag{8.2.5.2}$$

where $\boldsymbol{\xi}$ is the displacement during a small time-step τ. Furthermore, c denotes a constant and θ the angle between the vectors $\boldsymbol{\xi}$ and grad p.

Equation (8.2.5.2) is now to be averaged over the ensemble. The average $\boldsymbol{\xi}$ will be in the direction $-\mathrm{grad}\, p$ if the porous medium is assumed to be isotropic, as this is the only distinct direction. Its magnitude will be equal to the average of the component of $\boldsymbol{\xi}$ in the direction of $-\mathrm{grad}\, p$; i.e., equal to the average of $|\boldsymbol{\xi}| \cos\theta$. Thus we have

$$\overline{\boldsymbol{\xi}/\tau} = -\mathbf{n}\,c\sqrt{|\mathrm{grad}\, p|}\,\cos^{\frac{3}{2}}\theta/\sqrt{\rho} \tag{8.2.5.3}$$

or

$$\mathbf{V} = -\mathbf{n}\,c\sqrt{|\mathrm{grad}\, p|}\,\cos^{\frac{3}{2}}\theta/\sqrt{\rho}, \tag{8.2.5.4}$$

where $\mathbf{V}$ is the pore-velocity vector and $\mathbf{n}$ a unit vector in the direction grad p. Therefore, the probability distribution can now be expressed as follows:

$$\psi(\mathbf{x}, t) = (4\pi Dt)^{-\frac{3}{2}} \exp[-[\mathbf{x} + t\langle c\sqrt{|\mathrm{grad}\, p|}\,\cos^{\frac{3}{2}}\theta\rangle \mathbf{n}\rho^{-\frac{1}{2}}]^2/(4Dt)]. \tag{8.2.5.5}$$

The average position of the median $\bar{x}$ of this distribution is

$$\bar{x}(t) = -t\langle c\sqrt{|\operatorname{grad} p|}\cos^{\frac{3}{2}}\theta\rangle\mathbf{n}/\sqrt{\rho} \qquad (8.2.5.6)$$

and the average of the square-length deviation is, from the general theory of Gauss distributions,

$$\overline{x^2(t)} = 2Dt + \overline{\langle c\sqrt{|\operatorname{grad} p|}\cos^{\frac{3}{2}}\theta\rangle^2 t^2/\rho}. \qquad (8.2.5.7)$$

Thus, the mean pore velocity is

$$|\mathbf{V}| = -\overline{\langle c\sqrt{|\operatorname{grad} p|}\cos^{\frac{3}{2}}\theta\rangle}/\sqrt{\rho}. \qquad (8.2.5.8)$$

The above theory refers to homogeneous forces only. It is, however, easy to effect a generalization to inhomogeneous forces by observing, as was done in section 8.2.1, that the fundamental probability distribution is a solution of a differential equation as follows:

$$\partial\psi/\partial t = \operatorname{lap}(D\psi) + \operatorname{div}[\psi\langle c\sqrt{|\operatorname{grad} p|}\cos^{\frac{3}{2}}\theta\rangle\mathbf{n}\rho^{-\frac{1}{2}}], \qquad (8.2.5.9)$$

which is also valid for inhomogeneous forces. Finally, the continuity condition has to be formulated. Following the same procedure as in section 8.2.1, we end up with the following fundamental equation of turbulent flow:

$$\partial\rho/\partial t = \operatorname{lap}(D\rho) + \operatorname{div}[\rho^{\frac{1}{2}}\mathbf{n}\sqrt{|\operatorname{grad} p|}\,\bar{c}\cos^{\frac{3}{2}}\theta]. \qquad (8.2.5.10)$$

Taking everything together that depends on the porous medium only, this can be written as follows:

$$\partial\rho/\partial t = \operatorname{lap}(D\rho) + \operatorname{div}[m\rho^{\frac{1}{2}}\mathbf{n}\sqrt{|\operatorname{grad} p|}], \qquad (8.2.5.11)$$

where m is a constant of the porous medium and D a certain function that cannot be evaluated further without additional assumptions.

Equation (8.2.5.11) of turbulent flow shows that the flow is composed of two effects: first, one corresponding to the average turbulent flow through a set of small channels; and, second, a dispersivity effect. The heuristic equations mentioned earlier do not take the dispersivity into account. It is, however, well known that it does occur.

(b) *Molecular flow region* Finally, one can also apply statistical considerations to the molecular flow problem in porous media (Scheidegger 1955). Following a procedure analogous to that outlined above for laminar flow or for 'turbulent' flow, one ends up with the following equation:

$$\partial\rho/\partial t = \operatorname{lap}(D\rho) + \operatorname{div}[(c_1 + c_2/p)\rho\operatorname{grad} p), \qquad (8.2.5.12)$$

where D is a diffusivity function (cf. 8.2.5.7) and c_1 and c_2 are constants. This equation demonstrates that flow through porous media in the molecular regime can indeed be described by the flow through an assemblage of capillaries in each of which Knudsen flow occurs, but there is, in addition, a diffusivity effect which originates from the statistical considerations.

8.3
RANDOM-MEDIA MODELS

8.3.1 *Principles*

As already indicated, one can consider the flow of a particle through a porous medium not only as a 'random walk' through a homogeneous medium, but also as a deterministic process in a 'random' medium, i.e., in a medium whose possible configurations are only probabilistically known. Thus, the basic ensemble is a set of possible configurations (states) of the medium, the microdynamic equation of motion is a deterministic flow equation for the fluid, and 'tagged' fluid particles are again chosen as observables.

There are many ways by which random models of porous media can be constructed. However, only two basic procedures have been used: either a flow-net is built up by adding channels in a statistically prescribed fashion to progressively larger configurations (random-graph models), or flow channels are blocked (dammed) in a statistically prescribed fashion in an originally regular flow net (random-maze models). Evidently the two procedures *can* lead to the same ensembles. We shall discuss these two possibilities below.

8.3.2 *Random-graph models*

Instead of considering the dispersion of fluid particles during their passage through a porous medium as the result of a random walk, one can consider it as the outcome of deterministic flow through a random medium. This is best explained by using the visualization of the corresponding probability functions as the concentration of a pollutant, say of a 'spot of dye,' during its passage through a porous medium.

Thus, let us consider a small spot of dye injected into a porous medium. After a certain time, it will have spread in the medium owing to the successive branching processes imposed by the random complexities of the pore channels. The paths followed by the various fractions of the dye-spot will then have the shape of a topological 'tree,' the particles of dye being located at the free vertices. This leads to random-graph models as first considered by Liao and Scheidegger (1968, 1969). The ensemble is obtained by injecting the dye in macroscopically identical samples. According to the ergodic hypothesis, the expectation value for the position of the dye taken over the ensemble is equal to the time-expectation value for the position of the fluid caused by the stochastic branching process.

For simplicity's sake, one considers the branching process as a bifurcating

H

process. The ensemble then consists of all possible bifurcating arborescences with a given number of free ends. Each possible graph in this ensemble is considered as equally likely. An example of such a graph is shown in figure 31. For each graph one can calculate h_{mean} as the mean distance of the free vertices from the root of the tree and the variance σ_h^2 of the free vertices around this mean. The mean of h_{mean} and σ_h^2, i.e., $\bar{h}_{mean}$ and $\overline{\sigma_h^2}$ respectively, will give the expectation value for the mean position of the fluid and of the dispersion of the fluid over the ensemble of different arborescences.

As $\bar{h}_{mean}$ is the expected value of the mean fluid position, one has

$$\bar{h}_{mean} = vt, \tag{8.3.2.1}$$

v being the over-all pore velocity. From the general theory of dispersive processes one has

$$\overline{\sigma_h^2} = vt, \tag{8.3.2.2}$$

where D is the factor of dispersion. Thus if t is considered constant, a plot of $\bar{h}_{mean}$ versus $\overline{\sigma_h^2}$ corresponds to a plot of v versus D.

The task of calculating the required expectation values formulated above is a formidable one. The number N of bifurcating arborescences with n free ends was calculated long ago by Cayley (1859); it is

$$N = \frac{1}{2n - 1}\binom{2n - 1}{n}. \tag{8.3.2.3}$$

The binomial coefficient in this expression rapidly becomes very large, and it would be a hopeless task to enumerate all the graphs of an ensemble. Hence, one has recourse to a Monte Carlo procedure. One generates on a computer a *sample* of the ensemble in a random fashion. Since a computer cannot draw graphs, a representation which *can* be handled on a computer has to be found. For this purpose, Lukasiewicz's (see Berge 1958) representation of arborescences is especially convenient. One puts the root on top and reads the graph from top to

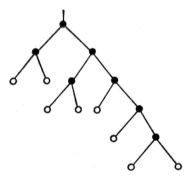

FIGURE 31　Example of a bifurcating arborescence.

bottom and from left to right. Each junction is represented by 'one' and each vertex by 'zero'. The graph in figure 31 can then be represented by the word

(110011001010100).

The number of junctions is always one less than the number of free vertices. Words of the above type can be random-generated on a computer.

By choosing the ensemble, one can simulate two- or three-dimensional models (a graph is always two-dimensional; however, if one counts all graphs which differ only by turning parts at the nodes 'left' and 'right' as identical, one obtains a 'three-dimensional' model). In all cases one finds, for longitudinal dispersion,

$$D \sim v^n \qquad\qquad (8.3.2.4)$$

with $n \sim 2.4$. For lateral dispersion, the result is similar, but with $n \sim 1$ (Liao and Scheidegger 1969).

In comparison with the random-walk theory, where relationships like (8.3.2.4) also occur, but with $1 \leq n \leq 2$, the value of $n = 2.4$ yielded by random-graph models for longitudinal dispersion appears very large. Therefore, several modifications of the random-graph model were tried by Torelli and Scheidegger (1972). In particular, ensembles of graphs were generated by actually modelling the stochastic branching processes that occur at every node, rather than by taking samples from an ensemble in which all graphs with a given number of free ends are considered as equally probable. One can thus obtain samples of graphs generated in a given number of cycles, and apply statistics over such samples of graphs. If *all* ends of the graphs are allowed to grow simultaneously, the relations between the coefficient of longitudinal dispersion D_{long} and the average pore velocity v, and between the coefficient of lateral dispersion D_{lat} and v, become

$$D_{\text{long}} \sim v^{1.2}, \qquad D_{\text{lat}} \sim v. \qquad\qquad (8.3.2.5)$$

These results compare favourably with those from the random-walk theory and, as we shall see later, with experimental evidence.

8.3.3 *Random-maze models*

For the construction of random-maze models, one starts with regular mazes called crystals. It was Broadbent and Hammersley (1957) who suggested this approach. One begins with a 'crystal' which is a structure composed of 'atoms' and of 'bonds' which connect the 'atoms'; the bonds represent the flow channels. A 'random maze' is then derived by assuming that each bond has a given probability of being obstructed. In an actual realization of a random maze, fluid injected at one atom, called a 'source atom,' will spread to all the atoms which are connected to the source atom by unobstructed paths.

It is apparent that the probability that a 'spot of dye' percolating with a fluid through a random maze will reach an infinite number of atoms cannot be 1 unless the maze reduces to a crystal and its random properties are lost.

Thus the original random-maze model of Broadbent and Hammersley (1957) does not seem suitable for describing the flow of a dye through saturated porous media very well, because the dye should be able to travel indefinitely far into the medium. For this reason, Torelli and Scheidegger (1971) modified the original random-maze model. They considered a crystal like that shown in figure 32 and derived a random maze from it by eliminating bonds at random, but with the condition that at least one of the two bonds originating at any one atom is left. In this way the possibility of infinite spreading of the dye is assured.

If an actual realization of the maze is given, if the pressure conditions at the upstream and downstream boundaries are specified, and if the bonds are assumed to be straight cylindrical capillaries of equal cross-section to which the Hagen-Poiseuille law applies, then the velocity of the flow in each bond can be calculated. In addition, the law for the dye dispersing at every bifurcation must be given. For this, one assumes that the ratio of the dye-masses which follow the two branches of the bifurcation equals the ratio of the fluid fluxes through the branches themselves.

Thus, if a unit mass of dye is released at time 0 at one source atom, one can at any time determine its distribution on a particular realization of the random maze. One then considers again the *ensemble* of all possible mazes for equal time and given boundary conditions, and calculates expectation values of the observables over such ensembles. Thus, one obtains relations between the mean longitudinal distance travelled by the dye and the expectation value of the longitudinal and lateral variances of the dye distributions. These correspond to relations between the mean pore velocity v and the longitudinal and lateral dispersion coefficients D_{long} and D_{lat}, which turn out to be linear.

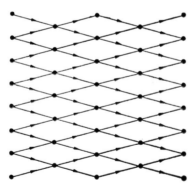

FIGURE 32   An ideal crystal for the description of the fluid flow through porous media.

Another way to obtain a random maze out of the crystal of figure 32 is to assign diameters to the bonds, at random, in lieu of considering them of equal cross-section and eliminating some of them at random. The result obtained from this model is again that the relations between D_{long} and D_{lat} and v are linear.

In effect, for all random-maze models the relation between D and v will be linear: evidently an increase in percolation velocity does not alter the geometrical configuration of the flow. Hence one will always find linearity in the relation between q and D (Torelli and Scheidegger 1972). The random-maze models thus yield the same result as that obtained with random-walk models in the 'geometrical' case.

8.4
SYSTEMIC APPROACH

8.4.1 *Principles*

Finally, a statistical model of flow through porous media can be constructed by omitting all direct references to microdynamical equations of motion, and by making instead only very generalized hypotheses about the statistical processes that affect the states of the ensemble. This approach is, in fact, based upon Gibbs' kinetic theory of gases (Scheidegger 1961b).

If a gas is considered as a large system which is described by the generalized position (p_n) and momentum (q_n) coordinates of all the molecules, then the motion of the system is described by a Hamiltonian function $H(p_n, q_n)$. The Hamiltonian in Gibbs' scheme represents the energy of the whole system, which must be a constant of the motion. Thus

$$\partial H/\partial t = 0. \tag{8.4.1.1}$$

One calls the space of all the p_ns and q_ns the 'phase space' of the system. Each possible configuration of the p_ns and q_ns is, therefore, represented by a point in phase space. A probability distribution $P(p_n, q_n, t)$ of points in phase space represents a probability distribution of states of the system. Thus, one can introduce ensembles of states over which expectation values can be calculated. In the kinetic theory of gases, one has the microcanonical ensemble for a thermally isolated system

$$p = \text{const.} \quad \text{for } E \leq H \leq E + dE. \tag{8.4.1.2}$$

Any small part of such a microcanonical ensemble is then described by a canonical ensemble

$$p = \frac{\exp[-\beta H(p,q)]}{\int \exp[-\beta H(p,q)] \, d^N p \, d^N q}, \tag{8.4.1.3}$$

where β is the parameter of the probability distribution. Then the thermodynamic functions are defined as follows: the partition function Z is

$$Z = \int \exp[-\beta H(p,q)]dpdq, \qquad (8.4.1.4)$$

the free energy F is

$$F = -(1/\beta)\ln Z, \qquad (8.4.1.5)$$

the temperature $\mathfrak{T}$ (where k is Boltzmann's constant) is

$$\mathfrak{T} = 1/(k\beta), \qquad (8.4.1.6)$$

and the entropy S is

$$S = -\partial F/\partial \mathfrak{T}. \qquad (8.4.1.7)$$

The above is a statement of Gibbs' theory. It can be extended to non-equilibrium thermodynamics if the following assumptions are made: (i) if one has non-equilibrium, the regression of the fluctuations is a linear function of the distance from the equilibrium case; (ii) the fluctuations are canonical; and (iii) one has microscopic reversibility (Onsager's relations). It is well known, that, under these conditions, one ends up with a diffusivity equation for the time-dependence of the temperature distribution $\mathfrak{T}(t)$ in a solid body (coordinates x_i) with a symmetric diffusivity tensor D_{ij}

$$\frac{\partial \mathfrak{T}}{\partial t} = \frac{\partial}{\partial x_i} D_{ij} \frac{\partial \mathfrak{T}}{\partial x_j}. \qquad (8.4.1.8)$$

The problem is now to carry Gibbs' kinetic theory of gases over to the case of flow of fluids through porous media. The way to do this is not immediately obvious because the existence of the Hamiltonian expression for energy as an integral of the motion is basic in Gibbs' theory. For porous media, one is faced with a dissipative system so that the energy is no longer a constant of the motion. One will therefore have to find some other constant of the motion which can play the role of the Hamiltonian function.

8.4.2 Thermodynamic analogy of flow through porous media

In order to solve the problem posed at the end of the last section, one notices that Gibbs' scheme does not necessarily require that the Hamiltonian function be the energy. One has the following basic structure of the theory: (a) a large system described by canonical variables p_n, q_n; (b) a Hamiltonian function $H(p_n, q_n)$ which is a constant of the motion and which is also positive-definite; (c) for the fluctuation, a linear regression and microscopic reversibility. These conditions can be satisfied by *mass* as well as by energy (Scheidegger 1961b).

Thus, let us consider a volume V travelling with the main stream of a fluid through a porous medium. We subdivide this volume into a large number of subvolumes (cells) of, say, unit volume ΔV. We now take a spot of dye and introduce it into a given cell of the volume V. After a certain time t has elapsed, the spot of dye will be found spread out over different cells, because of the statistical interaction between the cells. One wishes to calculate the dye-mass distribution over the cells with time. We thus have a system where the mass (of dye) is a constant of the motion, which results from the law of conservation of mass. This mass can then be called H, and identified with the Hamiltonian. It can be expressed as follows:

$$H = \sum_{i=1}^{N} m_i = \text{const.,} \tag{8.4.2.1}$$

where m_i is the mass in cell i. If one assumes that

$$p_i = + \sqrt{m_i}, \tag{8.4.2.2}$$

then the Hamiltonian can be written as

$$H = \sum_{i=1}^{N} p_i^2. \tag{8.4.2.3}$$

Then the q_i (the conjugates of p_i) are defined through the canonical equations of motion

$$\dot{q}_i = \partial H/\partial p_i \quad \text{and} \quad \dot{p}_i = - \partial H/\partial q_i. \tag{8.4.2.4}$$

The above Hamiltonian describes static conditions, i.e., the amount of dye in each cell remains constant. To obtain a change in the amount of dye in the individual cells, one has to modify the Hamiltonian expression by adding an interaction term U:

$$H = \sum_{i=1}^{N} p_i^2 + \varepsilon U(p_i, q_i, t), \tag{8.4.2.5}$$

where t is the time and ε is a small constant.

After defining the phase space as above, one can define an ensemble in this space. The ensemble will then attain a state of statistical equilibrium because of the presence of the interaction term in the Hamiltonian. The component systems are then distributed canonically, with a probability distribution P_i:

$$P_i(m_i) = C' \exp[-m_i/(k\mathfrak{T})], \tag{8.4.2.6}$$

where k is a constant, $\mathfrak{T}$ is the analogue of the temperature, and C' is a normalization constant. Thus (leaving out the index i)

$$P = (1/Z) \exp[-m/(k\mathfrak{T})], \tag{8.4.2.7}$$

where Z is the 'partition function' required for normalization:

$$Z = \int \exp[-m/(k\mathfrak{T})] \, dp \, dq = C \int \exp[-m/(k\mathfrak{T})] \, dp \tag{8.4.2.8}$$

with $C = \int dq$.

Based upon the partition function, one obtains an analogy between a thermodynamic field and flow through the porous medium, provided one sets $k = 2$, which maintains the following relations (Tomkoria and Scheidegger 1967):

temperature $(\mathfrak{T}) \to$ mass per unit cell $(m_0 = 1/(2\beta))$,

entropy $(S) \to \ln m_0 + 2K + 1$, where $K = \ln[c(\pi/2)^{\frac{1}{2}}]$,

internal energy $(U) \to 1/(2\beta)$,

work done $(W) \to -\alpha \int_V (m_0/V) dV$ (α is a constant),

energy potential (dU) (in differentials) $\to m_0 - (\alpha m_0/V) dV$,

Helmholtz's free energy $(F) \to -m_0(\ln m_0 + 2K)$,

Gibbs' free energy $(\Phi) \to -m_0(\ln m_0 + 2K) + \alpha m_0 \int dV/V$,

Gibbs' potential $(\Psi) \to -m_0(1 + \alpha \int dV/V)$,

heat capacity $\to 1$.

Extending this treatment to the non-equilibrium case, one ends up with a diffusivity equation for the mass-density m_i. For, it is well known that, for temperature conduction, a diffusivity equation is the outcome, which, by analogy, now also arises for dye-dispersion in a porous medium. The analogy is possible because, in both cases, a non-negative quantity (mass, energy) is transferred from cell to cell by a statistically fluctuating transfer process, whose exact nature is unspecified. It is only assumed that in both cases the fluctuations are very numerous and their effect is linearly additive, for then the central limit theorem of probability theory is applicable, which leads to a diffusivity equation.

It is, in fact, possible to give a much more detailed justification of the formal analogy between mass-transfer in porous media and energy-transfer in thermodynamics than was given above. The basis of the statistical theory of thermodynamics lies in the existence of a positive definite integral of the motion, i.e., the Hamiltonian expression for energy. It is then evident (Scheidegger 1961b) that the total mass of contaminant contained in a fluid volume percolating through a porous medium can be treated as a positive definite integral of the motion, just as the energy is treated in statistical thermodynamics. It can then be shown that in the equilibrium case the interaction function as used in the energy-based statistical mechanics of solids or weakly coupled gases entails a corresponding result in mass-dispersion systems (Chaudhari and Scheidegger 1964, 1965), which enables one immediately to set up an analogy between fluid dynamics in porous media and thermodynamics. This model has the advantage that no microdynamic assumptions (other than general statistical hypotheses) have to be made. On the other hand, it cannot make any specific predictions regarding the factor of dispersion occurring in the diffusivity equation. It has been useful, though, for the analysis of stability conditions in displacement processes.

8.5
APPLICATIONS OF THE STATISTICAL THEORY

8.5.1 *The phenomenon of dispersion*

The main achievement of the statistical theory of flow through porous media is that it predicts the phenomenon of dispersion. Owing to the complexities of the pore channels, neighbouring particles become mixed up with each other so that, for instance, a spot of dye should spread during percolation through a porous medium. The spread of the spot of dye should be given by a diffusivity equation, which reads in the linear case

$$\frac{\partial C}{\partial t} = D\frac{\partial^2 C}{\partial x^2} - v\frac{\partial C}{\partial x}, \tag{8.5.1.1}$$

where C is the concentration of dye and v is the pore velocity. The application of the above equation is mostly to miscible displacement, and specific solutions of it will be discussed in that context. However, from a theoretical standpoint, of greatest importance is the answer to the question whether this predicted dispersion does actually occur and, if so, which of the various statistical models that were introduced yield the correct connection between the factor of dispersion D and the pore velocity v (or the filtration velocity q).

In the laboratory, the predicted dispersion has indeed been found. Regarding longitudinal dispersion, i.e., dispersion in the direction of mean flow, a summary account has been given, for example, by De Wiest (1965, pp. 305ff). Accordingly, as a function of the percolation velocity q, the factor of dispersion D_{long} can be expressed by

$$D_{\text{long}} = D_m + aq^n, \tag{8.5.1.2}$$

where D_m and a are constants; the exponent n is close to 1 (Blackwell 1962; Harleman and Rumer 1963; Perkins and Johnston 1963; Bruch and Street 1967). The above empirical formula can be explained by assuming that the constant D_m represents mixing due to molecular diffusion which occurs also without any flow. This molecular diffusion is, of course, in *addition* to any dispersion and was neglected in all models discussed. On the other hand, the term dependent on q fits very well with those models that predict a linear relation between D_{long} and q. In this instance, the graph-theoretical models are the poorest, because for them n came out much too high.

Similar results have been obtained for lateral dispersion (Blackwell 1962; Harleman and Rumer 1963; Bruch and Street 1967) although the experiments are not quite as easy to perform and are therefore not quite as unequivocal. Most experiments yield again (for a discussion see, for example, Liao and Scheidegger 1969)

$$D_{lat} \sim q^n \tag{8.5.1.3}$$

with $n \sim 1$, but some yield much higher values for n. On the other hand, for lateral dispersion, all models discussed yield $n \sim 1$.

In summary, one can say that statistical models of flow through porous media correctly account for the phenomenon of dispersion, something which no capillaric model can do. It should be noted that statistical mechanics represents a procedural framework for constructing certain types of models; there is, naturally, still much room for further variations in the types of ensembles and microdynamic equations of motion that may be chosen.

8.5.2 Steady-state conditions

The thermodynamic analogy between mass flow in porous media and heat flow in a solid enables one to deduce the conditions for the establishment (or prevention) of a steady state in a displacement process in porous media.

In ordinary thermodynamics, there is a well-known condition which must be satisfied by a system in a steady state: the production rate of entropy throughout the system must be a minimum. To the extent that there is a complete thermodynamic analogy with mass transport, this principle of 'entropy' production must also be satisfied in the latter (Scheidegger 1968).

According to De Groot (1961), the entropy production rate σ in thermodynamics is given by

$$\sigma = \sum_i [(\text{grad } \mathfrak{T})/\mathfrak{T}^2] J, \tag{8.5.2.1}$$

where the sum is to be taken over all parts of the system, $\mathfrak{T}$ is the temperature, and J is the heat flux. Using now our analogy with flow through porous media, we obtain as stability condition in the linear continuous case

$$-\int_A^B J(C'/C^2)\, dx = \text{a minimum}, \tag{8.5.2.2}$$

where C is the mass concentration and J the mass flux.

Specific conditions for particular cases can then be deduced by deriving the Euler-Lagrange equations corresponding to the minimum principle given above. For this purpose, the form of the function $J(C)$ must be specified, which can only be done if one has specific applications in mind.

9
Elementary displacement theory

9.1
GENERAL REMARKS

We have discussed thus far in this book the flow of *homogeneous* fluids in porous media. On many occasions, however, the flow of *mixtures* of fluids is of considerable interest. The various constituents of the mixtures are usually referred to as 'phases,' and thus their flow is termed 'multiple phase flow in porous media.'

Contributions to the study of multiple phase flow are much fewer in number than those concerning single phase flow. This is owing to the somewhat restricted applicability of multiple phase flow, as compared with single phase flow, to practical cases. Nevertheless, in those fields where multiple phase flow plays any role at all it has a very important one. Contributions to the subject have been noted from the following fields:

(a) from the oil industry: the simultaneous flow of oil, water, and gas in porous rock strata is important in connection with the production of oil from oil fields;
(b) from the study of countercurrent towers in chemical engineering;
(c) from soil science: the flow of moisture in unsaturated soils (i.e., the simultaneous flow of water and air) is of importance;
(d) from groundwater hydrology: much attention has been paid to the encroachment of salt water into fresh water in reservoirs beneath oceanic islands.

The study of multiple phase flow in porous media could be split into sections similar to those for single phase flow. Thus one could characterize the various cases by the regime of flow prevalent in each; i.e., whether it is laminar, turbulent, molecular, etc.

However, it turns out that a much more important distinction than that of the flow regime is one concerning the fluids: viz., whether they are miscible or immiscible, leading to 'miscible' and 'immiscible' displacements in porous media. It will be seen that these two types of displacements are the two limit cases that are theoretically possible. The remainder of this book, therefore, is split into two chapters (10 and 11) dealing with immiscible and miscible flow, respectively. In addition, as a preliminary to the study of multiple phase flow where both phases are present simultaneously at certain points in the porous medium, we shall discuss various simplified treatments. These will be collected in the present chapter on elementary displacement theory.

9.2
THE MUSKAT MODEL

9.2.1 *Equal permeability-viscosity ratios*

It is possible to treat displacement processes in porous media in a very elementary fashion. The discussion of single phase flow, based on Darcy's law, may be extended to displacement processes if the following simplifying assumptions are made. If, for example, a fluid is displaced in a porous medium by another in such a manner that the 'input' and 'output' are conducted steadily and if, furthermore, the permeability-viscosity ratio is the same for both fluids, and if no mixing occurs at the interface, then one can argue that the flow of the two fluids would occur in the same manner as if they were one homogeneous fluid. The position of the interface at an arbitrary time t could be ascertained from a consideration of its position at the time t_0 and a study of the motion of each point of the interface during the time from t_0 to t.

It is thus possible to take over the solutions of Darcy's law discussed in chapter 5 and to use them for the description of displacement processes. The model of displacement processes thus envisaged is termed the 'Muskat model' after its originator.

9.2.2 *Unequal permeability-viscosity ratios*

It is not even necessary to assume that the permeability-viscosity ratio for the two fluids is identical, and it is then still possible to apply elementary steady-state flow considerations to multiple phase flow, provided no mixing occurs at the interface and provided the permeability-viscosity ratios are constant. Muskat (1934) has given an analytical formulation of this problem.

Let the interface be represented at the time t by the equation

$$F(x, y, z, t) = 0. \tag{9.2.2.1}$$

Then at the time $t + \delta t$ the new interface will be given by

$$F(x + v_x\delta t, y + v_y\delta t, z + v_z\delta t, t + \delta t) = 0 = F(x, y, z, t), \tag{9.2.2.2}$$

where v_i are the pore velocity components on the interface. We have at once:

$$\partial F/\partial t + \mathbf{v} \operatorname{grad} F = 0. \tag{9.2.2.3}$$

Applying Darcy's law (and the Dupuit-Forchheimer assumption $q = Pv$), we find that

$$\frac{\partial F}{\partial t} - \frac{k}{P\mu} \operatorname{grad} \Phi \operatorname{grad} F = 0 \quad \text{with } \Phi = p + \rho gz. \tag{9.2.2.4}$$

The problem can now be formulated as follows (Muskat 1934). Determine the potential functions Φ_1 between a surface S_w and $F(x, y, z, t) = 0$, and Φ_2 between a surface S_e and $F(x, y, z, t) = 0$, such that

$$
\left.
\begin{array}{l}
\left.
\begin{array}{l}
\Phi_1 = \Phi_w \text{ on } S_w \\
\Phi_2 = \Phi_e \text{ on } S_e
\end{array}
\right\} \text{boundary conditions,} \\[2em]
\left.
\begin{array}{l}
\Phi_1 = \Phi_2 \\[0.5em]
\dfrac{k_1}{\mu_1}\dfrac{\partial \Phi_1}{\partial n} = \dfrac{k_2}{\mu_2}\dfrac{\partial \Phi_2}{\partial n}
\end{array}
\right\} \text{on } F(x, y, z, t) = 0, \\[2em]
P\dfrac{\partial F}{\partial t} - \left(\dfrac{k}{\mu}\right)_{1,2} \operatorname{grad} \Phi_{1,2} \operatorname{grad} F = 0.
\end{array}
\right\} \tag{9.2.2.5}
$$

Here, n is the normal to the interface. The fact that Φ has been assumed to be a potential function implies, of course, that steady-state conditions (i.e., incompressibility) are maintained in each of the two flowing phases.

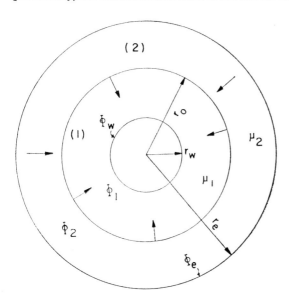

FIGURE 33   Geometrical layout in Muskat's approximate solution of a radial displacement problem (after Muskat 1934).

Even the very simplified problem outlined above is difficult to treat analytically. Muskat (1934) solved, for example, the case of radial encroachment (well) in two dimensions on this basis. The shape of the interface is known in advance in this case: it is always a circle. Let its position be denoted by r_0, and, furthermore, let the boundaries S_w and S_e also be circles of radius r_w and r_e ($r_w < r_0 < r_e$), respectively (see fig. 33). If the potentials are held constant upon the boundaries ($= \Phi_w$ and Φ_e, respectively), then the equation for the interface can be integrated readily and we obtain:

$$\frac{r_0^2}{r_e^2}\left(\ln\frac{r_e^2}{r_w^2} - (1 - \varepsilon) + (1 - \varepsilon)\ln\frac{r_0^2}{r_e^2}\right) = -4\frac{k_1}{\mu_1}t\frac{\Phi_e - \Phi_w}{Pr_e^2} + \ln\frac{r_e^2}{r_w^2} - (1 - \varepsilon),$$

$$(9.2.2.6)$$

with

$$\varepsilon = k_1\mu_2/(\mu_1 k_2). \tag{9.2.2.7}$$

Setting

$$4\frac{k_1}{\mu_1 P}(\Phi_e - \Phi_w)\frac{1}{r_e^2} = 1, \tag{9.2.2.8}$$

and adding the numerical assumption that

$$r_e/r_w = 2000, \tag{9.2.2.9}$$

one obtains curves representing equation (9.2.2.6) as plotted in figure 34.

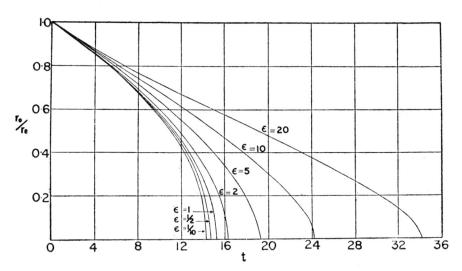

FIGURE 34   Muskat's solution of a radial displacement problem (after Muskat 1934).

More complicated geometrical patterns can be treated more easily by setting, in addition to the above simplifying assumptions,

$$k_1/\mu_1 = k_2/\mu_2, \qquad\qquad (9.2.2.10)$$

which implies that there is, in essence, a single-phase flow problem.

Muskat in his book (1937) discussed a variety of solutions for heterogeneous flow using this basis. Many solutions of the two-phase problem using this basis and relevant to the oil industry were also given in a later book by Muskat (1949). Hubbert (1940), Polubarinova-Kochina (1952), and Todd (1959) reviewed some cases relevant in groundwater motion.

As stated earlier, the calculations may become very involved for any but the simplest geometrical systems. This is particularly true if one of the flowing fluids is a gas whose compressibility is to be taken into account. Many of the above-mentioned investigations deal with this case. In general, advanced analytical procedures or high-speed electronic computers have to be employed to obtain the solutions (cf. Arthur and Metanomski 1966). Similarly, electrical analogues (Crawford and Collins 1955; Aronofsky and Ramey 1956; Burton and Crawford 1956; Karplus 1956; Meyer and Searcy 1956; Odeh et al. 1956; Crausse and Poirier 1957; Sinel'nikova 1957; Nobles and Janzen 1958; and others) and Hele-Shaw (cf. sec. 5.2.2) models (Collins and Gelhar 1971) have also been used.

We have given above Muskat's solution for the radial displacement from a well, which represents essentially a one-dimensional case. It may be useful to append the results obtained by Aronofsky (1952) for a particular two-dimensional system. Aronofsky studied (by a combined analogue and numerical procedure) the areal displacement caused by the injection of displacing fluid in a direct line drive into underground strata. The geometrical layout of the direct

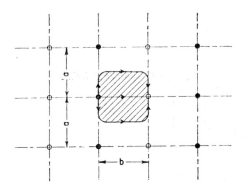

o   PRODUCING WELL
●   INJECTION WELL

FIGURE 35   Geometrical layout of a direct line drive.

line drive is shown in figure 35. Aronofsky obtained a series of solutions for various mobility ratios M,

$$M = (k/\mu)_{\mathrm{I}}/(k/\mu)_{\mathrm{II}}, \qquad\qquad (9.2.2.11)$$

of displacing (I) to displaced (II) fluids. The result at 'breakthrough' for three values of M, as obtained by Aronofsky, is shown in figure 36. Accordingly, with increasing mobility of the invading fluid, less and less of the fluid originally contained in the porous medium will be displaced.

Other solutions have been given by Meyer and Garder (1954) and Kawabata (1965), who included gravitational effects in a study of water coning in an oil well.

9.2.3 *Application of Muskat's model to groundwater hydrology*

An interesting application of Muskat's model to a special problem in groundwater hydrology has been reported. This is the behaviour of fresh water reservoirs beneath oceanic islands and in coastal areas (cf. Wentworth 1947, 1951). Thus it has been observed that rain water falling on the surface of, say, a circular island seeps downward and accumulates at the surface of the salt water at sea level. The fresh water builds up to a height above sea level determinated by the abundance of rain and the permeability of the island rock, thus forming a lens. At the lower boundary of the lens, the fresh water slowly displaces the sea water owing to the hydraulic head created by the build-up of water level at the top. Eventually a stationary state is reached in which the lower boundary extends downward to a depth z below sea level which corresponds to hydrostatic equilib-

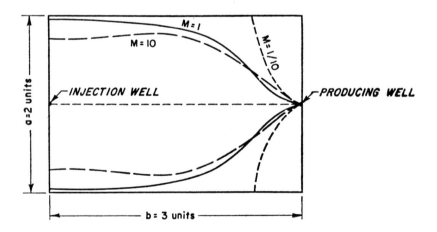

FIGURE 36   Comparison of flooded areas at the instant of breakthrough for various mobility ratios (after Aronofsky 1952).

rium of a body of fresh water floating upon sea water (cf. fig. 37). This depth z is
given by the equation (Ghyben-Herzberg equation)

$$z = \frac{\rho_f}{\rho_s - \rho_f} h_f, \tag{9.2.3.1}$$

where ρ_f is the density of fresh water, ρ_s the density of salt water, and h_f the
height of the water table above sea level. For $\rho_f = 1.000 \text{ g cm}^{-3}$ and $\rho_s = 1.025$
g cm^{-3}, one obtains

$$z = 40h_f. \tag{9.2.3.2}$$

The original displacement of sea water by fresh water must be assumed to have
taken place over a very long period of time; the displacement was almost without
mechanical mixing. The body of fresh water, therefore, has the shape of a doubly
convex lens with the circular edge coinciding with the circular coast of the island.
In groundwater hydrology, the principle of formation of a fresh water body
beneath an island in the above manner is referred to as the 'Ghyben-Herzberg
principle.' A drawing of a Ghyben-Herzberg lens is shown in figure 37.

The above theory is somewhat oversimplified, inasmuch as, in a *dynamic*
equilibrium, additional factors must be taken into account. What is important is
the *hydrodynamic* ('point') head, not the *hydrostatic* head. Accounting for the
difference between the two heads causes modifications in the Ghyben-Herzberg
equation. Such modifications have been proposed by Lusczinsky (1961), Perlmutter et al. (1959) and Goswami (1968a, b). The last author came up with the
relation

$$z = \frac{\rho_s}{\rho_s - \rho_f} h_{sp} - \frac{\rho_f}{\rho_s - \rho_f} h_{fp}, \tag{9.2.3.3}$$

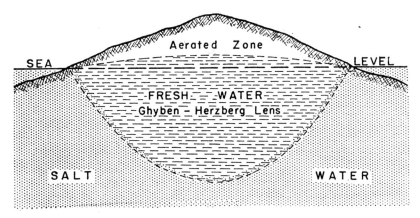

FIGURE 37   Drawing of a Ghyben-Herzberg lens beneath an oceanic island (after
Wentworth 1947).

where h_{sp} and h_{fp} are the saline and fresh water *point*-heads, respectively, i.e., the heads defined as the water-level, referred to sea level, found in a well filled with an amount of water of the *type* (salinity!) found at the point under consideration and in equilibrium at that point.

On the basis of equation (9.2.3.3), Goswami (1968a) found excellent agreement between theoretically predicted and actually measured data.

9.3
ELEMENTARY SYSTEMIC APPROACH

9.3.1 *Principles of solitary dispersion*

A further elementary treatment of displacement processes is obtained by considering the case of solitary dispersion (Scheidegger 1967).

Imagine that one has a mixture of a great number of substances percolating through a porous medium, but that one is interested in only one of these substances. One can then ignore all the other substances and consider the one of interest as 'solitary.' This yields the case of solitary dispersion.

Thus we characterize the amount of substance under consideration at the point $\mathbf{x}$ (Cartesian coordinates x_1, x_2, x_3) at time t by its concentration C (x_1, x_2, x_3, t). The continuity equation then states

$$\partial \rho C / \partial t = - \operatorname{div}(\rho \mathbf{v}), \tag{9.3.1.1}$$

where ρ is the density and $\mathbf{v}$ (components v_1, v_2, v_3) is the pore velocity vector of the substance under consideration. Often, it is convenient to introduce the overall pore velocity $\mathbf{V}$ (components V_i); one has then

$$v_i = r_{ik} V_k, \tag{9.3.1.2}$$

where r_{ik} is the fractional flow tensor (the summation convention is being applied). Then, the continuity equation becomes

$$\partial \rho C / \partial t = - \partial(\rho r_{ik} V_k) / \partial x_i. \tag{9.3.1.3}$$

Evidently one has in a first approximation

$$r_{ik} = C \delta_{ik} + \gamma_{ik}, \tag{9.3.1.4}$$

where δ_{ik} is the Kronecker tensor and γ_{ik} is a small deviation-tensor. This is so simply because the flow velocity of the substance under consideration must be close to the over-all flow velocity of the mixture. The deviation tensor indicates how much the flow velocity of the substance under consideration deviates from that of the mixture.

Introducing (9.3.1.4) into the continuity equation, we obtain

$$\partial \rho C / \partial t = - \partial[\rho(C \delta_{ik} + \gamma_{ik}) V_k] / \partial x_i, \tag{9.3.1.5}$$

which yields for an incompressible liquid

$$\partial C/\partial t = - \partial[(C\delta_{ik} + \gamma_{ik})V_k]/\partial x_i. \tag{9.3.1.6}$$

In the linear case this is further simplified to

$$\partial C/\partial t = - \partial[(C + \gamma)V]/\partial x. \tag{9.3.1.7}$$

Introducing travelling coordinates $x' = x - Vt, t' = t$, we obtain

$$\frac{\partial C}{\partial t'} + V\frac{\partial \gamma}{\partial x'} = 0. \tag{9.3.1.8}$$

9.3.2 Limiting cases

One can carry the above approach further by making educated guesses regarding the form of the deviation-flow γ. Two limiting cases are obviously possible:

(a) γ depends on C only   Thus we set for the linear case

$$\gamma = \gamma(C). \tag{9.3.2.1}$$

Introducing this into (9.2.1.7) we obtain, assuming V = const.,

$$\frac{\partial C}{\partial t} + V\left(1 + \frac{d\gamma}{dC}\right)\frac{\partial C}{\partial x} = 0 \tag{9.3.2.2}$$

or, in travelling coordinates,

$$\frac{\partial C}{\partial t'} + V\frac{d\gamma}{dC}\frac{\partial C}{\partial x'} = 0. \tag{9.3.2.3}$$

This is a well-known equation ('Buckley-Leverett equation'), which will be encountered again in immiscible displacement theory (cf. 10.3.1.10). The systemic approach shows that it must always appear when the deviation of a fractional flow from the main flow depends only on the concentration of the substance in question.

(b) γ depends on $\partial C/\partial x$ only   Thus, we set for the linear case

$$\gamma = - \alpha \partial C/\partial x. \tag{9.3.2.4}$$

Then the continuity equation becomes

$$\frac{\partial C}{\partial t} = - \frac{\partial(CV)}{\partial x} + \frac{\partial}{\partial x} D \frac{\partial C}{\partial x} \tag{9.3.2.5}$$

with $D = \alpha V$. This is again a well-known equation, the diffusivity equation with a mass-transport term. For V = const., one has

$$\frac{\partial C}{\partial t} = - V\frac{\partial C}{\partial x} + D \frac{\partial^2 C}{\partial x^2}. \tag{9.3.2.6}$$

It is thus seen that the general principles of systemic analysis yield the result that the behaviour of fluids that displace each other must lie between that described by the Buckley-Leverett equation and that described by a diffusivity equation. It will be seen later that these two equations describe immiscible and miscible displacements, respectively.

The extension of the treatment of this section (9.3.2) to three dimensions is obvious. The details are given in Scheidegger (1967).

9.3.3 Stability analysis

The elementary systemic approach lends itself to a stability analysis according to section 8.5.2, based on an analogue of the minimum-entropy principle (Scheidegger 1969). In the language of that section the quantity C is simply the mass-concentration per unit cell and V_j is the mass flux J. Hence, one has again two limiting cases (Tomkoria and Scheidegger 1967):

The first is obtained by setting

$$J = J(C). \tag{9.3.3.1}$$

The variational principle requires (in travelling coordinates x') that

$$\delta \int \frac{C'}{C^2} J(C) \, dx' = 0. \tag{9.3.3.2}$$

It becomes evident that it is not possible to find a solution for this case, i.e., the Euler equation corresponding to the minimum entropy principle becomes an identity since the expression under the integral sign is a total differential. This result indicates that, in the case under consideration, the general statistical principles pertaining to the steady state do not lead to any statement of the C-profile.

We try next the second limiting case

$$J = a \, dC/dx' \tag{9.3.3.3}$$

as a first approximation to J being a function of grad C; in this case, the dynamics of the displacement is governed by a diffusivity equation. It turns out that the minimization can then be performed; the solution is

$$C = c_1 e^{c_2 x'}, \tag{9.3.3.4}$$

where c_1, c_2 are constants of integration. Thus, a solution can be found in this case.

We thus have the result that in one of the possible theoretical limits of solitary dispersion $(J = J(C)$; Buckley-Leverett limit) no steady state is possible, and in the other $(J = aC'$; diffusivity limit), a steady state is possible. However, for the achievement of such a steady state, very special boundary conditions are required:

in the *moving* (with pore velocity) coordinate system, a constant C (i.e. concentration) must be maintained at two (moving!) points, say x_1' and x_2'. It is clear that, in practical cases, no such conditions can occur. Thus, in practice neither case allows a steady state to occur.

9.4
VISCOSITY-INSTABILITY

9.4.1 *The phenomenon of fingering*

The Muskat model, as discussed above, seems to yield reasonable results in many cases. However, it should be noted that a very stringent assumption in its applicability is that the front between the two fluids displaces itself in a regular manner. This assumption is by no means assured.

It has been observed that, under certain circumstances, instead of a regular displacement of the whole front, protuberances can occur which may shoot through the porous medium at relatively great speed. This effect is referred to as 'fingering.' It has been described, for instance, by Engelberts and Klinkenberg (1951), by Van Meurs (1957), by Terry et al. (1958), and by Perkins and Johnston (1969). A typical example of the development of fingers in a line drive experiment is shown in figure 38.

Thus, whenever instability phenomena occur, the Muskat model should no longer be employed. It will be shown below that there are indications that this should *always* be the case if the mobility ratio of invading to original fluid exceeds the value of 1.

9.4.2 *Instability criterion based on Muskat's model*

Let us now investigate the stability of a displacement front in the light of the

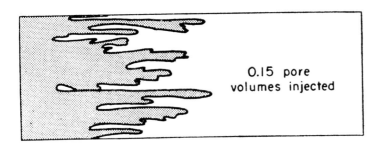

0.15 pore
volumes injected

FIGURE 38   Displacement front for a mobility ratio of 20, showing the development of fingers (after Terry et al. 1958).

model of fluid displacement in sands which was considered by Muskat (after Scheidegger 1960a). Accordingly, we assume the front as sharp, the displacement as complete, and Darcy's law as valid on both sides of the front. The Muskat procedure is to assume an initial front, and then to calculate the progress of that front numerically. This is a method which is ovbiously insufficient for the setting-up of general criteria for the stability of the front. What one would like to do is to assume a Muskat-type solution, then to introduce a perturbation into it, and, finally, to determine a criterion as to whether the perturbation will decay or grow in the course of time. Unfortunately, this is, in general terms, a very difficult problem, as it contains a floating boundary condition which is not easy to treat analytically.

It is, therefore, suggested that certain simplifications be made, which, however, will still preserve the basic physical features of the Muskat model. Thus, let us assume that a linear displacement experiment is being made (see fig. 39) where the pore velocity of fluid injection is v. If the displacement process is total and stable, then the displacement front will also move with the velocity v. During the time dt, the displacement of the front will therefore be

$$dx = vdt. \tag{9.4.2.1}$$

The quantity of fluid injection is then $Padx$, a being the cross-sectional area.

In order to have the front moving with the velocity v there must be a pressure gradient present in the fluids. If we denote the displacing fluid by 1 and the displaced fluid by 2, the pressure gradient in fluid 1 is given by

$$\text{grad } p_1 = \frac{\mu_1 P}{k_1} v \tag{9.4.2.2a}$$

and in fluid 2 by

$$\text{grad } p_2 = \frac{\mu_2 P}{k_2} v. \tag{9.4.2.2b}$$

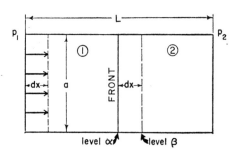

FIGURE 39   Stable displacement front.

The case represented in figure 39 corresponds to stable displacement. We shall refer to it as 'case A.'

We shall now consider instead of the above case of a stable displacement, the case where a 'finger' might develop ('case B'). This possibility is shown in figure 40. During the time dt the same quantity adx as before is being injected, and the problem is to determine whether the stable (case A) or the fingered (case B) displacement is more likely to occur under given external circumstances.

The possibility of the existence of a finger, as compared with the stable displacement process, depends on various conditions:

(a) *Volumetric condition*   The volume of the fluid contained in the finger and the small displacement of the front must be equal to the total fluid injected. Hence

$$Padx = (a'dx' + a''dx'')P,\tag{9.4.2.3}$$

where $a' = a - a''$.

(b) *Equilibrium condition*   A finger can be maintained only if there is no tendency for it either to spread out or to get pinched to nothing. This entails that the pressure gradient in the finger must be the same as that in the surrounding fluid. Since the mobility of the fluid contained within the finger is different from that contained in the surrounding part of the porous medium, this requires a different velocity in the finger and in the surrounding fluid. Equating (9.4.2.2a) and (9.4.2.2b) yields

$$P\frac{\mu_1}{k_1}v_1 = P\frac{\mu_2}{k_2}v_2,$$

and, since $v = dx/dt$ (with $m_i = k_i/\mu_i$),

$$m_1dx' = m_2dx'',$$

or

$$dx''/dx' = m_1/m_2.\tag{9.4.2.4}$$

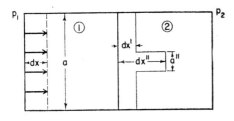

FIGURE 40   Model of a finger,

(c) *Energy condition*   A finger will develop only if, by its formation, it consumes less energy than the corresponding stable displacement of the whole front.

Let us, therefore, calculate the energy that is required to obtain the corresponding displacements during the time dt in the two cases (A and B) that are under consideration.

Case A. In this case, the displacement front is shifted steadily by the amount dx during the time dt.

On the side containing fluid 1, the pressure drop is grad p_1. Hence the work W_1^A required to move a slug of fluid 1 into the space between level α and level β (see fig. 39) is

$$W_1^A = a \operatorname{grad} p_1 dx^2. \tag{9.4.2.5}$$

Similarly, the work W_2^A required to move a slug of fluid *out* of the space between level α and level β is

$$W_2^A = a \operatorname{grad} p_2 dx^2. \tag{9.4.2.6}$$

However, one can now express the pressure gradients in terms of the mobility. One has

$$\operatorname{grad} p_i = \frac{\mu_i}{k_i} P \frac{dx}{dt}, \tag{9.4.2.7}$$

where k_i is the permeability and P the porosity of the medium. Thus, the total work required to move the quantity of fluid under consideration in case A is

$$W^A = Padx^2 \left(\frac{\mu_1}{k_1} + \frac{\mu_2}{k_2} \right) \frac{dx}{dt}. \tag{9.4.2.8}$$

Case B. We shall now consider this case, where there is a finger.

On the side containing fluid 1 the work required is

$$W_1^B = a' \operatorname{grad} p_1 (dx')^2 + a'' \operatorname{grad} p_2 (dx'')^2, \tag{9.4.2.9}$$

where use has been made of the fact that the pressure gradient in the finger must be the same as that in the surrounding *displaced* fluid (condition (b)).

Similarly, the work required to move the corresponding slug of displaced fluid is

$$W_2^B = a' \operatorname{grad} p_2 (dx')^2 + a'' \operatorname{grad} p_2 (dx'')^2. \tag{9.4.2.10}$$

Again, it is possible to express the pressure gradients in terms of the mobilities. One has

$$\operatorname{grad} p_i = \frac{P}{k_i} \mu. \frac{dx'}{dt}. \tag{9.4.2.11}$$

Hence

$$W^{\mathrm{B}} = \frac{dx'}{dt} P\left(\frac{\mu_1}{k_1} a'(dx')^2 + 2\frac{\mu_2}{k} a''(dx'')^2 + \frac{\mu_2}{k_2} a'(dx')^2\right).$$

(9.4.2.12)

The energy condition, as stated earlier, is

$$W^{\mathrm{B}} \le W^{\mathrm{A}};$$

(9.4.2.13)

hence

$$dx'\left(\frac{\mu_1}{k_1} a'(dx')^2 + 2\frac{\mu_2}{k_2} a''(dx'')^2 + \frac{\mu_2}{k_2} a'(dx')^2\right) \le a\,dx^3 \left(\frac{\mu_1}{k_1} + \frac{\mu_2}{k_2}\right).$$

(9.4.2.14)

It is now possible to combine the three conditions (a), (b) and (c) (i.e., equations 9.4.2.3, 9.4.2.4, and 9.4.2.14) to yield a condition for the critical mobility ratio at which fingering must develop.

For the sake of abbreviation, we write

$$a'/a = \alpha', \qquad a''/a = \alpha'',$$

(9.4.2.15)

$$dx'/dx = \xi', \qquad dx''/dx = \xi'',$$

(9.4.2.16)

$$k_1\mu_2/(k_2\mu_1) \equiv m_1/m_2 = \eta.$$

(9.4.2.17)

Then the three relevant conditions can be simplified to read

(a) $1 = \alpha'\xi' + \alpha''\xi''$,

(9.4.2.18)

$\qquad \alpha' = 1 - \alpha''$,

(9.4.2.19)

(b) $\xi'' = \eta\xi'$,

(9.4.2.20)

(c) $\xi'[\alpha'\xi'^2 + 2\eta\alpha''\xi''^2 + \eta\alpha'\xi'^2] \le 1 + \eta$.

(9.4.2.21)

The three conditions can be treated so as to yield a critical condition for the (inverse) mobility ratio η, which can be achieved by a procedure of elimination of the other variables from the system of equations.

From the two equations (a) one has

$$\alpha' = \frac{\xi'' - 1}{\xi'' - \xi'},$$

(9.4.2.22)

$$\alpha'' = \frac{1 - \xi'}{\xi'' - \xi'}.$$

(9.4.2.23)

If these values are inserted into the third condition, one obtains

$$\xi'\left(\frac{\xi'' - 1}{\xi'' - \xi'}\xi'^2 + 2\eta\frac{1 - \xi'}{\xi'' - \xi'}\xi''^2 + \eta\frac{\xi'' - 1}{\xi'' - \xi'}\xi'^2\right) \le 1 + \eta$$

(9.4.2.24)

and, using condition (b) for the elimination of ξ',

$$\frac{1}{\eta}\left[\frac{\xi''-1}{1-\frac{1}{\eta}}\frac{1}{\eta^2}\xi''^2 + 2\eta\frac{1-\frac{1}{\eta}\xi''}{1-\frac{1}{\eta}}\xi''^2 + \eta\frac{\xi''-1}{1-\frac{1}{\eta}}\frac{1}{\eta^2}\xi''^2\right] \le 1 + \eta, \qquad (9.4.2.25)$$

$$\frac{1}{\eta-1}\xi''^2\left(\frac{1}{\eta^2}(\xi''-1) + 2(\eta-\xi'') + \frac{1}{\eta}(\xi''-1)\right) \le 1 + \eta, \qquad (9.4.2.26)$$

$$\xi''^2\left(\frac{1}{\eta^2}(\xi''-1) + 2(\eta-\xi'') + \frac{1}{\eta}(\xi''-1)\right) \le \eta^2 - 1, \qquad (9.4.2.27)$$

$$\xi''^2\left(\frac{\xi''}{\eta^2} - \frac{1}{\eta^2} + 2\eta - 2\xi'' + \frac{1}{\eta}\xi'' - \frac{1}{\eta}\right) \le \eta^2 - 1. \qquad (9.4.2.28)$$

A finger exists only if $\xi'' > 1$, as this is, indeed, the very definition of a finger. Looking at the limiting case $\xi'' = 1$, one obtains a condition for η:

$$1 - 2\eta^2 + \eta - 1 + 2\eta^3 - \eta - \eta^4 + \eta^2 \le 0 \qquad (9.4.2.29)$$

or

$$\eta^2 - 2\eta + 1 > 0 \qquad (9.4.2.30)$$

which is always satisfied.

However, condition (b) (i.e., 9.4.2.20) implies that, *geometrically*, a finger can exist only if

$$\eta > 1 \qquad (9.4.2.31)$$

since the very definition of a finger implies that

$$\xi'' > \xi'. \qquad (9.4.2.32)$$

The last result shows that the Muskat model automatically implies that fingering should and will occur as soon as the mobility ratio of displacing vs. displaced fluid exceeds 1.

9.4.3 Incipient fingers

The model of a finger discussed above is somewhat oversimplified. What one would like to do is to introduce a perturbation into a general solution of a displacement problem as obtained by the use of the Muskat model, and then to follow the growth of this perturbation. As has already been stated, this is a very difficult problem involving a floating boundary condition. Using the simplified approach given in section 9.4.2 is one possibility to get around this difficulty;

another is to treat the perturbations as small. The latter leads to a description of the *initial* dynamics of fingering, i.e., of the dynamics applicable just when the instability begins to take effect.

Thus we consider a simple displacement experiment corresponding to that shown in figure 39. Formulating the conditions of the Muskat model, one can write down a series of differential equations describing the displacement. In a coordinate system x', y, where x' is parallel to the displacement (cf. fig. 39) and such that $x' = 0$ would correspond to the instantaneous position of the front if the displacement were stable, taking place with the velocity v, one can give the integrated form of the equation of motion for the disturbed interface if the perturbation is of the form of a sine-wave of wavelength λ. The solution is

$$x' = \frac{k}{P(\mu_f - \mu_D)v} e^{\alpha_K t} \varepsilon_f \left(1 - \frac{\mu_D}{\mu_f}\right) \sin Ky \qquad (9.4.3.1)$$

with

$$\alpha_K = vK \frac{\mu_f - \mu_D}{\mu_f + \mu_D}, \qquad (9.4.3.2)$$

$$K = 2\pi/\lambda, \qquad (9.4.3.3)$$

where μ_D, μ_f are the viscosities ($k/\mu_{D,f}$ the mobilities) of the displacing and original fluids, respectively. The equation (9.4.3.1) has been given (in a somewhat more elaborate form, including capillary pressure and gravity terms) by Chuoke et al. (1959). A deduction of the form as given in (9.4.3.1) has been given by the writer (Scheidegger 1960b), and the reader is referred to that paper for the details.

Equation (9.4.3.1) is not an exact solution of the displacement equation, but it can be shown that it is a valid approximation as long as the perturbations are small.

The structure of equation (9.4.3.1) shows that perturbations of all wavelengths should grow indefinitely as soon as the mobility ratio exceeds 1. This corresponds to the criterion of instability deduced in the last section from a different model. It may be interesting to note that if capillary pressure terms are introduced, instability should still occur as soon as the mobility ratio exceeds 1, but that only perturbations with wavelengths above a certain critical value (which depends on the capillary pressure term) will become unstable (cf. Chuoke et al. 1959). If, in addition, a gravity term is introduced, stability will not occur below a certain critical displacement velocity (cf. Chuoke et al. 1959).

Reverting to the case where the capillary pressure term vanishes and gravity is negligible, it is possible to prove that, as long as the perturbations are small, one can make a superposition of various solutions of the form (9.4.3.1). This yields the possibility of describing a general perturbation of the front in terms of a Fourier integral. The total displacement of the front is then given by the sum of the displacements of all its spectral components. Since all spectral components, if excited, will produce fingers in time, each of which is growing with its own speed, the rate of production of fingers is determined by the rate of introduction of velocity perturbations. The latter, however, is connected with the velocity-dispersion introduced in statistical theories of flow through porous media. It is then possible to calculate the spectrum of the perturbation in terms of the dispersive properties of the porous medium. The calculations to do this are rather involved and the reader is referred for details to the paper cited above (Scheidegger 1960b).

9.4.4 Statistical theory of fingering

The calculation of the actual interface in an unstable displacement process evidently meets with great analytical difficulties. Therefore, Scheidegger and Johnson (1961) have proposed a macroscopic treatment of the problem.

Thus, let us again consider an unstable linear displacement experiment, parallel, say, to the $+x$ direction. The fingers present in the porous medium may then be drawn schematically as shown in figure 41. In a cross-section, the intruding fingers will take up the fraction s_1 of the total area, and the problem is to set up a determining equation for the statistical behaviour of the function

$$s_1 = s_1(x, t). \tag{9.4.4.1}$$

The further discussion is based on Muskat's elementary displacement model (cf. sec. 9.2). Darcy's law for the intruding fluid (index 1) is

$$v_1 = \frac{k}{\mu_1 P} \frac{dp}{dx}, \tag{9.4.4.2}$$

where, as usual, v is the pore velocity, k the permeability, μ the viscosity, P the porosity, and p the pressure. A similar equation holds for the displaced fluid (replace index 1 by index 2 in (9.4.4.2)).

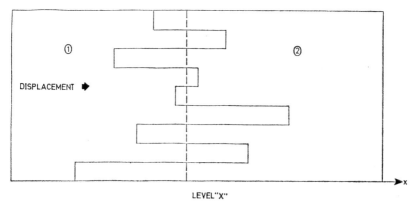

FIGURE 41 Schematic representation of fingers.

The theory cannot be carried further at this stage without making an additional assumption. Thus, we assume that at the level x the pressure does not vary within a cross-section. One has then for the flow of phase i ($i = 1, 2$) across a unit area at the position x

$$q_i = \frac{ks_i}{\mu_i}\frac{dp}{dx}.$$
(9.4.4.3)

The problem becomes determined if the continuity conditions

$$P\,\partial s_i/\partial t = -\operatorname{div} q_i,$$
(9.4.4.4)

$$s_1 + s_2 = 1$$
(9.4.4.5)

are added, provided the two fluids are assumed to be incompressible.

The above equations have an eliminant

$$P\frac{\partial s_1}{\partial t} + r'(s_1)q_x\frac{\partial s_1}{\partial x} = 0,$$
(9.4.4.6)

with q_x denoting the over-all filtration velocity in the x-direction and $r(s_1)$ the fractional flow

$$r(s_1) = \frac{s_1/\mu_1}{s_1/\mu_1 + (1 - s_1)/\mu_2} = \frac{m}{m - 1 + 1/s_1},$$
(9.4.4.7)

where

$$m = \mu_2/\mu_1.$$
(9.4.4.8)

The behaviour of the fingers can now be determined if appropriate solutions of (9.4.4.6) are found. This is best achieved by the method of characteristics. The partial differential equation (9.4.4.6) can be replaced by the following pair of total (characteristic) differential equations:

$$dx/dt = r'(s_1)q_x(t)/P,$$
(9.4.4.9)

$$ds_1/dt = 0.$$
(9.4.4.10)

The first of these describes the path (in the x–t plane) along which a constant saturation moves. Inserting (9.3.4.7) one finds that

$$\frac{dx}{dt} = \frac{m}{[s_1(m - 1) + 1]^2}\frac{q_x(t)}{P},$$
(9.4.4.11)

which yields, upon integration and assuming a constant over-all injection-velocity $q_x(t)$,

$$x = \frac{mt}{[s_1(m - 1) + 1]^2}\text{ const.,}$$
(9.4.4.12)

where the constant depends only on the injection pore velocity. It thus turns out that the characteristics are straight lines in the x–t plane; their slope gives the velocity $c(s_1)$ with which each s_1-value spreads into the porous medium:

$$c(s_1) = \text{const.} \frac{m}{[s_1(m-1)+1]^2} \cdot \qquad (9.4.4.13)$$

This is the solution of equation (9.3.4.6). A drawing is given in figure 42.

The manner in which time occurs in equation (9.4.4.12) shows that no stability is possible for the fingers; they simply grow ad infinitum by stretching.

Verma (1969) has elaborated upon the above theory and shown that for porous media with a specially chosen porosity P as a function of x (a heterogeneous porous medium) a stabilization of fingers is possible. The same is true if a capillary pressure differential is taken into consideration in a homogeneous medium. However, stabilization is not possible in a homogeneous porous medium for fluids between which no capillary pressure differential exists.

9.4.5   Thermodynamic analogy for fingers

The thermodynamic analogy between transport processes of mass and heat enables one to discuss the stability problem of viscous fingering in terms of the minimum entropy production principle.

The equations for $s_1(x, t)$ given in the last section describe nothing but a case of 'solitary dispersion' for the quantity s_1. Correspondingly, the stability analysis of section 9.3.3 can be applied directly to the fingering problem at hand (Scheidegger 1969). For this, it is necessary to write the pertinent formulas of the

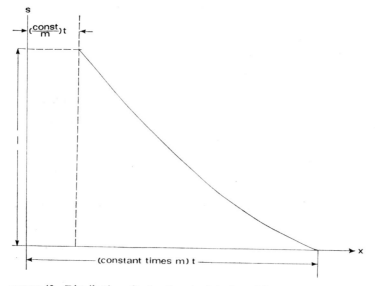

FIGURE 42   Distribution of saturation as a function of distance.

fingering theory of the last section compatibly with the formulas of solitary dispersion theory. One has complete correspondence if one identifies C in (9.3.1.7) with s_1 and sets

$$\gamma = \frac{\mu_2/\mu_1}{(\mu_2/\mu_1 - 1) + 1/s_1} - s_1. \tag{9.4.5.1}$$

Since γ is a function of s_1 only, one has the 'Buckley-Leverett' limit of solitary dispersion theory. It has been shown in section 9.3.3 that no stabilization is possible in this case.

The above thermodynamic analogy was further extended by Chow and Scheidegger (1972), who showed that, if capillary pressure is taken into account, stable fingers may be obtained under certain conditions. This confirms, from a thermodynamic-analogue standpoint, the direct calculations of Verma discussed in section 9.4.4.

9.5
GRAVITATIONAL INSTABILITY

9.5.1 *Principles*

In addition to hydrodynamic viscous instability, a gravitational instability may arise between fluids of different densities in a porous medium. Often, such a density difference may be caused by temperature differences; this possibility will be treated separately under the heading 'thermal effects' (sec. 9.6).

Purely gravitational effects can be exemplified by the following experiment. Let a long, vertical column filled with a homogeneous, isotropic porous medium be saturated with a fluid of density ρ_0. At the top of the tube, assume there is a connection to an open reservoir filled with a fluid of density $\rho_1 > \rho_0$, the two fluids being miscible. Under these conditions, convection (i.e., an instability) will set in until an equilibrium state is reached.

Investigations of this process have been undertaken by experimentation and by theoretical analysis. The possibilities will be discussed below.

9.5.2 *Heuristic investigations*

The classic experiment of vertical instability, as described at the beginning of the last section, using water as one fluid and salt solution as the other, was carried out by Wooding (1959). It turned out that the effective diffusivity in the porous column, D', of the solute through the saturated porous material is of importance. The process is then governed by the non-dimensional Rayleigh number Ra, defined as follows

$$\text{Ra} = \frac{d\rho}{dz} \frac{gkb^2}{\mu D'}, \tag{9.5.2.1}$$

where z is the vertical coordinate, ρ the density (assumed to be a function of z), g the gravity acceleration, μ the mean viscosity of the fluid, k the permeability (in cm^2) of the porous material, and b the radius of the (circular) column. For Ra $<$ Ra$_0$ the equilibrium of the column is stable, and for Ra $>$ Ra$_0$ it is unstable. One finds that

$$Ra_0 = 3390. \tag{9.5.2.2}$$

Dagan (1966) has extended the above analysis to the case in which there is a viscosity difference in addition to the density difference between two fluids.

The actual *process* of gravitational convection, in addition to the question of the *onset* of instabilities, has been studied experimentally by Bachmat and Elrick (1970). These authors showed that, in analogy with the case of viscous instability, the convection occurs in the form of *fingers*.

9.5.3 *Theoretical investigations*

Theoretical studies of the gravitational instability problem have been made by Wooding (1962), Dagan (1966), and Bachmat and Elrick (1970). Accordingly, the heuristic criterion for instability of a (circular) porous column given by equation (9.5.2.1–2) has been confirmed by theoretical calculations.

Bachmat and Elrick gave also an estimate of the time-variation of the concentration of a salt solution in a porous column. They based their work on a consideration of the average flux across a cross-section, for which an equation of motion, equations of state, and a continuity equation were formulated. The flux is then the sum of the fluxes due to ordinary molecular diffusion, mechanical dispersion, convective dispersion, and average convection. It was then shown that the time-variation of the concentration for a fixed volume of a uniformly concentrated solution on top of the porous material, a fixed total amount of salt, and a porous material of an unbounded depth is

$$C_{salt}(t) \approx C_0/(1 + Ft), \tag{9.5.3.1}$$

where F is a constant.

9.6
THERMAL EFFECTS

9.6.1 *The mechanism of heat transfer*

A further instance in which elementary displacement theory (i.e., Muskat's model) has been successfully employed is with regard to the displacement of a mass of fluid by a mass of identical fluid at a different temperature. The ensuing treatment of the displacement process neglects the mixing effects that occur at the interface of, say, hot and cold fluids. Such temperature gradations as may occur are assumed to be due to heat transfer within the various substances involved. Each elemental volume of fluid (at any temperature) is assumed to obey Darcy's law with its own characteristic density and, possibly, viscosity. The mechanism of heat transfer was studied by Jenkins and Aronofsky (1955), Adivarahan et al. (1962), and Verruijt (1969). Jenkins and Aronofsky (1955) have given a review of the mechanisms that can effect a heat transfer in the system. Accordingly, heat transfer from one part of the system to another is influenced by any of the following three mechanisms:

(a)   physical movement of the fluid which carries its own heat content with it;
(b)   thermal conduction from the colder portions of the system to the warmer;
(c)   the resistance to transfer of heat from the fluid to the solid (the transfer being assumed proportional to the temperature difference between fluid and solid).

In general, all three mechanisms are operative. However, it is customary to consider models in which either the second ('model A') or the third (temperature of fluid and solid always identical; 'model B') mechanism is neglected, since this greatly simplifies the calculations.

Of particular interest for applications are, first, the straightforward displacement of hot fluid by cold fluid or vice versa and, second, the possibility of onset of thermal convection in a porous medium in which a temperature gradient is being maintained (i.e., which is being heated from below). The latter case is of particular interest with regard to the behaviour of water and fluids in underground strata since a rather high temperature gradient is known to exist in the earth's crust.

We shall discuss briefly some of these problems below.

9.6.2 *Linear displacement*

The problem of linear displacement of a cold fluid by a hot one can be described as follows. Let a linear porous medium (extending from $x = 0$ to ∞) be filled with a fluid at initial temperature $\mathfrak{T}_i$, the fluid and porous medium being initially at the same temperature. Then at time 0 more of the same fluid is injected at temperature $\mathfrak{T}_f$, moving with pore velocity v. The question is: what is the temperature distribution $\mathfrak{T}_s(x, t)$ in the porous medium at the position x and time t, or, alternatively, what is the temperature in the fluid $\mathfrak{T}_F(x, t)$ at the position x and time t?

The two models introduced in section 9.6.1 and termed there 'model A' and 'model B' (after Jenkins and Aronofsky) have been used to solve the problem.

Solutions based upon model A have been given by Anzelius (1926), Nusselt

(1927), Schumann (1929), Furnas (1930), and Hausen (1931). These solutions contain integrals of Bessel functions and must be evaluated numerically. However, Klinkenberg (1948) has shown that a reasonable approximation to the solutions is represented by the following expression containing error functions:

$$\frac{\mathfrak{T}_F - \mathfrak{T}_i}{\mathfrak{T}_f - \mathfrak{T}_i} = \frac{1}{2}\left[1 + \mathrm{erf}\left(\sqrt{Z} - \sqrt{Y} + \frac{1}{8\sqrt{Z}} + \frac{1}{8\sqrt{Y}}\right)\right], \tag{9.6.2.1}$$

where

$$Z = \frac{hS}{C_S \rho_S (1 - P)}\,(t - x/v), \tag{9.6.2.2}$$

$$Y = \frac{hxS}{C_F \rho_F v P}, \tag{9.6.2.3}$$

with h denoting the fluid-solid heat transfer coefficient, ρ_F the density of the fluid, ρ_S the density of the solid (taken as constants), C_F the specific heat of the fluid, C_S that of the solid, S the specific surface of the porous medium, and P (as usual) the porosity.

'Model B' mentioned in section 9.6.1 has also been applied to the problem of a linear flood at constant velocity. Jenkins and Aronofsky (1955) give the following solution (note that here $\mathfrak{T} = \mathfrak{T}_S = \mathfrak{T}_F$; erfc denotes the error function complement):

$$\frac{\mathfrak{T} - \mathfrak{T}_i}{\mathfrak{T}_f - \mathfrak{T}_i} = \frac{1}{2}\left[\mathrm{erfc}\left(\frac{x}{2\sqrt{Kt}} - \sigma\sqrt{Kt}\right) + e^{2x\sigma}\,\mathrm{erfc}\left(\frac{x}{2\sqrt{Kt}} + \sigma\sqrt{Kt}\right)\right], \tag{9.6.2.4}$$

with

$$\sigma \equiv \frac{PvC_F\rho_F}{2K[PC_F\rho_F + (1 - P)C_S\rho_S]}, \tag{9.6.2.5}$$

where, in addition to the symbols defined above, K denotes the thermal diffusivity of the porous medium.

Many experimental tests of thermal effects in linear displacement experiments have been reported (Gamson et al. 1943; Wilhelm et al. 1948; Preston and Hazen 1954; Hadidi et al. 1956; Preston 1955, 1956). In a critical evaluation of these tests, Jenkins and Aronofsky (1955) show that 'model B' appears to be more applicable, although for most calculations 'model A' has been used.

The theory discussed above applies only to displacement at a constant rate in a linear system. It has been generalized to gas flow by Bland (1954) and Dejak and Trevissoi (1954), to unsteady-state conditions by Reilly (1957), and to complicated geometrical layouts by Landrum et al. (1958). The problem has found some practical applications in connection with hot-water injections made to stimulate the production from viscous-crude oil fields. Such applications have been discussed by, for example, Breston and Pearman (1953), Garbus (1956), Schild (1957), and Bernhard et al. (1958). The reader is referred to the original papers for the details.

9.6.3 *Thermal convection currents in porous media*

If a porous medium containing a fluid is being heated from below, the possibility exists that convection currents will arise. The reason for this is that hot fluid is less dense than the same fluid cold and, if the former be situated below the latter, it will have the tendency to rise, which may cause convection currents. The energy to maintain convection is taken from the heat supplied from below. While the convection is going on, heat is being transferred from the high-temperature region (the bottom) to the low-temperature region (the top). This is as it should be according to the second principle of thermodynamics.

From a study of the analogous problem with regard to convection in bulk masses of viscous fluids which are heated from below, one infers that instability need not always occur (Bénard 1901). Rayleigh (1916) and Jeffreys (1926, 1928) have deduced criteria which indicate that in bulk fluids no thermal convection can arise unless a certain dimensionless parameter has a minimum value. If convection does occur, it will establish itself in cylindrical (columnar) cells which have the shape of irregular hexagons or pentagons and whose axis is vertical. It stands to reason that similar phenomena will be observed in porous media.

The calculations of thermal convection in porous media that have been reported in the literature are all based on 'model B' of section 9.6.1 (this corresponds to slow flow so that $\mathfrak{T}_S = \mathfrak{T}_F$ always). The simplest case that can be treated is that where the density of the fluid is a linear function of the temperature $\mathfrak{T}$ ($\mathfrak{T}_0$ being some reference temperature):

$$\rho = 1 - \lambda(\mathfrak{T} - \mathfrak{T}_0) \tag{9.6.3.1}$$

and where the equilibrium gradient just prior to the onset of convection is also constant:

$$\mathfrak{T} = \mathfrak{T}_0 + mz. \tag{9.6.3.2}$$

Here z denotes the vertical coordinate (positive downward). The boundaries ($z = \pm l/2$) are assumed to be at fixed temperature. The problem of calculating the critical thermal gradient necessary for the onset of convection has been solved by Horton and Rogers (1945), Lapwood (1948), Goguel (1953), Elder (1967), and Katto (1967). We shall follow here the exposition of Goguel (1953).

Thus, of interest is the flow in the porous medium that is over and above the (possibly unstable) equilibrium state. Denoting the deviation of the pressure from the equilibrium state by Δp, the deviation of temperature by $\Delta\mathfrak{T}$, and the deviation of the fluid density by $\Delta\rho$, the hydrodynamic equations are

$$\left.\begin{aligned}
\partial\Delta p/\partial x &= -(\mu/k)q_x, \\
\partial\Delta p/\partial y &= -(\mu/k)q_y, \\
\partial\Delta p/\partial z &= g\Delta\rho - (\mu/k)q_z,
\end{aligned}\right\} \tag{9.6.3.3}$$

where all the symbols have the customary meaning. The continuity equation (assuming the fluid is incompressible) is

$$\text{div } \mathbf{q} = 0, \tag{9.6.3.4}$$

and the heat balance equation can be written as follows:

$$c\,\partial\Delta\mathfrak{T}/\partial t = \kappa_m \text{ lap } \Delta\mathfrak{T} - mq_z, \tag{9.6.3.5}$$

where κ_m is the thermal diffusivity of the system and c is the heat capacity. Both constants are normalized in such a fashion that the heat capacity and unperturbed density of the fluid equal 1.

The problem will be considered in two dimensions (x, z) only. Assuming that columnar convection may occur in a manner similar to that in which it is known to occur in bulk fluid, one can try the following solution, which satisfies the boundary conditions:

$$\left.\begin{aligned}
\Delta\mathfrak{T} &= \varepsilon \exp(t/\tau) \cos(\pi z/l) \cos(\pi x/l), \\
q_z &= -A \exp(t/\tau) \cos(\pi z/l) \cos(\pi x/l), \\
q_x &= -A \exp(t/\tau) \sin(\pi z/l) \sin(\pi x/l),
\end{aligned}\right\} \tag{9.6.3.6}$$

where ε is a small quantity and τ is a decay or growth time for the convection. Inserting the trial solution into the hydrodynamic equation (9.6.3.3) yields the equations

$$\left.\begin{aligned}
\frac{\partial\Delta p}{\partial z} &= \left(-g\lambda\varepsilon + \frac{\mu}{k}A\right)\exp\left(\frac{t}{\tau}\right)\cos\left(\frac{\pi z}{l}\right)\cos\left(\frac{\pi x}{l}\right), \\
\frac{\partial\Delta p}{\partial x} &= A\frac{\mu}{k}\exp\left(\frac{t}{\tau}\right)\sin\left(\frac{\pi z}{l}\right)\sin\left(\frac{\pi x}{l}\right),
\end{aligned}\right\} \tag{9.6.3.7}$$

which are compatible (requirement $\partial^2\Delta p/\partial x\partial z = \partial^2\Delta p/\partial z\partial x$) only if

$$A = \tfrac{1}{2}kg\lambda\varepsilon. \tag{9.6.3.8}$$

Thus

$$\Delta p = -\frac{\varepsilon\mu}{2\pi}g\lambda l\exp\left(\frac{t}{\tau}\right)\sin\left(\frac{\pi z}{l}\right)\cos\left(\frac{\pi x}{l}\right). \tag{9.6.3.9}$$

Inserting the calculated quantities into the heat balance equation yields

$$\frac{c}{\tau} = -2\frac{\pi^2}{l^2}\kappa_m + \frac{1}{2}\frac{mk}{\mu}g\lambda. \qquad (9.6.3.10)$$

Convection currents can arise only if this expression is positive; i.e., if

$$-2\frac{\pi^2}{l^2}\kappa_m + \frac{1}{2}\frac{mk}{\mu}g\lambda > 0, \qquad (9.6.3.11)$$

which yields the following critical condition for the temperature gradient m:

$$m > 4\pi^2\kappa_m\mu/(l^2kg\lambda), \qquad (9.6.3.12)$$

or

$$ml^2kg\lambda/(\mu\kappa_m) > 4\pi^2 \sim 40. \qquad (9.6.3.13)$$

The left-hand side in the above relation is commonly called the 'Rayleigh number' Ra in analogy with a similar number introduced for density gradients (cf. 9.5.2.1). In the above calculation, κ_m had been defined so that ρ and c are dimensionless quantities and equal to unity. If these quantities are reintroduced, the conditions and the definitions of the Rayleigh number become

$$Ra \equiv \frac{k\lambda gml^2\rho^2c}{K_m\mu} \gtrsim 40, \qquad (9.6.3.14)$$

where K_m is now the thermal conductivity of the saturated medium and ρc is the thermal capacity of fluid.

The above formulas refer to a special case only (as a constant viscosity of the fluid and a thermal environment linear in height are implied). Several attempts at discussing more general cases have been reported by Goguel (1953), Rogers and co-workers (Rogers and Morrison 1950; Rogers, Schilberg, and Morrison 1951; Rogers 1953), and, particularly, Elder (1967). The last author solved the system of basic equations (9.6.3.3–5) by a numerical method and arrived at the same condition (9.6.3.12) as Goguel.

Experiments to demonstrate thermal convection in porous media and to investigate its pattern have been reported by Rogers and co-workers (Morrison, Rogers and Horton 1949; Rogers and Schilberg 1951; Morrison and Rogers 1952), Wooding (1957, 1958, 1963, 1964), Elder (1967), Katto (1967), and Bories (1970a, b, c). The last author, in particular, published beautiful photographs of thermal convection cells in porous media. The onset criteria predicted from theory were confirmed by these experiments; the agreement of the various theories with experiments appears to be reasonable.

9.6.4 Geophysical applications

Finally, there might be some interest in applying the above results to geophysical problems. The question of prime importance is whether thermal convection of groundwater may occur as a result of the ordinary geothermal gradient (about $1/30$ deg/m) that is present in the earth's crust. A very crude estimate, using the criterion (9.6.3.14), shows at once that convection is not possible (Horton and Rogers 1945; Goguel 1953) unless the permeability of the strata is very high indeed. However, Goguel (1953) has shown that, if the real properties of water (temperature dependence of viscosity and compressibility) are taken into account, and if the calculation is made for a depth between 2500 m and 3000 m where a temperature gradient of $1/10$ deg/m may be assumed, the onset of thermal convection appears to be likely. Similarly, Wooding (1957) has applied the theory to the extraordinary conditions obtaining in the geothermal area of Wairakei, New Zealand. He claims that his results show features which are in fair agreement with temperature measurements made in the area.

More speculative applications are those of fluid convection in porous media to convection currents in the earth's mantle (assuming this to be porous) and to the theory of volcanism. A review of those possibilities has been given by Elder (1965).

10
Immiscible multiple phase flow

10.1
GENERAL REMARKS

The treatment of displacement processes presented in chapter 9 is based on extremely simplifying assumptions. In order to arrive at a more realistic theory, it will be necessary to take into account that the various phases are flowing *simultaneously* through the porous medium.

As has been stated earlier, a distinction has to be made as to whether the flowing fluids are miscible or not. The present chapter is devoted to immiscible multiple phase flow.

10.2
LAMINAR FLOW OF IMMISCIBLE FLUIDS

10.2.1 *Qualitative investigations*

In line with single phase flow experiments intended to test Darcy's law, similar experiments have been performed using two immiscible phases. In the early days it was assumed that Darcy's law should be valid for any fluid and thus also for a mixture of two immiscible fluids. Studies to test this assumption have been performed by Cloud (1930), Gardescu (1930), Versluys (1931), Bartell and Miller (1932), Uren and Bradshaw (1932), Garrison (1934), Uchida and Fujita (1937), Uren and Domerecq (1937), Plummer et al. (1937), Christiansen (1944), Pillsbury and Appleman (1945), Fletcher (1949), and Pirverdyan (1952). Experimental studies in this series soon rendered it evident that the presence of a second phase not only makes the permeability to the first much lower, but also greatly decreases the permeability to the mixture. Many qualitative arguments to explain this fact have been put forth. Gardescu (1930), for instance, ascribed it to an effect analogous to one noticed many years ago by Jamin (1860) in small tubes. Jamin observed that in a capillary tube containing a large number of detached drops of liquid interspersed with gas, a large difference in pressure between the ends of the tube without any appreciable movement of the drops along the tube may occur. The effect is ascribed to hysteresis of contact angle.

The Jamin effect gives a qualitative explanation of the observed 'threshold' pressures necessary before flow can occur. However, in fact, double phase flow in porous media probably does not occur in the manner observed by Jamin, but rather in the funicular channels postulated by Versluys (see chapter 3). The latter author also applied the concepts of funicular and pendular saturation regions to flow problems (Versluys 1931) by giving a series of qualitative microscopic arguments for the explanation of the observed qualitative results.

10.2.2 *Darcy's law*

The above qualitative investigations are not sufficient to provide a quantitative description of double phase flow in porous media. It is therefore necessary to extend the earlier investigations to obtain a quantitative description.

As a working assumption, one might be inclined to postulate that Darcy's law should be valid for *each* flowing phase. In this manner, it is possible to formulate a complete hydrodynamics for multiple phase flow.

In single phase flow, Darcy's law can be formulated as follows (cf. 4.3.1.2):

$$\mathbf{q} = - (k/\mu) (\operatorname{grad} p - \mathbf{g}\rho), \tag{10.2.2.1}$$

where $\mathbf{q}$ is the seepage velocity vector, k the permeability, μ the viscosity, p the pressure, $\mathbf{g}$ the gravity vector, and ρ the density of the fluid. Darcy's law is not sufficient to determine a given flow problem; in addition, there is the continuity equation

$$- P\, \partial\rho/\partial t = \operatorname{div}(\rho\mathbf{q}) \tag{10.2.2.2}$$

and a relationship between ρ and p:

$$\rho = \rho(p). \tag{10.2.2.3}$$

These three equations are the necessary and sufficient conditions, in conjunction with boundary conditions, to determine any given single phase flow problem.

The above conditions can be extended for double phase flow in an obvious way. Denoting the two phases by the subscripts 1 and 2 respectively, we have

$$\mathbf{q}_1 = - k\,(k_1/\mu_1)\,(\operatorname{grad} p_1 - \mathbf{g}\rho_1), \tag{10.2.2.4a}$$

$$\mathbf{q}_2 = - k\,(k_2/\mu_2)\,(\operatorname{grad} p_2 - \mathbf{g}\rho_2), \tag{10.2.2.4b}$$

$$P\, \partial(\rho_1 s_1)/\partial t = - \operatorname{div}(\rho_1\mathbf{q}_1), \tag{10.2.2.5a}$$

$$P\, \partial(\rho_2 s_2)/\partial t = - \operatorname{div}(\rho_2\mathbf{q}_2), \tag{10.2.2.5b}$$

$$s_1 + s_2 = 1, \tag{10.2.2.5c}$$

$$\rho_1 = \rho_1(p_1), \tag{10.2.2.6a}$$

$$\rho_2 = \rho_2(p_2), \tag{10.2.2.6b}$$

$$p_2 - p_1 = p_c(s_1). \tag{10.2.2.7}$$

Here the symbols are defined as follows: k is the total permeability, $k_{1,2}$ the relative permeability (in fractions of the total permeability), s the saturation, P the porosity, and p_c the capillary pressure. The above equations *define* the notion of 'relative permeability.' Sometimes the relative permeability is also given as a percentage of the total permeability, rather than as a fraction thereof.

The above extension of Darcy's law to double phase flow has been suggested by Muskat and co-workers (Muskat and Meres 1936; Muskat et al. 1937). The capillary pressure term seems to have been first introduced by Leverett (1939), who also postulated the validity of an obvious extension of the system of equations (10.2.2.4) to (10.2.2.7) for three phases (Leverett and Lewis 1941). Further discussions of the principles of the multiple phase flow equations have also been given in the general textbooks mentioned in the Introduction. Fatt (1953) showed that the compressibility of the porous medium could also be taken into account in the equations for multiple phase flow in a manner similar to that used for single phase flow.

The above deduction of Darcy's law for multiple phase flow is, so far, only theoretical speculation. It is, of course, always possible to *define* relative permeability as in equations (10.2.2.4) to (10.2.2.7), but these equations make sense only if that relative permeability is independent of the pressure and velocity, i.e., if it is a function of the saturation only. From a great number of investigations it appears that the relative permeabilities to *immiscible* fluids are indeed *within limits* such functions of saturation only. Experiments to substantiate this have been performed, first by the originators of the equations (10.2.2.4ff)., and also by Wyckoff and Botset (1936), Hassler et al. (1936), Reid and Huntington (1938), Botset (1940), and many others. Terwilliger and Yuster (1946, 1947) made a careful study of the possible effects of chemical agents. They, and later Calhoun (1951a, b), found that the chemical composition of the fluids does not matter much and that the relative permeability functions are approximately the same for any 'wetting–non-wetting' system. It should be noted, however, that accurate experiments show that this is, as indicated, only *approximately* true. The generalized Darcy law for multiple phase flow has, as ought to be expected, limitations. Some of these limitations are due to the same causes as the limitations of the Darcy law in single phase flow (such as adsorption, molecular slip, and turbulence) and will be dealt with later. However, the assumption that the relative permeability is a function of saturation only can possibly be only an approximation. Owing to the occurrence of hysteresis in wettability, the same phenomenon must occur in relative permeability to a greater or less extent.

The system of differential equations representing multiple phase flow can be generalized slightly if further parameters are introduced which may be associated with the two phases. Such parameters have to fulfil conservation laws, which will provide the additional equations necessary to make the problem determined. In practice, such a parameter will usually be an energy quantity (such as temperature), because energy satisfies a conservation law. However, it is conceivable that it could be some other constant of the motion.

10.2.3 *Measurement of relative permeability*

The measurement of relative permeability consists, essentially, of the determination of two flow rates under a given pressure drop and the determination of saturation. Seven methods are commonly applied, termed (i) the Pennsylvania State method, (ii) the single sample dynamic method, (iii) the gas drive method, (iv) the stationary liquid method, (v) the Hassler method, (vi) the Hafford technique, and (vii) the dispersed feed method. General discussions and reviews of these methods have been given by Brownscombe et al. (1949, 1950a, b), Rose (1951a, b), Osoba et al. (1951), and Richardson et al. (1952). A compilation of the virtues of the seven methods is given in table VII. The description of the seven methods given in this section is based very closely upon the reviews of Osoba et al. (1951) and Richardson et al. (1952).

The *Pennsylvania State method* (after Osoba et al. 1951) was developed by Yuster and co-workers (Morse et al. 1947a, b; Henderson and Yuster 1948a, b), and was modified by Caudle et al. (1951). The sample to be tested is mounted between two samples of similar material, the three being in capillary contact. This minimizes boundary effects and also effects a good mixing of the two phases

TABLE VII

A comparison of various techniques for relative permeability determination

Method	Reliability of results	Speed, hours/sample (permeability 10^{-9} cm^2)	Simplicity of operation	Remarks
Penn state	Excellent	8	Complicated	Uses three samples
Hassler	Excellent	40	Very complicated	Requires pressure gauges of very low displacement volume
Single sample dynamic	Questionable for short samples	6	Simple	For short samples the relative permeability to wetting phase is too high
Stationary liquid	Questionable at low gas saturations	4	Simple	Applicable only to measurement of relative permeability to gas
Gas drive	Good	2	Very simple	Can be operated with minimum amount of training and requires a minimum amount of equipment
Hafford	Excellent	7	Simple	Preferable to dispersed feed
Dispersed feed	Excellent	7	Simple	

before proceeding to the sample. The sample is first saturated with one phase and then this phase is allowed to flow through it at a rate that gives a predetermined pressure drop. The second phase (a gas) is then allowed to flow at a very low rate. After the system reaches equilibrium, the test sample is removed quickly, and the saturation is determined by weighing. The samples are then reassembled, and the flow of both phases is resumed at the rates that had existed previously. The rate of flow of the first phase is then decreased slightly, and the rate of flow of the gas is increased simultaneously to maintain the pressure drop across the sample at its previous value. After equilibrium has again been attained, the test section is removed and the saturation is determined. This stepwise procedure is repeated until a complete relative permeability curve is obtained at incremental changes in the fluid saturation. The Pennsylvania State method is very accurate, but rather complicated to use.

The *single sample dynamic method* (after Osoba et al. 1951) uses equipment similar to that employed in the Pennsylvania State technique, except that a single sample, mounted in plastic, is used. The two phases are introduced directly into the sample from a feed head. In making determinations by this method, the sample is initially saturated completely with the first phase. A rate of flow of this phase is then established which would correspond to a predetermined pressure drop across the sample. The second phase, a gas, is then admitted to the sample at a low rate. After equilibirium has been established, the saturation is determined by weighing. Successively higher gas flow rates, and correspondingly lower rates of the first phase, are employed in succeeding steps to determine complete relative permeability–saturation relations. The single sample dynamic method is very simple to apply but of questionable value for short samples.

In the *stationary liquid method* (Leas et al. 1950; Branson 1951) boundary effects are avoided by holding the wetting phase stationary within the sample by capillary forces. Only the non-wetting phase is permitted to flow. This is accomplished by the use of a very low pressure gradient across the sample. The method is applicable only to the determination of relative permeability to the non-wetting phase, which must be gaseous. With this method, relative permeability–saturation curves for gas are obtained by starting with the sample completely saturated with wetting phase. The wetting phase saturation is then reduced by blotting the sample. When the wetting phase saturation becomes less than the equilibrium saturation, the wetting phase is passed upward through the sample under a small pressure drop. The relative permeability to the non-wetting phase at that point is measured, and the wetting phase saturation is determined by weighing. This process is repeated until the complete relation of the relative permeability of the non-wetting phase to saturation is obtained. This method is very simple to apply but of rather questionable accuracy.

In the *gas drive method* the influence of the boundary effect is minimized by the use of a high rate of flow. The method differs from the stationary liquid method in that both phases flow simultaneously; it differs from the single sample dynamic method in that only the non-wetting phase, a gas, is admitted to the sample. The flow of both phases results from the displacement of portions of the wetting phase by the flowing gas: hence the name, 'gas drive.' This method was first suggested by Hassler et al. (1936). It is fairly good and very simple but occasionally gives faulty results.

Hassler (1944) also described another method for relative permeability measurements. The *Hassler method* has been modified by Gates and Lietz (1950), Brownscombe et al. (1950b), and Osoba et al. (1951). Its main features consist (after Osoba et al. 1951) in controlling the capillary pressure at both ends of the sample, which is accomplished by placing the sample between two discs permeable only to the wetting phase. This permits maintenance of a uniform saturation throughout the length of the sample, even at low rates of flow. The apparatus for measurements by the Hassler method is shown in figure 43. The semipermeable discs at each end of the sample allow the wetting phase (a liquid) but not the non-wetting phase (a gas) to pass. They also permit the pressure drop in the wetting phase to be measured independently of that in the gas phase. In the

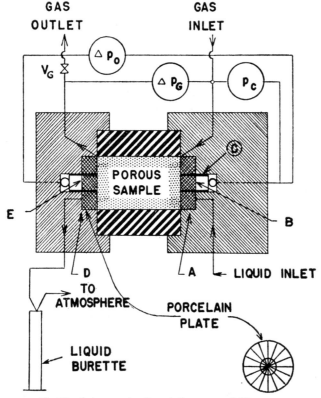

FIGURE 43   Hassler apparatus for relative permeability determination (after Osoba et al. 1951).

measurement of relative permeability, the sample to be tested is saturated with the wetting phase, and is then placed in the apparatus with a few layers of tissue paper at both ends to ensure good contact between the sample and semipermeable discs. The wetting phase enters the sample through disc A. Disc A is isolated from the small disc B by the metal sleeve C. The pressure in the wetting phase at the inflow end is measured through the disc B. Gas enters the sample through radial grooves in the face of disc A. The difference between the pressure in the wetting phase and in the gas at the inflow end is the capillary pressure and is measured by the gauge p_c. The wetting phase and gas pass through the sample at rates such that the pressure drop in the the two phases across the sample is the same. This is accomplished by adjusting valve v_G, which controls the rate of flow of gas through the sample, so that the pressure drop in the gas phase is equal to the pressure drop in the wetting phase. This is done so that the capillary pressure at the outflow end will be equal to the capillary pressure at the inflow end. The gas leaves the sample through radial grooves in disc D, and the pressure of the wetting phase at the outflow end is measured through disc E. After equilibrium is reached, the sample is removed and the saturation of the sample is determined by weighing. The Hassler method is exceedingly accurate but very complicated to use.

In the *Hafford technique* (after Osoba et al. 1951) the non-wetting fluid is fed directly into the sample and the wetting fluid is fed into the sample through a semipermeable disc that allows only the wetting fluid to pass. The central portion of the semi-permeable disc is isolated from the remainder of the disc by a small metal sleeve. The central portion is used to measure the pressure in the wetting fluid at the input end of the sample. The pressure in the non-wetting fluid is measured through a standard pressure tap machined into the lucite surrounding the sample. The pressure difference between the wetting and the non-wetting fluid is a measure of the capillary pressure of the sample at the inflow end. The Hafford technique gives excellent results and is fairly simple to apply.

The *dispersed feed method* (after Richardson et al. 1952) is similar to the Hafford and single sample dynamic methods. In this method the wetting fluid enters the test sample by first passing through a dispersing section, which is made of porous material similar to the test sample, but does not contain a device for measuring the pressure in the wetting fluid at the inflow end of the test sample as does the Hafford method. This porous material, which in some cases has been made of the same material as the test sample, serves to disperse the wetting fluid so that it enters the test sample more or less uniformly over the inflow face. Radial grooves are machined into the outflow face of the dispersing section. Gas is introduced into the test section through these radial grooves at the junction between the two sections. The dispersed feed method gives excellent results and is fairly simple to use.

Finally, the *seven methods* of relative permeability measurements are compared in table VII. For the convenience of the reader, the advantages and accuracy of each method are also listed there.

In addition to the methods mentioned above, attempts have been made to measure *directional* relative permeabilities (Corey and Rathjens 1957). Methods for *three-phase* relative permeability measurements have been described by Corey et al. (1956).

The various methods of measuring relative permeability have been applied on many occasions. Most such experiments are set up to find relative permeability curves for the flow of two or more phases in petroleum well cores. Such experiments and curves have been published (apart from the investigations already mentioned) for two-phase flow by Dunlap (1938), Krutter and Day (1943), Sen-Gupta (1943), and Henderson and Meldrum (1949), and for three-phase flow by Evinger and Muskat (1942) and Holmgren and Morse (1951). Richards and Moore (1952) made similar experiments on soil samples, and Fatt (1953) studied the effect of overburden pressure on relative permeability.

According to these investigations, the relative permeability curves are invariably S-shaped curves. A typical example which has been determined by

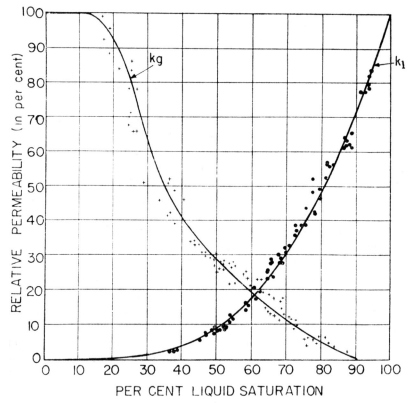

FIGURE 44 A typical example of relative permeability curves referring to a liquid (k_l) and a gas (k_g) (after Wyckoff and Botset 1936).

Wyckoff and Botset (1936) is supplied in figure 44. External parameters may, however, have an effect upon the relative permeability. Thus it has been found that overburden pressure, for instance, not only changes the total permeability, but also the shape of the relative permeability curves. A beautiful set of experiments to demonstrate this fact has been undertaken by Wilson (1956); an example of his results is shown in figure 45.

As the relative permeability curves are always S-shaped curves, attempts have been made to give an analytical expression for them. Thus, Jones (1946, 1949) found that the following expression represents experimental data reasonably well in most cases:

$$k_n = 1 - 1.11 s_w, \tag{10.2.3.1}$$
$$k_w = s_w{}^3, \tag{10.2.3.2}$$

where the subscript n stands for the non-wetting phase and the subscript w for the wetting phase. The above relation has been obtained from experiments involving oil as non-wetting phase and water as wetting phase.

10.3
SOLUTIONS OF DARCY'S LAW (IMMISCIBLE FLUIDS)

10.3.1 *The Buckley-Leverett case*

The system of equations (10.2.2.4) to (10.2.2.7) corresponding to Darcy's law is

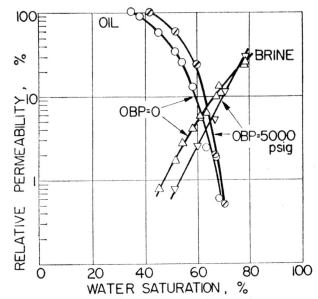

FIGURE 45   The effect of overburden pressure on relative permeability of an oil-brine system (after Wilson 1956).

extremely difficult to treat analytically, owing to its non-linear character. However, in some special cases solutions have been obtained.

An obvious solution has been given by Buckley and Leverett (1942). If gravity, capillarity, and variations in density are neglected ('Buckley-Leverette assumptions'), then it becomes evident that equations (10.2.2.5a–c) become an integral of the motion.

We set

$$r(s) = |\mathbf{q}_1|/|\mathbf{q}|, \tag{10.3.1.1}$$

i.e., $r(s)$ is the fractional flow of the 1-phase. Here, $\mathbf{q}$ is the total flow (equal to $\mathbf{q}_1 + \mathbf{q}_2$); the index '1' is omitted in the saturation as the equations are thought to pertain to the 1-phase only. It should be noted that $r(s)$ is indeed a function of saturation (and fluid viscosities) only, since one has, in virtue of (10.2.2.4), if the gravity term is neglected and p_1 set equal to p_2,

$$r(s) = \frac{|\mathbf{q}_1|}{|\mathbf{q}_1 + \mathbf{q}_2|} = \frac{k_1/\mu_1}{k_1/\mu_1 + k_2/\mu_2}. \tag{10.3.1.2}$$

As outlined in section 10.2.2, the relative permeabilities are functions of saturation only.

In view of the foregoing remarks, equation (10.2.2.5) can be written in the Buckley-Leverett case:

$$0 = P\frac{\partial s}{\partial t} + r'(s)\,(\mathbf{q}\,\text{grad}\,s), \tag{10.3.1.3}$$

$$\text{div}\,\mathbf{q} = 0, \tag{10.3.1.4}$$

where $r'(s) = dr/ds$.

In the one-dimensional case, equations (10.3.1.3–4) can be simplified to read

$$P\frac{\partial s}{\partial t} + r'(s)q_x\frac{\partial s}{\partial x} = 0, \tag{10.3.1.5}$$

$$\partial q_x/\partial x = 0. \tag{10.3.1.6}$$

Equation (10.3.1.6) gives at once:

$$q_x = q_x(t). \tag{10.3.1.7}$$

Thus we have

$$P\frac{\partial s}{\partial t} + r'(s)q_x(t)\frac{\partial s}{\partial x} = 0. \tag{10.3.1.8}$$

The last equation is a first-order differential equation and is amenable to treatment by the method of characteristics. A characteristic of a first-order partial differential equation is a line in the x–t plane along which the change of the unknown function (s) can be determined by a *total* differential equation. The change of s is

$$\frac{ds}{dt} = \frac{\partial s}{\partial x}\frac{dx}{dt} + \frac{\partial s}{\partial t}. \qquad (10.3.1.9)$$

If the ratio between dx and dt is chosen in such a manner that the right-hand side of equation (10.3.1.9) becomes equal to the left-hand side of equation (10.3.1.8), then the change ds is determined by a total differential equation. Obviously, dx/dt has to be chosen equal to the ratio of the coefficients in equation (10.3.1.8) associated with $\partial s/\partial x$ and $\partial s/\partial t$ respectively. We thus find that along the 'characteristic lines' determined by

$$dx/dt = r'(s)q_x(t)/P \qquad (10.3.1.10)$$

the change of s is given by the right-hand side of equation (10.3.1.8), viz.

$$ds/dt = 0. \qquad (10.3.1.11)$$

Thus we have succeeded in replacing the partial differential equation (10.3.1.8) by two total differential equations (10.3.1.10) and (10.3.1.11). The last two equations are the equations of Buckley and Leverett, which they obtained from (10.3.1.8) by 'dividing out' *partial* differentials, which is not always permissible as it may give a wrong result. In the present case, however, the rigorous derivation of (10.3.1.10) and (10.3.1.11) shows that the Buckley-Leverett equations are correct. It is quite immaterial whether in the partial differential equation (in our case equation (10.3.1.8)) the right-hand side is zero or not; the method of characteristics works in either case. In the present case, where the right-hand side of (10.3.1.8) *is* zero, we have the additional convenience that the right-hand side of (10.3.1.11) is also zero and the characteristics (equation (10.3.1.10)) are therefore lines in the x–t plane along which there is a constant saturation.

It is thus seen that any saturation proceeds with the speed given by the tangent to the $r(s)$-curve at that saturation (up to a factor which is constant for any given time, and constant *all* the time if the total flow is kept constant).

Brinkman (1948) has shown that equations (10.3.1.10) and (10.3.1.11) have solutions which take the form of a shock: a finite discontinuity of saturation is proceeding at a speed c. Although it is very difficult to prove that such a shock will always develop when the displacing fluid is injected at $x = 0$ at saturation 1 (thus one has saturation zero for the 1-phase to which our s refers), it is very easy to see that such solutions *do* exist and to give their explicit form.

Thus, let it be assumed that there is such a shock. To assist visualization, let it be assumed that the displacement takes place from left to right. To the right of the shock $s = 1$, and to the left of the shock there is some saturation s; to the right of the shock $r = 1$, and to the left some r (r and s behind the shock are to be determined). Thus, r and s refer to the *displaced* fluid.

We can formulate the continuity condition across the shock as follows:

$$Pc(1 - s) = q_x(1 - r); \qquad (10.3.1.12a)$$

this is termed the 'shock-condition' (Hugoniot equation). We obtain from it at once the speed of the shock

$$c = \frac{q_x}{P}\frac{1-r}{1-s}.$$

(10.3.1.12b)

Equation (10.3.1.10) gives the distance travelled by the various saturations of the displaced fluid in any given time. This distance is proportional to the $r'(s)$ function, the 'constant of proportionality' being equal to $\int q_x(t)dt/P$. Thus, a plot of s versus r', such as the one in figure 46 for a concrete case (see Welge 1952)

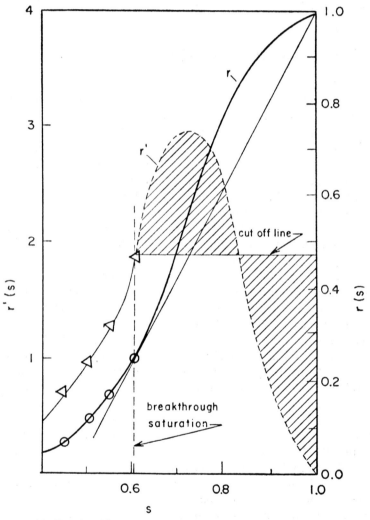

FIGURE 46   Fractional flow and its derivative in the Buckley-Leverett case (after Welge 1952).

also gives the saturation of the displaced fluid as a function of distance. It is only necessary to multiply the $r'(s)$ scale by the value $\int q_x(t)dt/P$ corresponding to the time at which the saturation distribution is desired. Thus, the curve describing saturation as a function of distance remains always similar except for horizontal 'stretching.'

In the event that the velocity and the saturation at the farthest point (i.e., the 'breakthrough saturation') is desired, Welge (1952) has shown that one can draw a tangent to the $r(s)$ curve from the point $r = 1$, $s = 1$, as shown in figure 46, for it is quite clear that if the saturation immediately behind the shock travels more slowly than the shock, it will take only a very short time until the size of the shock is decreased to a saturation which travels at least as fast as the shock. Once the saturation immediately behind the shock travels with the same speed as the shock, it will remain that way: the same saturation will always stay just behind the shock. This saturation, therefore, is the breakthrough saturation. The speed of any saturation is given by equation (10.3.1.10). The speed of the breakthrough saturation is c, as given by equation (10.3.1.12b). The condition for the breakthrough saturation is therefore obtained by equating the righthand-sides of equations (10.3.1.10) and (10.3.1.12b), which proves the validity of the tangent construction outlined above.

Instead of using the tangent construction of Welge, the breakthrough satura-tion can also be obtained by cutting off the $r'(s)$ curve (as indicated in figure 46) by making the two shaded areas of equal size (Brinkman 1948). Because of elementary properties of integrals, this construction is obviously equivalent to the tangent construction.

The problem of finding the saturations is now completely solved. If a displac-ing fluid is injected with (seepage) velocity q_x into a linear system, a bank forms where the saturation drops from 1 (of the displaced fluid) to the breakthrough saturation. The bank stays the same at all times and travels with speed c given by equation (10.3.1.12b). The saturations behind the bank travel with their charac-teristic speeds given by equation (10.3.1.10) and their advance can be found for any desired instant.

The fundamental solution obtained by Buckley and Leverett has been extended for more complicated occasions. Tarner (1944) applied it to gas flow in oil fields and calculated a few special cases. Nielsen (1949) used an analytical relative permeability formula (corresponding to equation 10.2.3.1–2) and obtained a complete analytical solution of the Buckley-Leverett case. Muskat (1950) investigated the problem of stratified layers of different permeability. Kern (1952) analysed the displacement mechanism in multiwell systems.

Some methods have been developed for the solution of special cases of the immiscible multiple phase flow problem. Thus, Oroveanu and co-workers (Oroveanu and Pascal 1956) studied the flow equation of mixtures of liquids and gases, the latter being treated as compressible.

Finally, the influence and possibility of not omitting the gravity term have been investigated by Lewis (1944), Brinkman (1948), Terwilliger et al. (1951), Philip (1957a), and Martin (1958). The fundamental equations are again obtained by writing down the two Darcy equations and the two continuity equations for the two phases.

10.3.2 *Other analytical solutions*

The solution of the linear flood of Buckley and Leverett, without accounting for capillarity, gravity, and density variations, is but a very particular solution of the general problem stated by equations (10.2.2.4–7). It cannot even be generalized to two dimensions. In two dimensions, the eliminant corresponding to equation (10.3.1.8) contains three independent variables; the characteristics therefore become surfaces. It can be seen that, rather than proceed in this manner, one might better treat the problem by a hodograph transformation. However, the real difficulty is to formulate the shock condition correctly. This now becomes a floating boundary condition, depending on the position of the shock *surface* (not on a point only).

The shock condition also makes it very difficult to obtain solutions of the basic equations by finite-difference methods on a computer: the derivatives become infinite at the shock-discontinuity, and a computer cannot handle this situation. The difficulty of a shock front can be avoided if in the fundamental equations (10.2.2.4–7) the capillary pressure p_c is not neglected but truly given as a function of the saturation. Then, no shock front develops. Numerical calculations for this case have been presented by Rachford and his co-workers (Douglas, Peaceman, and Rachford 1959, Blair and Peaceman 1963, Rachford 1964). The last paper gives a numerical calculation of instabilities that might arise in an immiscible displacement process.

10.3.3 *Imbibition*

A certain solution of the basic equations describes a peculiar phenomenon which is known as imbibition of a wetting fluid into a porous medium. This phenomenon has been discussed by Brownscombe and Dyes (1952), Enright (1954), Graham and Richardson (1959), and Verma (1969).

We assume, with Graham and Richardson (1959), that a cylindrical piece (linear coordinate x) of porous matrix is completely surrounded by an impermeable surface except for one end of the cylinder which is designated as the imbibition face. If the matrix is completely filled with some fluid, then it is observed that the presence of a more wetting fluid against the imbibition face gives rise to a flow of wetting fluid into the matrix and a counterflow of original fluid from the matrix.

If it is assumed that both the flow of wetting fluid (w) and the counterflow of original fluid (n) are governed by Darcy's law, we may write

$$q_w(x, t) = - [k_w(x, t)/\mu_w]k\, \partial p_w(x, t)/\partial x, \qquad (10.3.3.1)$$

$$q_n(x, t) = - [k_n(x, t)/\mu_n]k\, \partial p_n(x, t)/\partial x. \qquad (10.3.3.2)$$

Also, the capillary pressure is related to the non-wetting and wetting fluid pressures by

$$p_c = p_w - p_n. \qquad (10.3.3.3)$$

Combining (10.3.3.1), (10.3.3.2), and (10.3.3.3), and noting that $q_n(x, t) = - q_w(x, t)$, we obtain:

$$q_n(0, t) = k \left(\frac{k_w(0, t)k_n(0, t)}{\mu_w k_n(0, t) + \mu_n k_w(0, t)} \right) \left(\frac{\partial p_c}{\partial x} \right)_{x=0}. \qquad (10.3.3.4)$$

This equation gives the outflow rate of non-wetting fluid from the imbibition face as a function of time.

According to Leverett (cf. equation 3.4.1.20), a definite relation exists between the dimensionless quantity $(p_c/\gamma)(k/P)^{\frac{1}{2}}$ and s_w, the wetting fluid saturation. We may express this relationship by writing (cf. equation 3.4.1.20)

$$J(s_w) = \frac{p_c}{\gamma} \sqrt{\frac{k}{P}}. \qquad (10.3.3.5)$$

This equation, solved for p_c, yields

$$p_c = J(s_w)\gamma \sqrt{P/k}. \qquad (10.3.3.6)$$

Combining equations (10.3.3.4) and (10.3.3.6) gives

$$q_n(0, t) = \sqrt{kP} \left(\frac{k_w k_n}{\mu_w k_n + \mu_n k_w} \right)_{x=0} \left(\frac{\partial \gamma}{\partial x} J(s_w) + \gamma \frac{dJ}{ds_w} \frac{\partial s_w}{\partial x} \right)_{x=0} = 0. \qquad (10.3.3.7)$$

The last equation is the expression of Darcy's law for the case under consideration. The continuity condition for the same case has also to be formulated. We could proceed from the basic equations as listed in section 10.2.2, but it is just as easy to write down the balance for a slice of thickness dx of the cylinder:

$$q_w(x, t)dt - q_w(x + dx, t)dt = P[s_w(x, t + dt)dx - s_w(x, t)dx]. \qquad (10.3.3.8)$$

Dividing both sides of (10.3.3.8) by $(dxdt)$ and passing to the limit, we obtain

$$P(\partial s_w/\partial t)_x = - (\partial q_w/\partial x)_t, \qquad (10.3.3.9)$$

which is the equation of continuity. This equation also implies that

$$\left(\frac{\partial s_w}{\partial x} \right)_t = - \frac{(\partial q_n/\partial x)_t}{P(\partial x/\partial t)_{s_w}}. \qquad (10.3.3.10)$$

Combining equations (10.3.3.7) and (10.3.3.10), we obtain

$$q_n(0, t) = \sqrt{k} \left(\frac{k_w k_n}{\mu_w k_n + \mu_n k_w} \right)_{x=0} \left(\sqrt{P} \frac{\partial \gamma}{\partial x} J(s_w) - \frac{\gamma}{\sqrt{P}} \frac{\partial J}{\partial s_w} \frac{(\partial q_n/\partial x)_t}{(\partial x/\partial t)_{s_w}} \right)_{x=0}.$$

$$(10.3.3.11)$$

Equation (10.3.3.11) permits a comparison with data that can be obtained in the laboratory for the amount of original fluid produced (Q) by imbibition as a function of time, namely:

$$Q(t) = \int_0^t q_n(0, t) dt.$$

$$(10.3.3.12)$$

Although this integration is difficult to attempt, several important features of the imbibition phenomenon are brought to light:

1. The outflow rate of non-wetting fluid, $q_n(0, t)$, varies directly with $k^{\frac{1}{2}}$, all other factors remaining unaltered.

2. The term in the first brackets can be written as

$$\frac{k_w k_n}{\mu_w k_n + \mu_n k_w} = \frac{k_n}{\mu_w(k_n/k_w) + \mu_n}.$$

$$(10.3.3.13)$$

Since the region of matrix in the vicinity of the imbibition face soon acquires high wetting fluid saturation, it is a good approximation to write

$$\frac{k_w k_n}{\mu_w k_n + \mu_n k_w} = \frac{k_n}{\mu_n}.$$

$$(10.3.3.14)$$

Therefore, the outflow rate $q_n(0, t)$ varies nearly as $1/\mu_n$.

10.3.4 Solution subject to constraints

If there is an energy relationship between two flowing phases in a porous medium, the methods outlined above have to be modified. In general, such a relationship acts as a constraint in the solution.

Such a condition might occur if the liquid (index w) is displaced by its vapour (index G). The solution of the problem may be attempted for a linear system (linear coordinate x) as follows (suggested by J. von Neumann).

The continuity equation for mass yields

$$0 = P \frac{\partial}{\partial t} [s_w \rho_w + (1 - s_w) \rho_G] + \frac{\partial}{\partial x} [r_w \rho_w + (1 - r_w) \rho_G] q.$$

$$(10.3.4.1)$$

A similar continuity equation must hold for the heat flow:

$$\frac{\partial}{\partial t} [P(\zeta_w s + \zeta_G(1 - s) + \zeta_R)] + \text{HL} = - \frac{\partial}{\partial x} \{q[\zeta_w r + \zeta_G(1 - r)]\},$$

$$(10.3.4.2)$$

where ζ is the heat per unit mass for liquid, vapour, and porous matrix (corrected for porosity) denoted by the index w, G, R respectively, as a function of temperature; HL is a heat-loss term; and all the other symbols have the same meaning as before. If an index is omitted, it is understood that the corresponding term refers to *liquid*.

The problem can only be solved sensibly if there exists a front for the advance of the vapour (a bank in the usual sense). In that case the temperature in front of the bank will be $\mathfrak{T}_0$, i.e., the original temperature of the system; behind the bank it will be $\mathfrak{T}$, i.e., that of the vapour. Under these conditions, the heat-loss term is

$$\text{HL} = \frac{2C(\mathfrak{T} - \mathfrak{T}_0)}{h\sqrt{\pi}\sqrt{t - \tau}},$$ (10.3.4.3)

where C is some heat constant and h is the width of the system. If the velocity of the bank is denoted by c, then we have

$$\tau = x/c$$ (10.3.4.4)

if c is kept constant (the 'injection' of vapour can always be arranged in such a fashion). If c is not kept constant, we have to integrate in order to obtain τ.

The densities of all the phases involved are assumed to be constant. The heats per unit mass (ζ) refer now to the heat necessary to raise the temperature from $\mathfrak{T}_0$ to $\mathfrak{T}$ in the respective phase; the heat for the gas includes the heat of evaporation. These heats (ζ) are assumed constant.

Under these assumptions, we can write the two continuity conditions, valid anywhere *behind* the shock front (i.e., at temperature $\mathfrak{T}$) as follows:

$$P\frac{\partial}{\partial t}[\rho_w s + \rho_G(1 - s)] = -\frac{\partial}{\partial x}\{q[\rho_w r + \rho_G(1 - r)]\},$$ (10.3.4.5)

$$P\frac{\partial}{\partial t}[\zeta_w s + \zeta_G(1 - s) + \zeta_R] + \frac{2C(\mathfrak{T} - \mathfrak{T}_0)}{h\sqrt{\pi}\sqrt{t - x/c}} = -\frac{\partial}{\partial x}\{q[\zeta_w r + \zeta_G(1 - r)]\}.$$ (10.3.4.6)

Following the method of Buckley and Leverett, an eliminant of those two equations can be constructed. We shall use the following abbreviations:

$$\alpha = \frac{\rho_G}{\rho_w - \rho_G}, \qquad \beta = \frac{\zeta_G}{\zeta_w - \zeta_G},$$ (10.3.4.7)

$$\sigma = \frac{2C(\mathfrak{T} - \mathfrak{T}_0)}{h\sqrt{\pi}(\zeta_w - \zeta_G)}.$$ (10.3.4.8)

It is then obvious that the continuity equations can be written in the following manner:

$$P\frac{\partial s}{\partial t} = -\frac{\partial}{\partial x}[(r + \alpha)q],$$ (10.3.4.9)

$$P\frac{\partial s}{\partial t} = -\frac{\partial}{\partial x}[(r + \beta)q] - \frac{\sigma}{\sqrt{t - x/c}}.$$
(10.3.4.10)

This allows the construction of the following eliminant:

$$\frac{\partial q}{\partial x} = \frac{\sigma}{(\alpha - \beta)\sqrt{t - x/c}}$$
(10.3.4.11)

so that one obtains, with the correct constant of integration,

$$q = q(t) + \frac{2c\sigma}{\alpha - \beta}\left(\sqrt{t} - \sqrt{t - \frac{x}{c}}\right).$$
(10.3.4.12)

This can be inserted into the first continuity equation (10.3.4.9) to yield

$$P\frac{\partial s}{\partial t} = -qr'\frac{\partial s}{\partial x} - (r + \alpha)\frac{\sigma}{\alpha - \beta}\frac{1}{\sqrt{t - x/c}}$$
(10.3.4.13)

with

$$r' = dr/ds.$$
(10.3.4.14)

Using the method of characteristics, we obtain therefore

$$\frac{ds}{dt}\bigg|_{[dx/dt = qr'/P]} = -(r + \alpha)\frac{\sigma}{\alpha - \beta}\frac{1}{\sqrt{t - x/c}}.$$
(10.3.4.15)

An analysis of the actual numerical values of α and β shows that for all physically sensible cases the following conditions hold:

$$\alpha > 0, \quad \beta > 0, \quad \alpha - \beta < 0$$
(10.3.4.16)

so that the right-hand side of the characteristic equation is positive.

The equation for the characteristics gives the complete solution of the problem, except for the relationship between c and q, which is deduced from the 'shock condition,' i.e., the continuity condition around the bank.

Thus, the shock condition is obtained by formulating, for example, the heat-balance equation across the shock front. This procedure yields

$$cP[\zeta_w s + \zeta_G(1 - s) + \zeta_R] = q[\zeta_w r + \zeta_G(1 - r)].$$
(10.3.4.17)

After some algebraic transforming, this yields

$$c = \frac{q}{P}\frac{r + \beta}{s + \beta'},$$
(10.3.4.18)

with

$$\beta' = \beta + \frac{\zeta_R}{\zeta_w - \zeta_G}.$$
(10.3.4.19)

All this works only if c is kept constant, which is a working assumption. Otherwise, we have to introduce integrals for the calculation of τ.

The problem (with constant c) is now amenable to a graphiconumerical solution. The r versus s curve is experimentally known; it is an S-shaped curve, starting out from the origin. The speed of the front (up to the factor q/P) is given by the secant from the point $(-\beta, -\beta')$ to a point X on the curve, representing the instantaneous saturation at the front. If the tangent to the curve at the point X is steeper than this secant, then the saturations around the bank can be determined at the time $t + dt$ (where the time t corresponds to the point X). This is clear because the characteristics travel with the speed dr/ds (up to the factor q/P); thus, the information travels faster than the bank.

From the point $(-\beta, -\beta')$ it is possible to draw two tangents to the curve (the latter being S-shaped); call the point of contact of the lower tangent 'I', the point of contact of the upper tangent 'II.' We run into difficulty if X is not between the points I and II on the r versus s curve. It can be shown that if X is between 0 and I, it will go into I in an infinitesimal time; once at I it continues to ascend the curve. It is not at all clear, however, whether it will ever attain II; if it does, it would further ascend the curve and the problem would be undetermined. It is likely that in such a case a second bank would form.

A modification of the above problem has been considered by Bankoff (1969), who investigated the growth of a vapour bubble precipitating from a superheated liquid.

10.3.5 *Fractured media*

Particular problems arise when one is considering flow in highly fractured porous media. Bokserman et al. (1964) have shown that one can consider the medium as the combination of two systems, one comprising the pores and the other the fractures. Flow occurs through each system with an interaction taking place between them according to certain rules. This is expressed by introducing a continuous source ('impregnation') function in each system which depends on the pressure difference between the systems. One thus has, in essence, four interlocked equations of motion of the Darcy type, one for each of the two fluids in each of the two systems. The particular form of the impregnation function can be inferred from experiments performed by Mattax and Kyte (1962).

On the above basis, Bokserman et al. (1964) treated one-dimensional flow in a horizontal fractured reservoir without capillary pressure; Verma (1968) introduced the capillary pressure as well. As a special case, the problem of imbibition into a cracked medium (Verma 1969) may be treated in this fashion. Verma obtained solutions by means of an approximation procedure. Bear and Braester (1969) investigated what happens in a fissured reservoir during flow with pressure gradients which are not negligible in comparison with the pressure gradients that occur during imbibition, and set up the corresponding fundamental flow equations.

10.4
UNDERSATURATED FLOW

10.4.1 *The basic equation*

A special case of immiscible displacement in porous media is that in which the flow of only one phase is considered, the other being assumed a gas with a negligible pressure gradient. The pressure in the fluid is then equal to the capillary pressure at the corresponding saturation. Conditions of this type occur particularly when the flow of moisture through soil is treated; one fluid is then water, and the other air.

In the basic flow equations (10.2.2.4–7) one then retains only those terms which refer to the phase 1, and sets the air pressure equal to zero. Thus (dropping the index):

$$\mathbf{q} = - k\,(k_w/\mu)\,(\text{grad } p - \mathbf{g}\rho), \tag{10.4.1.1}$$

$$P\,\partial\rho s/\partial t = - \text{div}(\rho\mathbf{q}), \tag{10.4.1.2}$$

$$p = p_c(s). \tag{10.4.1.3}$$

Furthermore, if ρ is assumed constant (one assumes water to be of constant density), and using $p = p_c$, one can set

$$\mathbf{q} = k\,(k_w/\mu)\,(\text{grad } p_c - \mathbf{g}\rho), \tag{10.4.1.4}$$

$$P\,\partial s/\partial t = - \text{div } \mathbf{q}. \tag{10.4.1.5}$$

Inserting (10.4.1.4) into (10.4.1.5) yields

$$P\frac{\partial s}{\partial t} = \text{div}\left(\frac{kk_w}{\mu}\,(\text{grad } p_c - \mathbf{g}\rho)\right). \tag{10.4.1.6}$$

In soil physics it is customary to combine $gkk_w/\mu P$ into one factor K (μ, P are assumed constant), which, through the relative permeability k_w, is a function of s, termed the 'hydraulic conductivity':

$$K(s) = \rho g\, kk_w/(P\mu). \tag{10.4.1.7}$$

One can then write (10.4.1.7) as follows:

$$\frac{\partial s}{\partial t} = \text{div}(K\,\text{grad } \Phi) = \text{div}(K\,\text{grad } \Psi) - \frac{\partial K}{\partial z}, \tag{10.4.1.8}$$

where (if z is a vertical downward coordinate)

$$\Phi = \Psi - z \equiv \frac{p_c}{\rho g} - z. \tag{10.4.1.9}$$

If the relations between K, Ψ, and s are single-valued, equation (10.4.1.8) can be rewritten as follows:

$$\frac{\partial s}{\partial t} = \text{div}(D \text{ grad } s) - \frac{dK}{ds}\frac{\partial s}{\partial z},$$ (10.4.1.10)

where the 'moisture diffusivity' D is given by

$$D = K \, d\Psi/ds$$ (10.4.1.11)

and the differential quotients dK/ds are functions of s, which are assumed known. If, instead of s, Ψ is taken as the independent variable, equation (10.4.1.10) becomes

$$\frac{\partial \Psi}{\partial t}\frac{ds}{d\Psi} = \text{div}(K \text{ grad } \Psi) - \frac{dK}{d\Psi}\frac{\partial \Psi}{\partial z},$$ (10.4.1.12)

where K and $dK/d\Psi$ are assumed to be known functions of Ψ.

Equation (10.4.1.10) or (10.4.1.12) is the fundamental equation of under-saturated flow. It was deduced in an equivalent form by Richards (1931), and in the form of (10.4.1.10) by Klute (1952).

The same phenomenological equation (viz. a diffusivity equation) is obtained if the fluid transfer in the vapour and adsorbed phases is taken into account. The functions D and K then include these additional effects (Philip 1954, 1955a, b, 1957a, b).

10.4.2 Solutions

The theory of section 10.4.1 has been applied extensively to the undersaturated flow of water in soil. The basic equation (10.4.1.10 or 10.4.1.12) is of the dif-fusivity type, but, on account of the dependence of D and K on s, it is non-linear. Solutions are therefore difficult to obtain. Furthermore, specific solutions can be given only if the actual dependence of D and K on s is given. The factor K is essentially a relative permeability and can be measured by the corresponding methods. The quantity D is a 'soil moisture diffusivity' and methods have been developed to obtain it from laboratory outflow data (cf. e.g., Gardner 1956; Miller and Elrick 1958; Rijtema 1959; Butijn and Wesseling 1959; Kunze and Kirkham 1962; Bruce and Klute 1963; Peck 1966, 1969). Gardner and Mayhugh (1958) approximated D by the form

$$D(s) = D_0 \exp[\beta(s - s_0)],$$ (10.4.2.1)

where D_0 and β are empirical constants.

Solutions for filtration in one dimension are easiest to obtain, particularly if the gravity term (the second term in equation 10.4.1.10 or 10.4.1.12) is neglected. In the latter case, if the expression (10.4.2.1) is taken for D, the solution has the general property that the rate of advance of the position of any moisture content s is proportional to $t^{-\frac{1}{2}}$. A typical set of curves is shown in figure 47.

More complicated are the cases where gravity is accounted for. If a (dry) porous medium is brought into contact at the bottom with water, one obtains a capillary rise-effect which has been studied extensively (Rose et al. 1962; Philip 1966a). The opposite case is that of infiltration from rainfall, furrows, or ponded water (Gardner and Mayhugh 1958; Ferguson and Gardner 1963; Hanks and Bower 1962; Nielson et al. 1962; Rawlings and Gardner 1963). The theory of infiltration in a column was studied by Philip (1955a, b, 1957a, b, 1958). Further

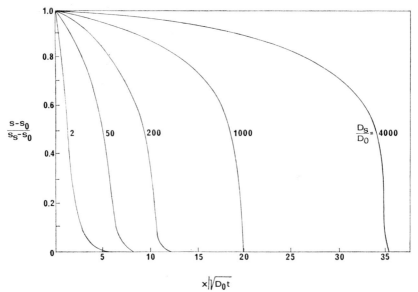

FIGURE 47 Solution of equation (10.4.1.10) for gravity-free infiltration into a porous medium (after Gardner and Mayhugh 1958).

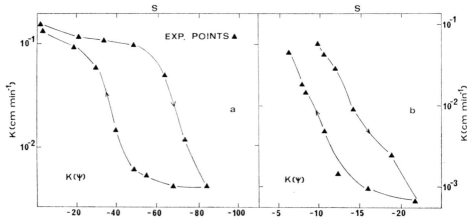

FIGURE 48 Hysteretic characteristics of Adelaide Dune Sand (left) and Molongo Sand (right) (after Talsma 1970a).

complications arise if the soil is inhomogeneous (Whisler and Klute 1967). Finally, attempts have been made to tackle three dimensional problems (Youngs and Towner 1963; Philip 1966b, 1969a, b; Raats and Gardner 1971; Freeze 1971).

10.4.3 *Limitations of the undersaturated flow theory*

The theory of undersaturated flow developed above assumes that there is a single-valued dependence between the saturation s and the quantities D, K, and p_c (or Ψ). This assumption is not always justified inasmuch as we have already seen in the chapter on hydrostatics that the relation between the capillary pressure p_c (or Ψ) and s is not univalued, but shows the phenomenon of hysteresis; the same is true in connection with dynamic phenomena (referring to the quantity K). The literature on this subject is large (cf. Philip 1970 for a review). Typical examples of such hysteresis phenomena are shown in figure 48 (after Talsma 1970a).

The existence of hysteresis in $p_c(s)$ or $\Psi(s)$ as well as in $K(s)$ represents a limitation of equation (10.4.2.10). Other limitations have been reviewed by Philip (1969b).

10.5
EXPERIMENTAL INVESTIGATIONS OF IMMISCIBLE DISPLACEMENT

10.5.1 *Principles*

Owing to the analytical difficulties of obtaining solutions to the Buckley-Leverett and similar equations, much effort has been devoted to experimental investigations, in which 'breakthrough' or 'recovery' curves have been obtained by displacing one substance by another. However, for such investigations to be meaningful, the appropriate scaling conditions have to be heeded. The general principles applicable in the deduction of scaling conditions have already been discussed in connection with gravity flow with a free surface (cf. sec. 5.4.4). The same principles, of course, have to be applied for immiscible displacement.

10.5.2 *Dynamical similarity of immiscible displacement processes*

The general principles of scaling have been applied to immiscible displacement processes in porous media by Rapoport and Leas (1953). Following these authors it is easy to see that, from the fundamental equations of multiple phase flow in porous media (10.2.2.4–7), the following eliminant can be formed, provided the two fluids are incompressible and gravity is neglected:

$$P\frac{\partial s_1}{\partial t} + r'(s_1)\mathbf{q}\,\mathrm{grad}\,s_1 - \frac{k}{\mu_2}\,\mathrm{div}[k_2 r(s_1)p_c'\,\mathrm{grad}\,s_1] = 0, \qquad (10.5.2.1)$$

where r is the fractional flow of the one-phase (as defined in 10.3.1.2) and the other symbols have the same meaning as in section 10.2.2. If an index is omitted, it is understood that the corresponding term refers to the 1-phase.

By means of the above eliminant, scaling problems can be studied. The eliminant equation can be brought into dimensionless form if the following substitutions of variables are made:

$$X = x/L, \tag{10.5.2.2}$$

$$T = t|\mathbf{q}|/LP, \tag{10.5.2.3}$$

$$P_c = p_c/p_0, \tag{10.5.2.4}$$

where p_0 is some characteristic reference pressure and L is some characteristic macroscopic length of the flow system. This yields

$$\frac{\partial s_1}{\partial T} + r'\mathbf{n}\,\mathrm{grad}_X\,s_1 - \frac{k\mu_1}{\mu_2}\frac{p_0}{L|\mathbf{q}|\mu_1}\,\mathrm{div}_X\,[k_2 r P_c'\,\mathrm{grad}_X\,s] = 0, \tag{10.5.2.5}$$

where $\mathbf{n} = \mathbf{q}/|\mathbf{q}|$. This equation being in dimensionless form, it is seen that dynamical similarity is obtained if the 'scaling' coefficient $\mathcal{3}$

$$\frac{1}{\mathcal{3}} = \frac{kp_0}{L|\mathbf{q}|\mu_1} \tag{10.5.2.6}$$

and the viscosity ratio μ_2/μ_1 have identical values in two flow systems. Naturally, the two systems must also be geometrically similar and must consist of identical porous media.

For small values of the scaling coefficient $\mathcal{3}$, examination of the dimensionless eliminant equation (10.5.2.5) shows that the second-order term, containing the capillary pressure, is relatively large in comparison with the two first-order terms. Under such circumstances, the flow behaviour is influenced by capillary pressure forces and is markedly dependent upon rate, length of flow system, and fluid viscosities. As the value of the scaling coefficient increases, it is, however, seen that the relative importance of the second-order term diminishes and that it will eventually become negligible. The eliminant equation then reduces to a simplified form, in which only first-order terms are retained:

$$\frac{\partial s_1}{\partial T} + r'\mathbf{n}\,\mathrm{grad}_X\,s_1 = 0. \tag{10.5.2.7}$$

Expression (10.5.2.7) is independent of L, q, and the viscosities, provided the viscosity *ratio* remains unchanged. Accordingly, displacement experiments ('floods') may be defined as 'stabilized' if they are independent of the rate of injection, the length of the flooded system, and the fluid viscosities. This is the case if the scaling coefficient $\mathcal{3}$ is large.

One can extend the above by starting from the general flow equations. The following eliminant can be formed from equations (10.2.2.4–7) (Rapoport 1955):

$$P\frac{\partial s_1}{\partial t} + r'(s_1)\mathbf{q}\,\text{grad}\,s_1 - \frac{k}{\mu_2}\,\text{div}(k_2 r p_c'\,\text{grad}\,s_1) - \frac{k}{\mu_2}g(k_2 r)'\Delta\rho\frac{\partial s_1}{\partial z} = 0,$$

$$(10.5.2.8)$$

where $\Delta\rho$ is the density difference between the two flowing fluids (defined as $\rho_1 - \rho_2$), which are again assumed to be incompressible. However, the gravity term is now not being neglected.

As before, the eliminant equation (10.5.2.8) makes possible a study of scaling problems. Denote the 'original' by capital letters, the 'model' by lower case ones. We can then introduce the following scaling factors:

$X/x = Y/y = Z/z = A$ (linear dimensions), $\qquad$ (10.5.2.9)

$U/u = V/v = W/w = C$ (seepage velocities), $\qquad$ (10.5.2.10)

$N/n = B$ (volume flow rates), $\qquad$ (10.5.2.11)

$K/k = D$ (permeabilities), $\qquad$ (10.5.2.12)

$P_c/p_c = E$ (capillary pressures), $\qquad$ (10.5.2.13)

$\Delta P/\Delta\rho = F$ (density differences), $\qquad$ (10.5.2.14)

$T/t = G$ (times), $\qquad$ (10.5.2.15)

$M/\mu = H$ (viscosities), $\qquad$ (10.5.2.16)

$P/p = R$ (porosities). $\qquad$ (10.5.2.17)

If, now, the eliminant equation (10.5.2.8) is used to determine scaling factors, we obtain at once:

$$ADE = BH = A^2 DF \qquad (10.5.2.18)$$

or, preferably written,

$$E = AF, \qquad (10.5.2.19)$$

$$BH = A^2 DF. \qquad (10.5.2.20)$$

It may be noted that the scaling of times and velocities is not arbitrary; we have

$$C = B/A^2, \qquad (10.5.2.21)$$

$$G = A^3 R/B. \qquad (10.5.2.22)$$

From a macroscopic viewpoint, the conditions expressed by equations (10.5.2.19–20) are sufficient to achieve proper scaling between an 'original' and a corresponding 'model.' However, it was simply assumed that the relative permeability curves of wetting and non-wetting fluid (not the total permeabilities) are identical in the model and the original. It seems desirable to investigate the conditions for which this is true, considering the microscopic aspects of flow through porous media. Thus, we must also discover what similarity conditions apply in a microscopic sense to properties such as pore geometry and capillary pressure curves.

Rapoport (1955) has given a discussion of this problem, but based upon the Kozeny equation (cf. sec. 6.3.2). As stated earlier, the Kozeny equation makes use of very definite assumptions about the constitution of porous media, which we have maintained earlier are somewhat doubtful. The same results, however, can be obtained on the basis of much less stringent ideas about the constitution of porous media as shown by Miller and Miller (1956).

Thus, assuming that one has two porous media which are geometrically similar, i.e., identical up to a scaling factor λ (this scaling factor being applied microscopically to the pore geometry), the dynamics within are characterized by two kinds of laws: (i) the surface tension law, and (ii) the (microscopically applied) Navier-Stokes equation.

The surface tension equation (3.4.1.9),

$$(4 \cos \theta)/ \delta_c = p_c/\gamma \qquad (10.5.2.23)$$

(see sec. 3.4.1 for the definition of symbols), indicates by its form that if the surface tension γ is doubled and the variable p_c is everywhere doubled, the interface shapes (i.e., their radii of curvature δ_c) consistent with (10.5.2.23) will be unchanged. The porosity does not enter into these considerations. Thus, in order to make interfaces between two phases in geometrically similar porous media similar also, we have to choose, assuming identical contact angles θ,

$$\frac{1}{\lambda} = \frac{P_c/p_c}{\Gamma/\gamma} = \frac{E}{S} \qquad (10.5.2.24)$$

or

$$S = E\lambda, \qquad (10.5.2.25)$$

where the various scaling factors are defined as follows:

$$\Delta_c/\delta_c = \lambda \,(\text{microscopic lengths, radii of curvature}), \qquad (10.5.2.26)$$

$$\Gamma/\gamma = S \,(\text{surface tensions}). \qquad (10.5.2.27)$$

Furthermore, the Navier-Stokes equation (2.2.2.1) yields that, for a given liquid geometry, the resistance to flow will vary with the viscosity μ and inversely with δ_c^2. Thus, averaging over the sample, the permeability to a phase will vary directly with δ_c^2, i.e.,

$$kk_1 = \text{const.}\, \delta_c^2, \qquad (10.5.2.28)$$

Thus:

$$D\kappa_1 = \lambda^2, \tag{10.5.2.29}$$

where the scaling factor κ_1 is defined by

$$K_1/k_1 = \kappa_1 \text{ (relative permeabilities).} \tag{10.5.2.30}$$

Since the macroscopic eliminant equation requires that κ_1 equal 1, we obtain as microscopic similarity conditions:

$$D = \lambda^2 = S^2/E^2. \tag{10.5.2.31}$$

This relation was also found by Rapoport (1954), thus demonstrating that in its deduction the particular ideas introduced by the application of the Kozeny equation would actually have been justified.

10.5.3 *Experimental displacement studies*

Owing to the analytical difficulties of solving the multiple phase flow Darcy equations, attention has been directed towards the possibility of performing experiments at a very early stage. Unfortunately, the idea of scaling outlined in section 10.5.2 has not been applied to such experiments until later, although the general principles of dynamical similarity have been well known in physics for about a century.

In fact, most experiments on two-fluid flow that have been performed were undertaken by the oil industry in order to determine the 'recovery' of oil under various conditions of 'gas-' or 'water-flooding.' The reports on the influence of the various variables on flooding experiments yielded, however, very erratic results – as was to be expected, since scaling had been neglected. The results obtained in the models, equally, are often not at all significant in relation to the condition in the 'field' which they are supposed to represent. There is one saving factor, however. This is that the 'scaling coefficient' β is often (inadvertently) very large in both the 'model' and in the 'field' so that the corresponding experiments are very often dynamically similar, i.e., the flood experiments are 'stabilized' in the sense described in section 10.5.2.

Thus, most experimental investigations are nothing but straightforward 'floods' (using water or gas) on petroleum well cores in order to determine the 'oil recovery' obtainable therefrom without any regard to scaling. Their number is very great. Reviews may be found in the pertinent books mentioned in the Introduction.

Finally, it should be mentioned that not only have porous media model experiments of displacement processes been made, but also analogy-model experiments using processes that are (or are hoped to be) *analogous* to flow in porous media. Of particular importance are, in this connection, models using Hele-Shaw (cf. sec. 5.2.2) cells (Thirrot and Aribert 1969).

10.6

PHYSICAL ASPECTS OF RELATIVE PERMEABILITY

10.6.1 *Visual studies*

The extension of Darcy's law to multiple phase flow as performed in section 10.2.2 is, in fact, a heuristic procedure suggested by the analogy with single phase flow. It does not provide an understanding of the physics of multiple phase flow. However, as the law has been substantiated by experiments, it must be assumed that it is at least partly correct and a physical understanding of it has to be sought.

Some direct investigations into the problem of double phase flow have been made. Versluys (1917, 1931) had already shown that, statically, two phases may be in three different states in a porous medium, which were termed fully saturated, pendular, and funicular respectively (see chap. 3). It must be assumed that the three states have different flow patterns.

If the porous medium is completely saturated with one phase, then a displacing phase must break a way into the displaced fluid. It is observed that this happens by establishing fingering into the porous medium. The fingers then branch out and eventually form a network.

This leads to a state where both the displacing and the displaced fluid form a network of funicular channels where it is possible to get from any point in either fluid to any other point in the same fluid by not leaving that fluid. The displacing fluid will flow along the established network of its own (and so will the displaced fluid), gradually increasing its hold, until the channels of the displaced fluid break up and this fluid then occurs in the pendular state. Eventually, depending on wettability characteristics, it may be possible that no more displaced fluid can be removed from the porous medium; the displacing fluid moves in its own network and the displaced phase may stay immobile as a 'connate' phase.

The prevailing regime of double phase flow is that where both phases are flowing along their own funicular channels. It is easy to envisage that these channels may be much more tortuous than the pore channels for single phase flow. Thus, we arrive at a qualitative explanation for the low relative permeabilities characteristic of multiple phase flow systems.

A substantiation of these ideas has been achieved by visual (i.e., microscopic) flow studies. Such studies have been made by Mahoney (1947), Ryder (1948), Wilson and Calhoun (1952), Chatenever (1952; also Chatenever and Calhoun 1952; and Chatenever, Indra, and Kyte 1959), and Muskat et al. (1953).

10.6.2 *Theoretical studies*

Ultimately, it will be theoretical studies that will provide the explanation of relative permeability. In qualitative terms, such studies were begun many years ago. Gardescu (1930) discussed the behaviour of gas bubbles in capillary spaces, Ceaglske and Kiesling (1940) studied the mechanics of capillary flow in solids, and Houpeurt (1949) investigated the effects of surface forces on the equilibrium and motion of oil deposits in soil.

A serious treatment of the theory of relative permeability will make use of the same concepts as those used in the treatment of single-phase permeability. The first possibility is thus to establish heuristic correlations. This has been done by Atkinson (1948), Childs and Collis-George (1948), Harrington (1949), Whiting and Guerrero (1951), and Collis-George (1953).

The next possibility is to use capillaric models. Before doing so, however, one has to investigate the flow of several phases in a single capillary. This has been done by Bergelin (1949), Porkhaev (1949), and Templeton (1953, 1954). These investigations are largely experimental, since techniques have been devised for the observation of gas-liquid or liquid-liquid displacements in uniform capillaries with diameters as small as four microns. Qualitative descriptions of the behaviour of moving interfaces were thus obtained. Quantitative observation of gas-liquid displacements with zero contact angle indicates the adequacy of a model based on Poiseuille's law, and the independence of capillary pressure from interfacial velocity.

The concepts of displacement in single capillaries have been applied to porous media, the media being represented as a bundle of such capillaries, by Gates and Lietz (1950), Fatt and Dykstra (1951), Burdine (1953), Hassan and Nielsen (1953), and Irmay (1954). A theory of relative permeability based upon such 'capillaric models' has been given by Fatt and Dykstra (1951). It is outlined below.

As usual in capillaric model theories, it is assumed that the sample can be represented by a bundle of capillary tubes in which the fluid path length is not the same as the bulk length. In addition, the fluid path length varies with saturation. For a bundle of N capillary tubes the flow through dN tubes will be

$$dQ = q_{av}dN, \qquad (10.6.2.1)$$

where Q is the total flow rate through all the tubes and q_{av} is the average flow rate through the tubes in the interval dN. If the interval dN is made small, the average flow rate through a tube in the interval dN is given by Poiseuille's Law

$$q_{av} = \pi r^4 \Delta p/(8\mu l), \qquad (10.6.2.2)$$

where Δp is the pressure drop across the tube of length l and radius r, and μ is the viscosity of the fluid. Substituting for q_{av} in equation (10.6.2.1), Fatt and Dykstra obtained

$$dQ = [\pi r^4 \Delta p/(8\mu l)]dN. \qquad (10.6.2.3)$$

Darcy's law for linear flow of an incompressible fluid in a porous medium is

$$Q = kA\,\Delta p/(\mu L), \qquad (10.6.2.4)$$

where k is the (total) permeability and Δp is the pressure drop across the bundle of tubes which has replaced the porous medium of length L and cross-sectional area A. Differentiating Darcy's law with A, Δp, μ and L constant, we obtain

$$dQ = [A\,\Delta p/(\mu L)]dk. \tag{10.6.2.5}$$

Equating (10.6.2.3) and (10.6.2.5) we have

$$dk = [\pi r^4 L/(8Al)]dN. \tag{10.6.2.6}$$

If the pores are assumed to be cylinders,

$$dV = \pi r^2 ldN, \tag{10.6.2.7}$$

where V is the volume of flowing fluid in the pores. Substituting dN in equation (10.6.2.6) Fatt and Dykstra obtained

$$dk = [r^2 L/(8Al^2)]dV. \tag{10.6.2.8}$$

By definition, the saturation, s, of the porous sample is given by

$$s = V/V_p = V/(PAL), \tag{10.6.2.9}$$

where V_p is the total pore volume and P is the porosity. Differentiating equation (10.6.2.9) and substituting for dV in equation (10.6.2.8) we get

$$dk = dsr^2 PL^2/(8l^2). \tag{10.6.2.10}$$

The ratio of the fluid path length l to the length L of the sample has earlier been termed the 'tortuosity,' T:

$$T = l/L. \tag{10.6.2.11}$$

The tortuosity, T, has now to be related directly to measurable properties. Fatt and Dykstra did this by considering what happens to the wetting phase in the pore spaces as the sample is desaturated. During desaturation, the wetting phase retreats into the smaller pores and the smallest crevices. Liquid in such crevices has a small radius of curvature and by the derivation given is considered to be in small pores. It can be reasoned then that liquid flowing in the crevices and small pores will travel a more tortuous path than liquid flowing through the large pores. As a first approximation, T can thus be assumed to vary inversely as r^b (with $b > 0$)

$$T = a/r^b, \tag{10.6.2.12}$$

where a and b are constants. Substituting equations (10.6.2.11) and (10.6.2.12) in (10.6.2.10) gives

$$dk = \frac{Pr^{2(1+b)}}{8a^2}\,ds. \tag{10.6.2.13}$$

The pressure, p_c, across a curved interface between two fluids is the capillary pressure given by

$$p_c = (2\gamma \cos \theta)/r, \tag{10.6.2.14}$$

where r is the radius of curvature, γ the interfacial tension at the interface, and θ the liquid-solid contact angle. When equation (10.6.2.14) is applied to porous media, r is the 'pore radius' at which a non-wetting phase just displaces a wetting phase out of the pore. Equation (10.6.2.14) thus gives r as a function of the capillary pressure. Substituting for r in equation (10.6.2.13), we obtain

$$dk = \frac{P(2\gamma \cos \theta)^{2(1+b)}}{8a^2 p_c^{2(1+b)}}\, ds. \tag{10.6.2.15}$$

If the constants, a and b, are known, the effective permeability k_e, at any saturation, to the corresponding phase can be calculated by integrating equation (10.6.2.15) from zero saturation to the desired saturation s. Thus, Fatt and Dykstra obtained:

$$k_e = \frac{P(2\gamma \cos \theta)^{2(1+b)}}{8a^2} \int_0^s \frac{ds}{p_c^{2(1+b)}}. \tag{10.6.2.16}$$

The relative permeability, k_r, however, can be calculated without knowing the constant a if the constant b is known or assumed, because by definition $k_r = k_e/k$, thus giving

$$k_r = \frac{\displaystyle\int_0^s ds/p_c^{2(1+b)}}{\displaystyle\int_0^1 ds/p_c^{2(1+b)}}. \tag{10.6.2.17}$$

Gates and Lietz (1950) assumed $b = 0$, and Fatt and Dykstra (1951) $b = \frac{1}{2}$. However, a comparison of measured and calculated relative permeability data indicates that this constant may not be the same for all types of materials. In this instance, the constant must be held to be one of the usual undetermined factors so common in hydraulic radius theories.

The capillaric model theory has been improved by Purcell (1950), who attempted to calculate relative permeabilities by assuming pores which were doughnut-shaped instead of cylindrical.

A further refinement of the above theory is obtained if, instead of capillaric models, the Kozeny theory is extended to multiple phase flow. This has been done by Rose and co-workers (Rose and Bruce 1949; Rose 1949; Rose and Wyllie 1949), Wyllie and co-workers (Wyllie 1951; Wyllie and Spangler 1952), and Rapoport and Leas (1951).

Following Rose (1949), the Kozeny equation for single phase flow can be expressed as:

$$k = P/(S_p^2 T), \tag{10.6.2.18}$$

where P is the fractional porosity, S_p is the specific internal surface area of the pores per unit *pore* volume, and T is again a dimensionless textural constant related to the shape and orientation (tortuosity) of the pores. Carman (1941), in studying the properties of unconsolidated sands, claimed empirically that

$$S_p = p_D/\gamma, \tag{10.6.2.19}$$

where p_D is the displacement pressure and γ is the interfacial tension. Equations (10.6.2.18) and (10.6.2.19) can be combined to yield

$$k = P\gamma^2/(p_D^2 T). \tag{10.6.2.20}$$

Now, considering the effective permeability to the wetting phase (k_{ew}) in a polyphase flow system, it can be argued by analogy with equation (10.6.2.20) that

$$k_{ew} = P_{ew}\gamma^2/(p_c^2 T_{ew}), \tag{10.6.2.21}$$

where the effective porosity P_{ew} is equal to $s_w P$, and where p_c, the capillary pressure, is accepted as being the effective displacement pressure characterizing the partially saturated system. The effective textural constant, T_{ew}, is an undetermined factor and may therefore, for the purposes of this analysis, be assumed equal to T throughout the saturation range of interest. In any event, simply dividing equation (10.6.2.21) by equation (10.6.2.20) yields an expression for the wetting phase relative permeability, k_{rw}, namely:

$$k_{rw} = \frac{k_{ew}}{k} = \frac{P_{ew}}{P} \frac{p_D^2}{p_c^2} \frac{T}{T_{ew}}, \tag{10.6.2.22}$$

which leads to the following equation, since $s_w = P_{ew}/P$:

$$k_{ew} = s_w p_D^2/p_c^2. \tag{10.6.2.23}$$

In order to test equation (10.6.2.23) it is necessary to have such pressure versus saturation data available as are obtained in capillary pressure (static) experiments or displacement (dynamic) experiments. Rose showed, however, that a more general relationship between the wetting phase relative permeability and saturation can be derived in which the relative permeability is described entirely in terms of explicitly known saturation parameters. This can be accomplished by treating the Kozeny equation (10.6.2.18) in a manner similar to the derivation of equation (10.6.2.23) from equation (10.6.2.20); that is, by assuming $T = T_{ew}$ through the saturation range of interest, and by recognizing that $P_{ew}/P = s_{ew}$. Thus, by analogy with equation (10.6.2.18), Rose set

$$k_{ew} = P_{ew}/(S_{pew}T_{ew}) \tag{10.6.2.24}$$

or

$$k_{ew} = s_w S_p^2/S_{pew}^2. \tag{10.6.2.25}$$

This is the desired relationship. Here S_{pew} may be regarded as the effective surface separating the wetting phase from all other elements of the system. Comparison of equation (10.6.2.25) and equation (10.6.2.23) suggests that

$$S_{pew} = p_c/\gamma \qquad (10.6.2.26)$$

is a reasonable expression for this effective surface area analogous to Carman's expression for specific surface area, equation (10.6.2.19).

An approach similar to that of Rose was chosen by Rapoport and Leas (1951) in order to give a theory of relative permeability. Following these authors, the total permeability is given by the Kozeny equation

$$k = cP^3/S^2, \qquad (10.6.2.27)$$

where S is now the specific surface area per unit *bulk* volume and c is the Kozeny constant, often set equal to $\frac{1}{5}$. In the case of double phase flow (phases 'liquid' and 'gas'), Rapoport and Leas represented the areas contacted by the two phases as follows:

$$\left.\begin{aligned}
S_L &= R_L + I &&= \text{specific surface area of the liquid system,} \\
S_G &= R_G + I &&= \text{specific surface area of the gas system,} \\
S &= R_G + R_L &&= \text{specific surface area of the solid,}
\end{aligned}\right\} \qquad (10.6.2.28)$$

where R_L represents the contact area between the solid and the liquid, R_G the contact area between the solid and the gas, and I the interfacial area between the two phases.

Referring to the conditions of applicability of Darcy's law, it is readily seen that the Kozeny equation can be applied to the liquid phase by (a) the replacement of the absolute porosity, P, by sP, the liquid filled pore volume per unit of bulk volume, i.e., by the 'effective porosity' with respect to the liquid; (b) the substitution of the specific pore area, or total specific surface S, by S_L, which represents the specific total boundary area of the liquid phase 'flow matrix.'

By substituting the above terms into equation (10.6.2.27) Rapoport and Leas obtained the effective permeability to the liquid phase:

$$k_L = cs^3P^3/S_L^2, \qquad (10.6.2.29)$$

and by division of equation (10.6.2.29) by equation (10.6.2.28), they obtained the following expression for relative permeability to liquid:

$$k_L = s^3/(S_L/S)^2. \qquad (10.6.2.30)$$

In order to obtain a more usable expression for the relative permeability to liquid, the following considerations must be taken into account. It is generally recognized that towards the end of a capillary displacement process an irreducible minimum saturation s_0 is reached, and no more liquid can be forced out, except by diffusion. At this saturation the effective permeability to liquid becomes zero. Therefore, it seems logical to treat, in first approximation, the irreducible liquid as an essentially stationary element which reduces the porosity of the porous medium as well as the volume of flowing liquid. Once this liquid is considered as part of the solid framework, it is necessary to use an *effective* area, S_E, representing the surface that separates the flowing fluid from the immobile flow matrix composed of solid and stationary fluid, instead of the specific *solid* surface S; furthermore, instead of using the total liquid saturation s, a *reduced saturation* s' has to be introduced:

$$s' = (s - s_0)/(1 - s_0). \tag{10.6.2.31}$$

On substitution into equation (10.6.2.30), Rapoport and Leas obtained the following expression for the relative permeability to liquid:

$$k = \frac{s^3}{(S_L/S_E)^2} \left(\frac{s - s_0}{1 - s_0}\right)^3 \left(\frac{S_E}{S_L}\right)^2. \tag{10.6.2.32}$$

It will be noted that the derivation of equation (10.6.2.32) implies the two following assumptions: (a) the molecular layers at the interfaces between solid liquid and gas liquid are stationary – if they were not, the basic Kozeny treatment should be slightly modified so as to take into account a certain 'slippage' of the fluid on part of its boundaries; (b) the Kozeny constant of the wetting fluid 'matrix' is the same as that of the total pore space – such an approximation seems to be justified since in general the possible variations of the Kozeny constant have been recognized to be comparatively small.

The calculation of k_L, the relative permeability to liquid, requires a knowledge of the surface areas, S_L and S_E. With the help of a thermodynamic approach the following expressions can be derived from a liquid–gas capillary pressure curve:

$$S_G = \Sigma_s \equiv - (P/\gamma) \int_1^s p_c ds, \tag{10.6.2.33}$$

$$S_E = \Sigma_{s_0} \equiv - (P/\gamma) \int_1^{s_0} p_c ds, \tag{10.6.2.34}$$

$$S_L = \Sigma_w + 2I \equiv - (P/\gamma) \int_s^{s_0} p_c ds + 2I, \tag{10.6.2.35}$$

in which the integrals indicate the areas under the p_c versus s curves measured between the indicated limits. It is seen that a system of three equations is obtained for the four unknowns S_G, S_E, S_L, and I. Consequently, while the values for S_G and S_E are directly indicated, separate solutions cannot be defined for S_L or I.

In order to evaluate S_L, Rapoport and Leas found it necessary to introduce an additional assumption concerning the distribution of the fluids. Such an assumption, expressed in terms of R_G, R_L, and I, will then represent one more independent equation, which together with the above relations will furnish a system from which all the unknowns can be deduced. It might be noted that on the basis of thermodynamic considerations only a more or less complex 'trend' can be indicated for the liquid distribution, namely maximum contact area with the solid compatible with a minimum amount of interfacial area. In order to express such a trend in a more exact manner, it would be necessary to consider in detail the geometry of the porous medium, which depends on many parameters such as grain size distribution, sphericity or shape factor of the grains, packing, and consolidation. Therefore, no exact additional relation can be established that would lead to unique solutions for S_L and I. The only logical approach that can be attempted consists of establishing the most general limiting conditions of fluid distribution that would be expected to apply to any, or at least to most, of the possible geometrical systems. According to such a principle, separate calculations for the minimum and for the maximum values of k_L can be made.

In accordance with the previously established relations, Rapoport and Leas considered the following system of equations:

$$\Sigma_w = S_L - 2I, \tag{10.6.2.36}$$

$$\Sigma_s = R_G + I, \tag{10.6.2.37}$$

where Σ_s and Σ_w have a well-defined value at any saturation and can, therefore, be considered as 'constants' relative to the unknowns S_L, I, and R_G. It is readily seen that, according to equation (10.6.2.36) for any set of Σ_s and Σ_w, i.e., for any saturation, a *maximum* value will be obtained for S_L if I is maximum, and that according to equation (10.6.2.16) the maximum value for I obtains if R_G is assumed to be zero. Consequently, Rapoport and Leas arrived at the following maximum value that S_L may take at any saturation:

$$S_L(\text{Max.}) = \Sigma_w + 2\Sigma_s = \Sigma_{s_0} + \Sigma_s. \tag{10.6.2.38}$$

According to equation (10.6.2.32) it is clear that at any saturation the smallest possible value will be obtained for the relative permeability to liquid k_L if a maximum area S_L is considered, so that

$$k_L(\text{Min.}) = s'^3[S_E/S_L(\text{Max.})]^2 = s'^3[\Sigma_{s_0}/(\Sigma_{s_0} + \Sigma_s)]^2. \tag{10.6.2.39}$$

Physically the derivation of $k_L(\text{Min.})$ means that the distribution of the fluids is such that it does not permit the existence of any contact between the gas phase and the solid surface. The gas can then be visualized as flowing inside a network of channels completely surrounded by liquid, and it might be noticed that such an assumption concerns only the disposition of the fluids and does not correspond to any hypothesis about the geometry of the porous medium itself.

Referring to the system of equations (10.6.2.36) and (10.6.2.37), Rapoport and Leas showed that the theoretical absolute *minimum* value for S_L would obtain under the assumption $I = 0$, i.e., $S_L = \Sigma_w$. Such a situation would correspond to $R_G = S_G$, i.e., to the assumption of having the gas phase (and equally the liquid phase) bounded exclusively by solid surfaces. In that case, the porous medium should be visualized as formed by non-interconnected capillaries. It is clear that such an assumption is by far too limiting and leads actually to k_L values which are too high. Thus, instead of considering (even for the purpose of limiting conditions) a porous medium as equivalent to a bundle of capillary tubes, it is reasonable to represent it as an isotropic random packing of grains. Under this more general assumption, interfaces must exist between the gas and liquid, and none of the three surface areas, I, R_G, R_L, is zero. The lowest possible value of the ratio I/R_G can then be evaluated. The use of this value in equation (10.6.2.37) leads to a minimum value of I, of S_L, and consequently to a maximum of k_L.

With the help of general capillary pressure relationships, it is possible to show that

$$I/R > [sP/(1 - P)] (p_m/p_c)^2, \tag{10.6.2.40}$$

where p_c represents the capillary pressure at any saturation s, and p_m the mean value of p_c corresponding to the saturation range s to s_0. The above inequality led Rapoport and Leas to postulate the following minimum value of S_L:

$$S_L(\text{Min.}) = \Sigma_w + \cfrac{2\Sigma_{s_0}}{1 + \cfrac{1 - P}{sP} \left(\cfrac{p_c}{p_m}\right)^2}, \tag{10.6.2.41}$$

and the final expression for the limiting maximum relative permeability to liquid is then obtained as

$$k_L(\text{Max.}) = s'^3 \cdot \cfrac{\Sigma_{s_0}{}^2}{\left\{\Sigma_w + \cfrac{2\Sigma_s}{1 + \cfrac{1 - P}{sP} \left(\cfrac{p_c}{p_m}\right)^2}\right\}^2}. \tag{10.6.2.42}$$

Equations (10.6.2.39) and (10.6.2.42) represent the minimum and maximum possible values of relative permeability, according to Rapoport and Leas.

Refinements in the hydraulic radius theories of immiscible multiple phase flow were reported by Fatt (1956) and Rose and Witherspoon (1956). Fatt extended his network model (cf. sec. (6.2.5) to multiple phase flow, and Rose and Witherspoon used the doublet model (cf. sec. 6.2.5) for the explanation of relative permeability.

The theories of relative permeability outlined above rely either on capillaric models or on the Kozeny theories. Therefore, they are subject to the usual limitations of the hydraulic radius theory. However, attempts to extend the drag theory or the statistical theory of hydrodynamics in porous media to double phase flow do not seem to have been attempted.

The papers on the theory of relative permeability mentioned above also report experiments which were performed to substantiate claims made. However, as there are a variety of undetermined factors involved in these theories, an agreement between experiments and theory is not necessarily meaningful.

10.7
GENERAL FLOW EQUATIONS FOR IMMISCIBLE FLUIDS

10.7.1 *Limitations of Darcy's law*

The limitations of Darcy's law in multiple phase flow of immiscible fluids are partly due to the same causes as in single phase flow. Thus, in the high-velocity regime, one has to expect non-linearity to play a role; and in the flow of gases, molecular effects and slippage phenomena might become important. Furthermore, there is also a series of 'other' effects such as boundary phenomena and ionic phenomena. In addition, there are effects due to hysteresis of contact angle, as has been mentioned earlier.

The limitations of Darcy's relative permeability law have been discussed in general terms, for example, by van Wingen (1938), Ivakin (1951), and Richardson et al. (1952). In particular, boundary and related effects (such as macroscopic channelling) have been discussed by O'Connor (1946), Krynine (1950), Geffen et al. (1951), and Hill (1952). The very concept of relative permeability is in doubt (Rose 1969) inasmuch as hysteresis and a dependence of the measured values of k_i on the dynamic conditions has been noted. Thus, Henderson and co-workers (Henderson and Yuster 1948a; Henderson and Meldrum 1949) and Calhoun (1951b) found some dependence of relative permeability not only on saturation, but also on pressure. Furthermore, Yuster (1951) stated that relative permeabilities also depend on the absolute viscosities of the fluids involved. He tried to explain this by the remark that there is a shear transmitted at the two-phase interface within the porous medium which would actually entail such a phenomenon. A further discussion of this 'Yuster effect' has been given by Scott and Rose (1953). In addition, Sen-Gupta (1943) found abnormal changes in air permeabilities with water content in oil sands, and Geffen et al. (1951) observed hysteresis. These effects all signify limitations of Darcy's law.

For a completely general formulation of linear flow laws encompassing dynamic effects one would have to set (Rose 1969) in one dimension

$$q_i = \sum_j D_{ij} \operatorname{grad} p_j, \tag{10.7.1.1}$$

where the indices i, j run through the values identifying the components of the mixture, and the D_{ij} may be arbitrary functions of the saturations, saturation gradients, etc. The case for

$$D_{ij} = D(s_i)\delta_{ik} \tag{10.7.1.2}$$

has already been treated by Scheidegger as 'solitary dispersion' (cf. sec. 9.3.1).

In the further exposition in this book, we shall discuss the usual deviations from Darcy's law in the high-velocity and molecular-flow regimes, assuming the classical description to be valid under ordinary circumstances.

10.7.2 *High-velocity regime*

It must be expected that non-linearity plays a significant role in the high-velocity regime of multiple phase flow in porous media. Unfortunately, analyses of non-linear effects are confined to the study of relatively coarse-grained porous media (pore diameters of the order of centimetres), such as are found in industrial towers. Non-linear flow of two immiscible phases in such towers has been investigated especially by chemical engineers. A great number of observations have been accumulated and various theories have been advanced to explain these observations.

The exposition of non-linear flow in coarse-grained porous media contained in the present section is essentially that given in a review article by Lerner and Grove (1951).*

Accordingly, most of the investigations on the behaviour of industrial towers are heuristic, corresponding to the heuristic investigations on single phase turbulent flow.

One of the earliest investigators of this problem, Peters (1922), observed that non-linear conditions of countercurrent flow for a liquid-gas system occur rather abruptly. White (1935) extended this observation by noting that in a logarithmic plot of pressure drop versus gas velocity two breaks ('critical' points) occur. With reference to the two breaks in the logarithmic plot, White defined the 'loading point' of a column as 'the gas velocity at which, for a given liquor rate, the logarithmic pressure drop-gas velocity curve first deviates from a slope of approximately two.' The 'flooding point' was defined as 'that velocity at which the same curve turns abruptly almost vertically upward.' This latter point was said to be accompanied by a marked spraying of the liquid. Further investigations along these lines were made by Simmons and Osborn (1934), Mayo et al. (1935), and Baker et al. (1935). The definition of loading point as given above was retained by Mach (1935) and again by White (1935).

Sherwood (1937) used the critical points of White's data interchangeably and changed the data of Baker et al. (1935) on loading velocities to flooding data. This may be readily seen by a comparison of Sherwood's work with the original data of Baker et al. A further anomaly arises from the fact that, although Baker et al. used the term 'loading,' they provided no definition of it.

Similar investigations were made by Uchida and Fujita (1936, 1937), Verschoor (1938), Vilbrandt et al. (1938), and Furnas and Bellinger (1938). Sherwood (1937) revised his earlier work by using White's definitions for his own determinations

---

*Permission to use the article by Lerner and Grove so extensively has kindly been granted by the American Chemical Society.

of the loading and flooding points. Although he defined the flooding point as a graphical flood point, the flooding condition was actually taken 'by visual observation of the liquid flowing over the packing and down the walls of the tower.' Inasmuch as Sarchet (1942) has since reported an appreciable discrepancy between visual and graphical flood points, this would seem to render Sherwood's data inconsistent with his own definition, as the visual points were found by Sarchet to be from 15 to 20 per cent above or below the graphical flood points, the magnitude and direction of the deviation being a function of packing size. Sarchet further concluded that the graphical determination of the critical flow velocities was more dependable than visual observation. The basic correlation of flooding velocity used for design purposes in most instances is that of Sherwood (1937). While originally accurate only within 40 per cent, the correlation has been improved by further work on packing characteristics (see Lobo et al. 1949).

Considerably fewer data have been published on loading conditions than on flooding velocities. The data that have been published are somewhat contradictory. Elgin and Weiss (1939), measuring both *hold-up* (i.e., *saturation*) and pressure drop, found no abrupt break point but rather a gradual transition, and suggested that the loading point be properly represented as a zone. On the other hand, Piret, Mann, and Wall (1940) concluded from their data on a column 2.5 feet in diameter, packed with round gravel stones 1.75 inches in diameter, that a definite break in the hold-up occurs at a point corresponding to the loading point (see Lerner and Gove 1951).

Further experimental investigations with attempts at empirical correlations have been made by many people. Most of these investigations result in the plotting of some 'correlation curves,' but some of them end up in an analytical equation representing these curves. Particularly notable is the work of Brownell and Katz (1947, see also Dombrowski and Brownell 1954), who extended their correlation between the friction factor and Reynolds number for single phase flow to double phase flow by including functions of saturation in the two correlated quantities.

It is difficult to judge how accurate the correlations advanced in the cited papers are if they are considered from a general standpoint. However, in view of the criticisms pointed out during the discussion of single-phase flow correlations, it must be held that the present ones are subject to the same limitations. There seem to have been no fundamental improvements achieved over the basic methods used for single-phase flow correlations, which have been shown to be inadequate.

Among the theoretical attempts at correlations, Zenz (1947) has advanced a mechanism and a correlation for the limiting velocities of flow in packings based on an analogy with fluid flow through valves and orifices. Zenz smoothed the normal log-log plots into continuous curves, and by application of thermodynamic relationships defined the flood point as the gas velocity corresponding to a constant 'critical pressure drop' above which the log-log pressure drop–gas velocity curves become vertical. The existence of a definite critical velocity for two-phase flow through orifices – i.e., a break point in the log-log plot – would obviate the basic premise of this theory (see Lerner and Grove 1951).

Similar attempts at correlation based on the mechanisms of discontinuity were suggested by Sarchet (1942), Bertetti (1942), and Lerner and Grove (1951). The last authors concluded that the variables under consideration, i.e., the superficial rates of flow based on the empty cross-sectional area of the bed, could not be directly related with the desired degree of accuracy. It was felt that, if the superficial rates could be converted to the actual flow rates through the interstices, then these latter variables would represent a better picture of the actual flow conditions, and would probably be more amenable to correlation.

In the case of countercurrent liquid-gas flow in vertical packed columns, qualitative analysis of the problem of converting superficial to actual flow rates leads directly to a general form of the quantitative relationships. If, as a first approximation, it is assumed that the liquid flows through the packing under a constant head, then any increase in the mass liquid throughput will result in an increase in the cross-sectional area through which the liquid flows, rather than an increase in the linear liquid velocity. It is implicitly assumed that the liquid-packing interaction forces remain constant, and that the velocity of the liquid is simply related to the free fall velocity under the given gravity head. Thus, if the increase in cross-sectional area through which the liquid flows is proportional to the increase in liquid rate, the mass velocity of the liquid phase remains substantially constant. On the other hand, the total flow area for both gas and liquid must remain constant so that the area pre-empted by the increase in liquid rate diminishes the area available for gas flow. Thus, the actual mass rate of flow of the gas increases with an increase in liquid rate, although the superficial gas flow rate is unchanged.

If now the limiting criterion for continuous gas flow be the velocity of the gas phase, it should be possible to achieve the loading and flooding conditions by an increase in liquid rate alone, at constant gas rate. Evidence that the actual gas flow rate is the limiting variable has been advanced by Furnas and Bellinger (1938), Piret et al. (1940), and Elgin and Weiss (1939), who found that the liquid saturation is independent of the gas rate up to the loading point, at which point it increases sharply. Furthermore, the fact that Elgin and Weiss were able to obtain flood points with zero gas flow is confirming evidence for the expectation of a point of discontinuity with increasing liquid rate.

The above reasoning may be placed on a quantitative basis by the use of the data of Jesser and Elgin (1943) on the relation between liquid rate and liquid saturation in packed columns. The objective of the analysis is to derive an expression relating what appears to be the critical variable, the actual gas flow rate, to the superficial liquid and gas rates.

Following the definition of Jesser and Elgin, the term 'hold-up,' i.e., 'liquid saturation,' refers only to the dynamic hold-up, which is that portion of the total saturation which varies with liquid rate. The static saturation, which is independent of liquid rate, is therefore included in the wet-drained fractional voids. Essentially the same procedure was utilized by Cooper et al. (1941) in their computation of the linear velocities of gas flow in packed columns.

Considering now a differential height in a vertical tower, we have, from the definition of hold-up (after Lerner and Grove 1951),

$$A_L = as_0, \qquad (10.7.2.1)$$

where A_L is the cross-sectional area through which liquid is flowing, s_0 is the dynamic hold-up (a fractional saturation), and a is a proportionality constant. The critical hold-up (i.e., saturation) corresponding to the point of transition (flow discontinuity) at zero gas rate may be designated as $(s_0)_m$. Inasmuch as the critical flow point is marked by a break in the hold-up, $(s_0)_m$ is not the hold-up that completely occupies the free void space in the packing, but rather the hold-up immediately preceding the attainment of the critical point at zero gas rate. In terms of relative velocity, zero gas throughput is obviously not equivalent to zero gas velocity relative to the falling liquid, but is nevertheless a reproducible reference flow for a given apparatus with a fixed free fall of liquid. Therefore, taking the zero gas throughput condition as the primary reference point, $(s_0)_m$ is a constant, and will result in liquid closure of the flow channel. Equation (10.7.2.1) then becomes:

$$A_T = a(s_0)_m, \qquad (10.7.2.2)$$

where A_T is the total area available for flow and may be defined as

$$A_T = A_g + A_L = P_{wd}A_0, \qquad (10.7.2.3)$$

where A_g and A_L are the gas and liquid flow areas, respectively, A_0 is the cross-sectional area of the empty tower, and P_{wd} is the wet-drained fractional voids (a porosity value).

Equation (10.7.2.3) may be transposed to the form

$$A_g = A_T(1 - A_L/A_T). \qquad (10.7.2.4)$$

Substituting equations (10.7.2.1) and (10.7.2.2) into (10.7.2.4), we obtain

$$A_g = A_T[1 - s_0/(s_0)_m] \qquad (10.7.2.5)$$

or

$$A_g = P_{wd}A_0[1 - s_0/(s_0)_m]. \qquad (10.7.2.6)$$

By a simple material balance for the gas stream, the relation between the actual and superficial mass gas rates (denoted by G with appropriate subscripts) is given by

$$G_a = A_0 G_0/A_g, \qquad (10.7.2.7)$$

which is the Dupuit-Forchheimer assumption. Substituting equation (10.7.2.6) into equation (10.7.2.7), we obtain:

$$G_a = \frac{G_0}{P_{wd}[1 - s_0/(s_0)_m]}.$$
(10.7.2.8)

For the purpose of visualization it is advantageous to change from a mass flow rate to a pore velocity v:

$$v = \frac{G_0}{\rho_G P_{wd}[1 - s_0/(s_0)_m]}.$$
(10.7.2.9)

With water as the liquid, the data of Jesser and Elgin (1943) show that

$$s_0 = bL_0^c,$$
(10.7.2.10)

where b and c are constants for any given packing and L_0 is the superficial liquid mass rate. Substituting equation (10.7.2.10) in equation (10.7.2.9) and solving for G_0 gives

$$G_0 = v\rho_G P_{wd}[1 - bL_0^c/(s_0)_m].$$
(10.7.2.11)

Because $(s_0)_m$ is a constant, the superficial liquid and gas rates are now related through the actual pore velocity or, converse ly, the pore velocity may be expressed as a function of L_0 and G_0.

Thus far, the actual mechanism of channel closure has not been considered. However, the literature contains considerable material on two-phase flow transition in pipes which is relevant to the problem of the mechanism of closure in packed-column channels. Boelter and Kepner (1939), investigating two-phase flow in horizontal pipes, found that with increasing gas velocity, waves are generated on the surface of the liquid. As the gas velocity was further increased, the amplitude of the waves was observed to increase, until it became sufficiently large to cause the gas flow to become discontinuous. Although Boelter and Kepner employed parallel *horizontal* flow and an oil-air system, the same observations were made by O'Bannon (1924) and Houghten et al. (1924) for *vertical* and inclined countercurrent flow for air and water and steam and water. Later, Martinelli and co-workers (1944, 1946) and Gazley (1949) published extensive observations of the role of wave formation in channel closure.

The fact that a minimum gas velocity is necessary for inducing and sustaining waves on a liquid surface leads to the conclusion that the limiting velocity of two-phase flow is directly related to the gas velocity which is causing wave-formation. For small channel diameters, of the order of magnitude of twice the gravitational wave amplitude, the limiting two-phase flow gas velocity and the critical velocity of wave formation may be assumed identical. Inasmuch as wave formation is primarily an interfacial phenomenon, the shape of the channel should have little effect on the inception velocity of the waves. For channel diameters larger than the critical minimum (twice the wave amplitude), the initiation of waves due to attainment of the critical gas velocity decreases the area available for gas flow and the velocity decreases so that an unstable situation results, terminating in channel closure. Therefore, the period between wave initiation and channel closure is necessarily short (Gazley 1949). Within the limits of channel size ordinarily encountered in packed columns, the size and shape of the channel should have only a minor effect on the mechanism of closure, or the critical gas velocity of closure.

In view of the arguments cited, Lerner and Grove (1951) advanced the following postulates about flooding in packed columns in countercurrent two-phase flow:

(a) The mechanism of flooding in packed columns is wave formation on the liquid surface of sufficient amplitude to close the gas flow channel.

(b) Because the gas velocity appears to be the determining factor, there exists a definite velocity of gas flow at which waves will close the flow channel. This gas velocity is defined as the velocity of wave closure, $(v)_c$. One can also postulate that for any given liquid-gas system in countercurrent flow the velocity of wave closure is a constant, dependent primarily on the physical properties of the liquid and gas, and to a minor extent on the size and shape of the channel. On this basis, equation (10.7.2.11) becomes the correlating equation for limiting flow in packed columns.

Lerner and Grove (1951) substantiated their theory, outlined above, by a number of experiments. Although it is based upon semi-empirical arguments, and contains a number of undetermined factors, it seems to be at least qualitatively correct. It should not be forgotten, however, that it was advanced essentially in order to explain non-linear multiphase flow phenomena in rather *coarse* porous media such as are found in industrial towers. It should, therefore, not be expected to apply to substances with significantly finer pores.

As outlined earlier, systematic applications of either capillaric models or statistical mechanics to non-linear multiple phase flow phenomena, which would be applicable to media with fine pores, do not seem to have been undertaken.

10.7.3 *Molecular effects*

As in single phase flow, the molecular effects can be split into two groups: first, if one of the phases is a gas, slippage will be expected; and second, if one of the phases is strongly adsorbed by the porous medium, adsorption and capillary condensation might play a role.

Gas slippage in double phase flow has been investigated by Rose (1948), Fulton (1951), and Estes and Fulton (1956). The essential results of these investigations are (i) that the value of relative gas permeability, extrapolated in the manner of Klinkenberg for infinite mean pressure, is the same as the non-wetting liquid relative permeability; and (ii) that the effect of gas slippage decreases as the wetting liquid saturation increases.

Multiple phase flow through adsorptive materials has been investigated essentially by soil scientists, textile researchers, and building engineers, who are concerned with the motion of moisture through soil, textiles, or building materials, respectively, air being the second phase. It is generally assumed that the flow of moisture follows heuristically a diffusivity equation where the partial water vapour pressure is the driving agent. Experimental investigations of such flow phenomena have been made by Babbitt (1939, 1940), Christensen (1944), Peirce et al. (1945), Hauser and McLaren (1948), Johansson et al. (1949), Wink and Dearth (1949), and Pfalzner (1950). Klute (1952) developed a numerical method for solving the flow equations. Kirkham and Feng (1949) formulated imbibition equations that contradict the usual diffusivity equation.

Understanding of diffusion of moisture through various materials is presumably to be sought in some application of the principles of adsorption. Babbitt (1948) observed that the permeability of hygroscopic materials decreases steadily at decreasing humidity, which he ascribes to the disappearance of sorbed layers. Accordingly, it seems that surface flow is the chief contributor to the total flow through the porous medium and thus that the phenomenon can be explained by an effect analogous to capillary condensation in the flow of condensable vapours.

However, the whole problem of the movement of moisture or other vapours through solids is an extremely complex one. The literature bearing upon it is, accordingly, very large, and cannot be listed here in detail. A substantial monograph on the subject has been written by Lykow (1958). The reader is referred to it for further details.

11
Miscible displacement

11.1
GENERAL REMARKS

The final chapter of this book will be devoted to the discussion of miscible displacement. This type of displacement has been of great concern to hydrologists who have been studying the problem of displacement of fresh water by sea water in coastal areas. Recently, a modern note has been struck as the problem has become important to people who are trying to dispose safely underground of ever-increasing amounts of atomic waste products from nuclear reactors. The oil industry has also become involved in miscible displacement studies in connection with the possibility of flushing oil from reservoirs by solvents.

The all-important phenomenon affecting miscible displacement is that of *mechanical dispersion*. Miscible displacement in porous media is a type of two-phase flow in which the two phases are completely soluble in each other. Therefore, capillary forces between the two fluids do not occur. The progressive mixing of the phases is then caused by the flow of the individual particles through the complexities of the pore channels. This is 'dispersion.'

Several attempts at describing dispersion have been made. A first attempt based on a capillaric model does not seem to lead to adequate results. It turns out that the most successful description of dispersion must be based on the statistical theories of flow through porous media as presented in chapter 8 of this book. Thus, the present chapter will be found to be, in many ways, an extension and application of that chapter on statistical hydrodynamics in porous media, which represents the key to the understanding of miscible displacement.

11.2
DARCY'S LAW

11.2.1 *Formulation of Darcy's law*

At first it might be thought that miscible displacement could be described in a very simple fashion. The mixture, under conditions of complete miscibility, could be thought to behave, locally at least, as a single-phase fluid which would obey Darcy's law. The change of concentration, in turn, would be caused by diffusion along the flow channels and thus be governed by the bulk coefficient of diffusion of the one fluid in the other. In this fashion, one arrives at a heuristic description of miscible displacement which looks, at a first glance at least, very plausible.

In the case of incompressible fluids, the equations corresponding to the description given above, are as follows (Offeringa and Van der Poel 1954):

(i) Darcy's law (cf. 4.3.1.2):

$$\mathbf{q} = - \frac{k}{\mu(C_1)} [\text{grad } p - \rho(C_1)\mathbf{g}]; \qquad (11.2.1.1)$$

(ii) Continuity-diffusivity condition:

$$- P \frac{\partial C_1}{\partial t} - \text{div}(C_1\mathbf{q}) + PED_1 \text{ lap } C_1 = 0, \qquad (11.2.1.2)$$

where the indices 1, 2 refer to the two miscible phases, and C, the concentration, corresponds to the saturation s in immiscible displacement. The quantity D_1 is the usual diffusion coefficient of phase 1 in phase 2 in bulk masses of the liquids, and E is a dimensionless factor introduced to account for the fact that pore channels are tortuous and therefore longer than the corresponding macroscopic stream line intervals. Diffusion should take place along the tortuous channels and the factor E, therefore, should always be smaller than 1.

11.2.2 *Criticism*

The flow equations (11.2.1.1–2) can be tested by experiment. If this is done, it soon becomes evident that they are inadequate to describe the observed phenomena correctly. The diffusion factor D_1 is allegedly the molecular diffusion factor; the diffusion should thus depend on the time only, irrespective of the flow, but this does not appear to be the case. According to experimental investigations, it appears that during miscible displacement a fairly sharp concentration front becomes established, and that the front is the sharper the more slowly the displacement takes place. It also appears that below a finite displacement rate a definite, special pattern of flow occurs. Thus it seems that a finite, characteristic time is necessary for the displacing fluid to diffuse sideways within the individual flow channels, thereby completely 'mopping up' the original fluid. If the forward motion of the displacing fluid is slow, i.e., of the order of a pore diameter during the characteristic time, the displacement may be called *molecular*; if the displacement is fast (i.e., of the order of many pore diameters during the characteristic diffusion time), the displacement is *invasive*. In both cases, the degree of mixing seems to be governed by mechanical dispersion.

The phenomenological flow equations suggested by Darcy's law as a generalization of single phase flow to miscible multiple phase flow do not allow for any characteristic times. A satisfactory heuristic theory of such phenomena does not yet exist. It is doubtful whether the heuristic approach will ever yield any suitable results in the present context, since it appears preferable to develop a theory of miscible displacement starting from microscopic and statistical considerations.

The above remarks pertain to what corresponds to the laminar flow regime in single phase flow. It is, of course, to be expected that modifications will have to be introduced in flow regimes where turbulence, molecular streaming, or other effects occur.

11.3
DISPERSION EFFECTS

11.3.1 *Description*

It has been stated above that the theories based on Darcy's law of miscible displacement in porous media are inadequate. This fact is not really surprising in the light of earlier discussions of the physical aspects of permeability.

First, the factor E in (11.2.1.2) is tied up with the tortuosity of the pore channels, and it has been shown earlier that the concept of tortuosity is doubtful at best. Second, the diffusion of the one fluid into the other is certainly not taking place in the manner envisaged by the simple theory of section 11.2.1, but rather by a mechanism of mechanical dispersion of one fluid into the other owing to the interconnections of flow channels. This effect must be connected with the dispersivity of porous media introduced in the statistical theories of single phase flow. (For a confirmation of this thought, see Handy 1959.)

Heuristically, it has been found that the process represents a 'smearing out' (dispersion) of the displacement front (concentration C of invading fluid) while the front is in convective motion with the mean pore velocity $\mathbf{v}$. Phenomenologically, the process can therefore be described by the equation

$$\frac{\partial C}{\partial t} = \frac{\partial}{\partial x_i} D_{ij} \frac{\partial C}{\partial x_j} - \frac{\partial (v_i C)}{\partial x_i} , \tag{11.3.1.1}$$

where D_{ij} is some tensor of dispersion and the summation convention has been used.

Two attempts seem to have been made to account for this heuristic result from a theoretical standpoint. The first is an extension of the idea of capillaric models, and the second is an extension of the statistical methods of chapter 8 of this book.

11.3.2 *Capillaric models*

It is fairly straightforward to set up a capillaric model theory in which Taylor's displacement process would obtain in each capillary. This has been done by Von

Rosenberg (1956). Thus, if the porous medium is envisaged as a body consisting of straight, parallel capillaries of uniform size and equal length, then exactly the same displacement law as was found by Taylor for each of the capillaries should hold. Thus, according to section 2.5.2, the 'length' L of the saturation front in a linear displacement experiment should be directly proportional to the square root of the seepage velocity, and at a given velocity the length L of the front should increase as the square root of the distance traversed. In mathematical symbols, one thus has for the length of the front:

$$L \sim \text{const. } t^{\frac{1}{2}}q \sim \text{const. } (qx)^{\frac{1}{2}}, \tag{11.3.2.1}$$

where $x \sim qt$; x is the mean linear position coordinate of the front, q the seepage velocity, and t the time. Aris (1956) has refined the above theory and given an expression for the factor of dispersion, based on a capillaric model. However, the specific assumptions that have to be made are not very satisfactory; it is therefore much more desirable to use the perfectly general methods of statistical mechanics for the explanation of the heuristic equation (11.3.1.1).

11.3.3 *Statistical models*

The application of the statistical theory of single phase flow of chapter 8 can be extended to multiple phase flow if it is assumed that the displacing and the displaced fluids have similar physical characteristics. Then it is possible to identify the probability that a 'particle' is at a certain spot at a certain time with the 'concentration' of the invading fluid within that originally contained in the porous medium. As has been shown in section 8.5.1, the statistical models lead to equations essentially equivalent to (11.3.1.1). The various models, however, lead to different dependences of the factor of dispersion on the pore velocity v (or on the filtration velocity $q = Pv$). This dependence is of the form

$$D = \text{const. } q^n \tag{11.3.3.1}$$

with values of n ranging from 1 to 2.4. The result (11.3.2.1) from the capillaric model theory implies that $n = 2$, which is, as will be seen below, too restrictive. It thus turns out that the only reasonable way to account for the observed phenomenon of dispersion in miscible displacement is by means of the statistical theory of hydrodynamics in porous media.

11.4
SOLUTIONS OF THE DISPERSION EQUATION

11.4.1 *General remarks*

In order to evaluate and apply the various theories of miscible displacement in porous media, it is necessary to have solutions of the fundamental equation

(11.3.1.1), which has the form of a diffusivity equation. If the tensor of diffusion D_{ik} is a function of C, and v is a function of x, the equation is non-linear. However, in the light of the statistical models, one can assume that D is a function of q (or v) only; if v is independent of position, then D as well as v can be taken out of the differential sign and we have

$$\frac{\partial C}{\partial t} = D_{ij} \frac{\partial^2 C}{\partial x_i \partial x_j} - v_i \frac{\partial C}{\partial x_i}. \tag{11.4.1.1}$$

This is, in fact, still a rather complicated partial differential equation. The value of v must be found first, using Darcy's law or some equivalent equation of motion. Subsequently, the dispersion has to be superposed. Needless to say, this is not an easy task, and except in the simplest cases, the finding of solutions is a difficult undertaking.

11.4.2 The one-dimensional case

The simplest case of the dispersion problem is the linear case. The fundamental equation becomes, for constant v,

$$\frac{\partial C}{\partial t} = D \frac{\partial^2 C}{\partial x^2} - v \frac{\partial C}{\partial x}, \tag{11.4.2.1}$$

solutions of which have to be found. One can introduce moving coordinates

$$x' = x - vt, \tag{11.4.2.2}$$

$$t' = t, \tag{11.4.2.3}$$

and then the fundamental equation becomes

$$\partial C/\partial t' = D \, \partial^2 C/\partial x'^2. \tag{11.4.2.4}$$

The solution of this equation must be found for appropriate initial conditions.
    We shall start with a consideration of the progress of a thin slug. This requires that the fundamental differential equation be solved for the initial condition

$$C(x', 0) = m_0 \delta(x'), \tag{11.4.2.5}$$

where δ denotes the Dirac delta function and m_0 (which we shall set equal to 1 in the following lines) is the total amount of invading fluid contained in the slug. Then C is a linear density (amount per unit length).
    The solution of equation (11.4.2.4) for the initial condition (11.4.2.5) is well known; it is (see e.g., Crank 1956)

$$C = \frac{1}{(4\pi Dt')^{\frac{1}{2}}} \exp\left(-\frac{x'^2}{4Dt'}\right). \tag{11.4.2.6}$$

The shape of the curves representing $C(x')$ for various values of $t'(=t)$ is shown in figure 49.

We now turn our attention to the progress of an initially sharp front through the porous medium. This case can immediately be calculated from the solution for the progress of a slug by superposition.

Denote the solution for a slug by $\sigma(x', t')$ (note that we are always referring to the moving coordinate system x', t'); then the progress of a front is simply given by

$$C(x', t') = \int_{-\infty}^{0} \sigma([x' - \zeta], t')\, d\zeta \qquad (11.4.2.7)$$

if originally (at $t' = 0$) all space $x' < 0$ (this is also $x < 0$) is filled with that kind of fluid to which C refers. Setting

$$x' - \zeta = \eta \qquad (11.4.2.8)$$

yields

$$C(x', t') = -\int_{\infty}^{x'} \sigma(\eta, t')\, d\eta = \int_{x'}^{\infty} \sigma(\eta, t')\, d\eta. \qquad (11.4.2.9)$$

This can be written as follows because σ is normalized and symmetrical with respect to $x' = 0$:

$$\left. \begin{aligned} C(x', t') &= \frac{1}{2} - \int_{0}^{x'} \sigma(\eta, t')\, d\eta \quad \text{for } x' > 0, \\[2mm] &= \frac{1}{2} + \int_{x'}^{0} \sigma(\eta, t')\, d\eta \quad \text{for } x' < 0. \end{aligned} \right\} \qquad (11.4.2.10)$$

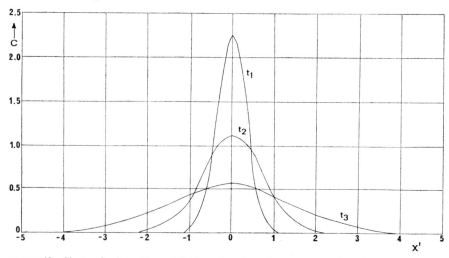

FIGURE 49   Shape of a slug of tagged fluid passing through a porous medium for $t_1 < t_2 < t_3$.

Inserting (11.4.2.6) for σ, we find (see, e.g., Crank 1956)

$$C(x', t') = \frac{1}{2} - \int_0^{x'} \frac{1}{(4\pi Dt')^{\frac{1}{2}}} \exp\left(\frac{-\eta^2}{4Dt'}\right) d\eta$$

$$= \frac{1}{2} - \int_0^{x'/\sqrt{4Dt'}} \frac{1}{\sqrt{\pi}} e^{-z} dz = \frac{1}{2} - \frac{1}{2} \operatorname{erf} \frac{x'}{\sqrt{4Dt'}}, \qquad (11.4.2.11)$$

where erf denotes the well-known error function, for which tables are available. The general appearance of the solution (11.2.4.11) is shown in figure 50.

The above solutions represent but the simplest cases of one-dimensional dispersion. Ebach and White (1958) have given a solution for an input concentration which is a periodic function of time. Other one-dimensional cases are obtained if axially symmetric or radial flows are considered; in these cases, of course, v is no longer independent of position. Thus Hoopes and Harleman (1965) used various coordinate systems and, in this fashion, treated the problem of radial flow from a well fully penetrating an aquifer. The reader is referred to the original paper for the details.

11.4.3 Simple two-dimensional cases

Solutions of the linear case of miscible displacement completely neglect the fact that D_{ik} is a *tensor*. As a matter of fact, it has already been stated (cf. sec. 8.2.2) that the dispersion even in an isotropic medium is anisotropic, the lateral dispersion (D_{lat}) being different from the longitudinal dispersion (D_{long}).

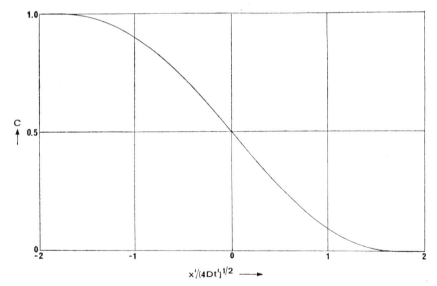

FIGURE 50 Shape of a concentration front during miscible displacement.

The simplest conditions which will account for this fact arise in one-dimensional convective flow in the x-direction, with dispersion in both the x and y directions, in an isotropic medium. For this case, equation (11.4.1.1) becomes

$$\frac{\partial C}{\partial t} = D_{\text{long}} \frac{\partial^2 C}{\partial x^2} + D_{\text{lat}} \frac{\partial^2 C}{\partial y^2} - v \frac{\partial C}{\partial x}. \tag{11.4.3.1}$$

Solutions of this equation have been calculated for various boundary conditions by Harleman and Rumer (1963), Ogata (1961), and Bruch and Street (1967). Thus, Harleman and Rumer (1963) considered a steady uniform flow in the x-direction with a pore velocity v through a porous medium contained between impermeable boundaries at $y = \pm\infty$, with the following boundary conditions:

$$\left. \begin{array}{ll} C(0, y) = C_0 & \text{for } -\infty < y \le 0, \\ C(0, y) = 0 & \text{for } 0 < y < +\infty, \\ \partial C/\partial y = 0 & \text{for } y = \pm\infty \text{ for all } x. \end{array} \right\} \tag{11.4.3.2}$$

Then equation (11.4.3.1) reduces to

$$v \, \partial C/\partial x = D_{\text{lat}} \, \partial^2 C/\partial y^2, \tag{11.4.3.3}$$

whose solution is

$$\frac{C}{C_0} = \frac{1}{2} \operatorname{erfc}\left(\frac{y}{2(D_{\text{lat}}x/v)^{\frac{1}{2}}}\right). \tag{11.4.3.4}$$

Bruch and Street (1967) solved equation (11.4.3.1) for somewhat more complicated boundary and initial conditions in the domain $x > 0, 0 \le y < n_0$, $t > 0$, viz.

$$\begin{array}{ll} C(0, y, t) = C_0 & \text{for } 0 \le |y| \le \varepsilon, \\ = 0 & \text{for } \varepsilon < |y| < n_0. \end{array} \tag{11.4.3.5}$$

$$\left. \begin{array}{l} \partial C(x, 0, t)/\partial y = 0, \\ \partial C(x, n_0, t)/\partial y = 0, \\ |C(\infty, y, t)| \text{ bounded,} \end{array} \right\} t > 0 \tag{11.4.3.6}$$

$$C(x, y, 0) = 0 \quad \text{for } x > 0, 0 \le y < n_0. \tag{11.4.3.7}$$

The solution was then

$$C(x, y, t) = \frac{\varepsilon C_0}{2n_0} \operatorname{erfc}\left(\frac{x - vt}{2(D_{\text{long}}t)^{\frac{1}{2}}}\right) + \frac{C_0\varepsilon}{2n_0} \exp\frac{v}{D_{\text{long}}} x \left(\operatorname{erfc}\frac{x + vt}{2(D_{\text{long}}t)^{\frac{1}{2}}}\right)$$

$$+ \frac{1}{2} \sum_{n=1}^{\infty} \left(F_n \cos\frac{n\pi}{n_0} y\right) \exp\left[\frac{1}{2}\left(\frac{v}{D_{\text{long}}} - J_n\right)x\right] \operatorname{erfc}\left(\frac{x - J_n D_{\text{long}}t}{2(D_{\text{long}}t)^{\frac{1}{2}}}\right)$$

$$+ \frac{1}{2} \sum_{n=1}^{\infty} \left(F_n \cos \frac{n\pi}{n_0} y \right) \exp \left[\frac{1}{2} \left(\frac{v}{D_{\text{long}}} + J_n \right) x \right] \text{erfc} \left(\frac{x + J_n D_{\text{long}} t}{2(D_{\text{long}})^{\frac{1}{2}}} \right), \quad (11.4.3.8)$$

where

$$J_n = \left[\left(\frac{v}{D_{\text{long}}} \right)^2 + \frac{4n^2\pi^2}{n_0^2} \frac{D_{\text{lat}}}{D_{\text{long}}} \right]^{\frac{1}{2}} \quad (11.4.3.9)$$

and

$$F_n = \frac{2C_0}{\pi n} \sin \frac{n\pi}{n_0} \varepsilon \quad \text{with } n = 1, 2, \ldots \quad (11.4.3.10)$$

11.4.4 Numerical methods

In general, it cannot be assumed that v in the fundamental equation (11.4.1.1) is constant, so that D as well as v cannot be taken out from under the differential signs and the fundamental equation becomes non-linear. Still, the technique is to determine the flow-field (v) from some equation of motion such as Darcy's law, and then to superpose the dispersion.

This can only be done by numerical methods. Thus, Peaceman and Rachford (1962) solved equation (11.4.3.1) by a numerical technique based on finite difference approximations, in which at each time step the distribution of pressure in the fluid and the flow field is calculated on a grid from Darcy's law. The new concentration distribution can then be calculated and the procedure continued. The solution can be streamlined by using the method of characteristics (Garder et al. 1964).

A thorough analysis and review of the numerical possibilities of solving equation (11.4.3.1) has been given by Stone and Brian (1963). A further review has been given by Shamir and Harleman (1967), who also proposed numerical techniques for the solution of equations (11.4.1.1) in two dimensions where D_{long} and D_{lat} are linear functions of v. Various cases were discussed, such as the two-well problem in an aquifer, for which breakthrough curves were calculated. An extension to three dimensions was proposed but not tested.

A specific application of the theory to salt water intrusion into a coastal aquifer was calculated by Pinder and Cooper (1970), using the method of characteristics in two dimensions.

It is clear that numerical methods offer the only hope for the solution of complicated two- or three-dimensional problems. However, the demands on machine capacity and time become rapidly very large.

11.4.5 Scale model experiments

As always when it is difficult to obtain solutions of a set of basic equations of motion, recourse is had to scale model experiments. For such experiments to be useful, the appropriate scaling relations have to be applied. The general procedure to deduce such relations has already been explained in this book in connection with the scaling of gravity flow (sec. 5.4.4). Using the analogous procedure, Bachmat (1967) deduced the scaling relations for dispersive flow.

As before, we denote again the ratio of a variable x in prototype and model by $R(x)$. Then the scaling conditions required the following:

(i) From geometrical similarity

$$R(L_1/L_2) = 1, \tag{11.4.5.1}$$

where L_1 and L_2 are any two lengths.

(ii) From kinematic similarity

$$R\left(\frac{qt}{PL}\right) = 1, \tag{11.4.5.2}$$

where all symbols have their usual meaning.

(iii) From dynamic similarity

$$R(p/(\rho gz)) = 1, \tag{11.4.5.3}$$

where z is the vertical coordinate;

$$R(\text{Re}^*) = 1, \tag{11.4.5.4}$$

where Re^* is a sort of macroscopic Reynolds number:

$$\text{Re}^* = \frac{q}{TP\mu} \frac{k}{Pa} \tag{11.4.5.5}$$

with a denoting the (lateral or longitudinal) dispersivity (cf. eq. 8.2.2.12) and the other symbols having their usual meaning; and

$$R\left((\mu/k)q/g\right) = 1. \tag{11.4.5.6}$$

(iv) From physicochemical similarity

$$R(\mu\beta/\rho) = 1, \tag{11.4.5.7}$$

$$R\left(\frac{\mu\rho_0}{\rho\mu_0}\right) = 1, \tag{11.4.5.8}$$

$$R(C/C_0) = 1, \tag{11.4.5.9}$$

where the index 0 denotes reference values and β is defined from the equation of state:

$$\mu = \mu_0 + \beta(C - C_0). \tag{11.4.5.10}$$

(v) From dispersive similarity

$$R(\text{Pe}^*) = 1, \tag{11.4.5.11}$$

where Pe* is a Peclet number:

$$\text{Pe}^* = qa/(PTD_m). \tag{11.4.5.12}$$

Here a is again the (longitudinal or lateral) dispersivity and D_m the molecular diffusivity of the solute in the solvent.

As is seen, the scaling relations constitute a severe limitation on modelling possibilities.

11.5
COMPARISON OF THE DISPERSION THEORY WITH EXPERIMENTS

11.5.1 *Principal remarks*

With the knowledge of typical solutions of the dispersion equations, the theory can be compared with experiments. There are several levels at which this can be done. The simplest measurements are those of longitudinal dispersion, and the first question which arises is then that of the validity of the dispersion theory. Experiments to test this will therefore be discussed first (sec. 11.5.2).

Next, it has been noted that dispersion, even in isotropic media, is not isotropic, so that one must distinguish between longitudinal and lateral dispersion. Experiments on this problem will be discussed next (sec. 11.5.3).

Finally, the dispersion theory predicts that D should depend on the pore (or seepage) velocity. This has also been tested by experiments, and the results will be given in section 11.5.4.

11.5.2 *Longitudinal dispersion*

It was noted very early (cf. section 11.3.1 on heuristic equations) that the dispersion equation describes the process of longitudinal dispersion correctly qualitatively. To test this, one generally uses a porous column in which the position of the outflow end is kept fixed. As input, either a slug or, more commonly, a step-displacement is used; the concentration is then measured at the outflow end as a function of time (Von Rosenberg 1956; Day 1956).

Although the dispersion equation does seem to give generally correct results if used to describe flow through porous media, Aronofsky and Heller (1957) noted in their discussion of the equation that published experimental tests appear to indicate that there are, in fact, minor deviations from the predicted values. Therefore, the writer (Scheidegger 1959) examined a series of experiments in order to test the validity of the dispersion equation.

These experiments were conducted on a linear system using fluids carefully matched as to viscosity and density. At the outflow end of the system, the capacity due to the dielectric property of the effluent fluids was monitored from which the composition could be calculated. In order to avoid any interference from a possible dependence of the factor of dispersion on the over-all displacement velocity, the velocity was carefully held constant. Assuming that the diffusivity equation *does* describe the displacement process correctly, the appropriate integral was obtained and the one unknown parameter in this integral (the factor of dispersion D) was calculated for each run by a least squares method.

After the best-fitting factor of dispersion had been determined for each displacement experiment, it was possible to use this value to calculate the *expected* value of concentration at each time step. This expected (or *calculated*) value for the effluent concentration can then be compared with the actually *observed* value.

It is to be anticipated that there will, in general, be deviations of the 'calculated' values from the 'observed' ones. The mean of the root-mean-square deviations turned out to be 0.0337, which must be considered as the standard error in displacement experiments of the type performed, if these are compared with the diffusivity equation. In turn, the value of 0.0337 can be considered as the standard error to be expected if the diffusivity equation is applied to a miscible displacement process.

However, a closer inspection of the deviations between calculated and observed values of the effluent concentration shows that the deviations are systematic. The signs of the deviations are of such a nature that the calculated values lie below the measured ones for small times; at the half-way mark the calculated curve always lies above the measured values, and at the end of the process it lies again below the observed concentrations. A typical curve is shown in figure 51.

It is difficult to speculate on the cause of these systematic deviations. It stands to reason that there is an effect occurring in the experiments which is not properly accounted for by the integral of the diffusivity equation on which the comparison is based. One possibility is that the boundary conditions used are not those that were prevalent in the experiments; for the calculations, it was assumed that the linear porous medium extends to plus and minus infinity, whereas in reality a constant concentration is injected at one point. However, this possibility has been discussed by Aronofsky and Heller (1957) and by Scheidegger and Larson (1958) and it has been shown that the deviation from the usual solution should be exceedingly small for average factors of dispersion. Moreover, the deviation should become the smaller, the smaller the factor of dispersion is, and it should also decrease with increasing length of the flow system. There was a lack of correlation between the observed deviations and either the factor of dispersion or the system length which, therefore, speaks against this possibility. It should be noted, though, that the *signs* of the deviations are in the proper direction.

All that can be said, therefore, is that experiments to date confirm the diffusivity equation up to a standard error of 0.03.

In consequence of the observation that the diffusivity equation describes longitudinal miscible displacement correctly qualitatively, various experiments have been performed to determine the relevant dispersion factors. Most of these experiments assume a priori that the diffusivity equation is the proper equation.

Thus, frequency-response techniques (Kramers and Alberda 1953), pulse-function methods (Yagi and Miyauchi 1955), and step-function methods (Lapidus and Amundson 1952) have been employed. A summary of these techniques has been given by Ebach and White (1958), and an evaluation of a set of published displacement data in terms of the corresponding dispersion factors has been provided by Aronofsky and Heller (1957). Table VIII lists some of the factors of dispersion as determined by Aronofsky and Heller.

TABLE VIII

Some factors of dispersion (after Aronofsky and Heller 1957)

v (ft/day)	D (cm²/sec)	k (darcy)
81	4×10^{-3}	0.4
500	2×10^{-2}	7.7
2	8.1×10^{-5}	7.2
16	7.7×10^{-4}	7.2
120	7.2×10^{-3}	7.2

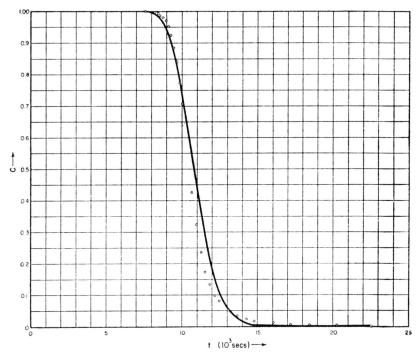

FIGURE 51   Comparison of calculated and measured values of the concentration C of the effluent in an experiment. The best-fitting calculated curve is shown as a black line, and the measured values as circles. (After Scheidegger 1959)

11.5.3 *Lateral dispersion*

The dispersion theory predicts that lateral dispersion may be different from longitudinal dispersion even in isotropic porous media. In fact, it was noted quite early by de Josselin de Jong (1956, 1958) that the transverse dispersion may be six to eight times smaller than the longitudinal dispersion.

Experiments on lateral dispersion are much more difficult to perform than those on longitudinal dispersion, because one is now dealing with a two-dimensional flow system. Studies have been made, for instance, by Ogata and Banks (1961), Simpson (1962), Harleman and Rumer (1963), Li and Lai (1966), List and Brooks (1967), and Bruch (1970). A typical set-up is shown in figure 52 (after Harleman and Rumer 1963) to which the mathematical solution given in equation (11.4.3.4) corresponds. The concentration, as a function of time, may be measured by conductivity probes, using a conducting and a non-conducting fluid as the two fluids. A typical result comparing the experimental data with the theoretical prediction (solid lines) is shown in figure 53.

11.5.4 *The factor of dispersion*

The dispersion theory predicts that the factor of dispersion should depend on the pore (or seepage) velocity v of the flow through the porous medium. As noted earlier, the various models yield (cf. sec. 8.5.1)

$$D \sim q^n \qquad (11.5.4.1)$$

with n between 1 and 2.4.

Some experiments to test this relation have already been mentioned in section 8.5.1. We append here, in figure 54, the results of a typical test run for the

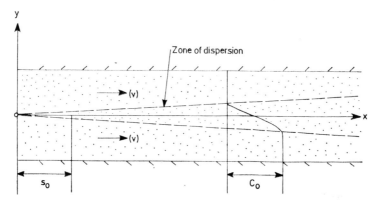

FIGURE 52   Sketch of a lateral dispersion experiment (after Harleman and Rumer 1963).

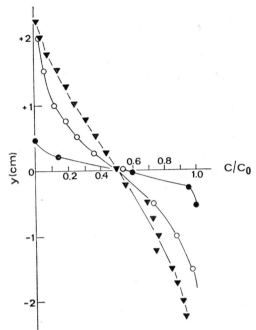

FIGURE 53   Concentration profiles for a lateral
dispersion test with $v=0.0635$ cm/sec. Full circles,
$x=5$ cm; open circles, $x=151.5$ cm; triangles,
$x=273.5$ cm. (After Harleman and Rumer 1963)

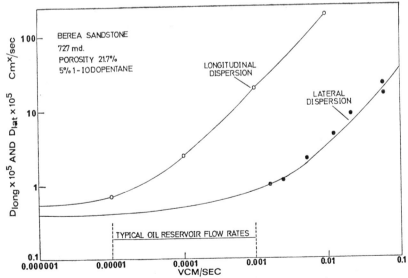

FIGURE 54   Coefficients of longitudinal and lateral dispersion for a sandstone at
various flow rates (after Crane and Gardner 1961).

determination of D_{lat} and D_{long} as a function of pore velocity for Berea Sandstone (after Crane and Gardner 1961). Similar results have been found for glass beads and other unconsolidated media.

At very low velocities the longitudinal and lateral dispersion factors are almost equal to each other and to the apparent molecular diffusion coefficient (at $v = 0$, $D_{long} = D_{lat} = D_m$, where D_m is the apparent molecular diffusion coefficient). Thus at low velocities the mixing is controlled by diffusion. Lateral mixing is controlled by diffusion up to greater velocities than is longitudinal mixing. At high velocities, D_{long} is much greater than D_{lat} and the straight-line sections indicate that D_{long} and D_{lat} are approximately proportional to v. This supports the statistical models of flow through porous media for which n in (11.5.4.1) is close to 1.

In consequence of the above, one makes the hypotheses

$$D_{long} = D_m + a_{long} v, \tag{11.5.4.2}$$

$$D_{lat} = D_m + a_{lat} v, \tag{11.5.4.3}$$

where a_{long} and a_{lat} are dispersivity coefficients with the dimension of a length that depend on the pore geometry. It was further shown (Raimondi et al. 1959) that

$$a_{long} = \beta_{long} \delta, \tag{11.5.4.4}$$

$$a_{lat} = \beta_{lat} \delta, \tag{11.5.4.5}$$

where δ is the average pore diameter. From a survey of the literature, Perkins and Johnston (1963) subsequently chose $\beta_{long} = 1.75$ and $\beta_{lat} = 0.055$ as representative values. Naturally, these values cannot be universally valid inasmuch as the shape and size distribution of the particles have been shown to affect the factor of dispersion. In general, the factor of dispersion increases as the particle sphericity decreases and wide particle size distributions will lead to increased dispersion. Graphical correlations of this have also been presented by Perkins and Johnston (1963).

Equations (11.5.4.2–3) imply that D_m is the residual diffusion for zero flow velocity. However, there is in fact a range of low flow velocities for which the molecular diffusion and convective dispersion have a varying degree of influence (curved stretches in fig. 53). This is the transition zone. Although the dispersion theory appears to predict factors of dispersion adequately at high velocities, (where convective dispersion predominates) or at low velocities (where diffusion effects predominate), there is as yet no satisfactory explanation applicable for the transition zone. From the data available in the literature, longitudinal dispersion appears to be controlled by diffusion for values of $v\delta/D_0 < 10^{-2}$ and by convection for values greater than 1, where D_0 is the true molecular diffusion coefficient (as opposed to the apparent D_m in the porous medium) and δ is again the pore diameter. The corresponding values for transverse dispersion are 1 and 100.

11.6
GENERALIZATIONS OF THE DISPERSION THEORY

11.6.1 *General remarks*

The dispersion theory, as it has been presented thus far, assumes laminar displacement in equal-viscosity miscible fluids. The basic theory can be generalized in
various ways.

First, there is the possibility of changing the fundamental statistical assumptions to allow for autocorrelation between individual time steps. Second, flow
regimes other than laminar ones can be considered. Finally, the possibility of
having different mobilities in the two miscible fluids leads to peculiar problems;
in particular, the possibility exists that instabilities arise at the displacement front.

We shall discuss these various generalizations in their turn below.

11.6.2 *The effect of autocorrelation*

We have seen in section 8.2.3 that the velocity dispersion theory of flow through
porous media can be modified by allowing for autocorrelation between the
individual time steps. The fundamental 'dispersion' equation then becomes, in
the linear case, in moving coordinates x', t' (cf. 8.2.3.38)

$$\frac{\partial C}{\partial t'} = D\frac{\partial^2 C}{\partial x'^2} - A\frac{\partial^2 C}{\partial t'^2}.$$ (11.6.2.1)

In terms of fixed coordinates this yields

$$\frac{\partial C}{\partial t} + v\frac{\partial C}{\partial x} = D\frac{\partial^2 C}{\partial x^2} - A\left(v^2\frac{\partial^2 C}{\partial x^2} + 2v\frac{\partial^2 C}{\partial x\partial t} + \frac{\partial^2 C}{\partial t^2}\right).$$ (11.6.2.2)

The quantities A, D, v depend on other dynamic parameters such as the pressure
drop according to the statistical model that is chosen. However, if an 'over-all
steady state' is considered, i.e., a state in which the total flow velocity is steady,
then these various parameters are constants.

We turn first to the linear progress of a slug through the porous medium. In
this case, one has to solve equation (11.6.2.1) for the boundary condition

$$C(x', 0) = m_0\delta(x')$$ (11.6.2.3)

(cf. discussion in connection with equation 11.4.2.5). The solution for this
problem has been given by Goldstein (1951); it is

$$C(x_0't) = \frac{e^{-t/(2A)}}{4Au}\left(I_0(Y) + \frac{t}{2A}\frac{I_1(Y)}{Y}\right) \quad \text{for } |x'| < ut$$
$$= 0 \qquad\qquad\qquad\qquad\qquad\qquad \text{for } |x'| > ut$$ (11.6.2.4)

with

$$Y = (u^2 t^2 - x'^2)^{\frac{1}{2}}/(2Au),$$
(11.6.2.5)

$$D = Au^2, \quad u = \sqrt{D/A},$$
(11.6.2.6)

$$x' = x - vt.$$
(11.6.2.7)

Here I_0, I_1 are Bessel functions with imaginary argument:

$$I_0(x) \equiv J_0(ix) = \sum_{m=0}^{\infty} \frac{(\frac{1}{2}x)^{2m}}{(m!)^2},$$
(11.6.2.8)

$$I_1(x) \equiv -iJ_1(ix) = \sum_{m=0}^{\infty} \frac{(\frac{1}{2}x)^{2m+1}}{m!(m+1)!}.$$
(11.6.2.9)

In order to evaluate the above formulas, let us set

$$\xi = x'/(2Au) = x'/(2\sqrt{AD})$$
(11.6.2.10)

and

$$\tau = t/(2A).$$
(11.6.2.11)

Then the solution becomes

$$\left.\begin{array}{ll} C = \frac{1}{2}e^{-\tau}[I_0(\sqrt{\tau^2 - \xi^2}) + \tau\{I_1(\sqrt{\tau^2 - \xi^2})\}/\sqrt{\tau^2 - \xi^2}] & \text{for } \xi < \tau \\ = 0 & \text{for } \xi > \tau \end{array}\right\}, \quad (11.6.2.12)$$

which is also properly normalized. The values of C have been calculated for a representative number of arguments, using the tables of functions of Jahnke and Emde (1945). The corresponding curves are shown in figure 55. The chief feature of the present case is the sharp cut-off of the distribution curve at early times. The cut-off travels with the 'wave velocity' u (referred to the mean position $x' = 0$) through the porous medium.

We turn our attention now to the progress of an initially sharp front through the porous medium. This case can immediately be calculated from the solution for the progress of a slug by superposition (cf. eq. 11.4.2 7)

Thus, inserting the expression (11.6.2.4) for σ into (11.4.2.7) yields

$$C(x', t') = \frac{1}{2} - \int_0^{x'} \frac{e^{-t'/(2A)}}{4Au} \left[I_0\left(\frac{\sqrt{u^2 t'^2 - \eta^2}}{2Au}\right) + \frac{t'}{2A} \frac{I_1\left(\frac{\sqrt{u^2 t'^2 - \eta^2}}{2Au}\right)}{\frac{\sqrt{u^2 t'^2 - \eta^2}}{2Au}} \right] d\eta,$$
(11.6.2.13)

where all the symbols have the meaning defined earlier. This integral has been evaluated by Goldstein (1951); the result is

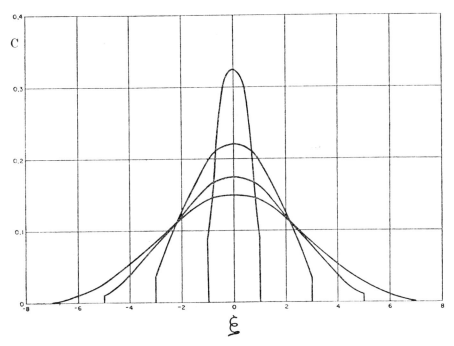

FIGURE 55   Shape of a slug passing through a porous medium if autocorrelation is taken into account (after Scheidegger 1958a).

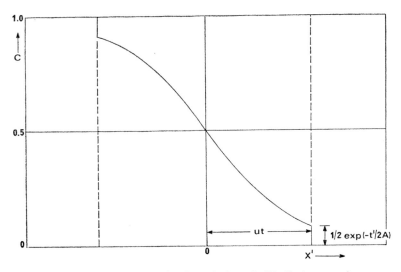

FIGURE 56   Shape of a concentration front during miscible displacement in a porous medium with autocorrelation (after Scheidegger 1958a).

$$C(x', t') = \tfrac{1}{2} e^{-t'/(2A)} \left[I_0 \left(\frac{\sqrt{u^2 t'^2 - x'^2}}{2Au} \right) + 2 \sum_{n=1}^{\infty} \left(\frac{ut' - x'}{ut' + x'} \right)^{n/2} \right.$$

$$\left. \times I_n \left(\frac{\sqrt{u^2 t'^2 - x'^2}}{2Au} \right) \right] \quad \text{for } 0 < x' < ut',$$

$$= 0 \qquad\qquad\qquad \text{for } x' > ut', \qquad (11.6.2.14)$$

and corresponding values for $x' < 0$.

The characteristic feature of this solution is again the sharp cut-off at $x' = ut'$. Thus, the cut-off is again travelling with the 'wave velocity' u into the porous medium. The general picture of the solution is as shown in figure 56.

If the above solutions are compared with the experimental results adduced in section 11.5, it turns out that autocorrelation, under ordinary circumstances, appears to be unimportant.

11.6.3 *General flow equations*

As was the case in single phase flow and immiscible multiple phase flow, it seems desirable to investigate flow regimes other than the laminar one also in miscible multiple phase flow.

It stands to reason that the extension of the dispersion theory from the laminar flow regime to other regimes is very simple. In the construction of any statistical theory, we have noted (cf. sec. 8.1.2) that a choice of the microdynamics is involved. The concept of dispersion is independent of the microdynamics so that the latter can be changed at will. The various flow regimes are expressed primarily by the choice of the microdynamics, which determines only the mean flow behaviour of the displacement upon which the dispersion is superposed. The *value* (but not the occurrence) of the factor of dispersion may also depend on the microdynamic assumptions as has already been found in the laminar flow regime (cf. the possible occurrence of 'geometrical' or 'dynamic' dispersivity).

With regard to various possible flow regimes we may note the following:

(*a*) *High velocities* The relevant equations may be taken over from section 7.2.3(*e*). These equations simply show that dispersion should be superimposed upon the mean flow which obeys a suitable high-velocity (non-linear) flow equation.

The description of mixing in turbulent flow in a porous medium by means of a diffusivity equation was proposed by Bernard and Wilhelm (1950) simply because such an equation applies to the mixing process in turbulent bulk masses of fluids. Later, experiments were reported, for example, by Carberry and co-workers (Carberry 1958; Carberry and Bretton 1958), Klinkenberg and Sjenitzer (1956), Aris and Amundson (1957), and McHenry and Wilhelm (1957), which seem to confirm the theory. In the last two papers the authors report that in the high flow regime the dispersion may also be anisotropic so that the factor of dispersion may have to be treated as a tensor rather than as a scalar.

(b) *Molecular effects*   There are generally two types of molecular effects which one considers as relevant in flow through porous media: gas slippage and adsorption. With regard to gas slippage, the relevant equation combining dispersion with the Knudsen flow regime is (Scheidegger 1955)

$$\partial C/\partial t = \text{lap}(DC) + \text{div}[(c_1 + c_2/p) \, C \, \text{grad} \, p], \tag{11.6.3.1}$$

where D is a dispersion factor and c_1 and c_2 are constants. This equation denomstrates that flow through porous media in the molecular regime can indeed be described by the flow through an assemblage of capillaries in each of which Knudsen flow occurs, but there is, in addition, a dispersion effect which originates from the statistical considerations.

A corresponding equation can be set up to account for the superposition of adsorption effects during flow dispersion (Ogata 1958). A series of breakthrough curves which exhibit these effects is shown in figure 57.

11.6.4 Unequal mobilities

In the dispersion theory discussed so far it has been supposed that the two flowing fluids have the same mobility. It is natural to ask how the theory would have to be modified if the two fluids were to have *unequal* mobility.

From the structure of the statistical theory it is obvious that only geometrical concepts enter into the argument. The dispersion is caused mainly by the interconnections of the flow channels in the porous medium, and only to a very minor

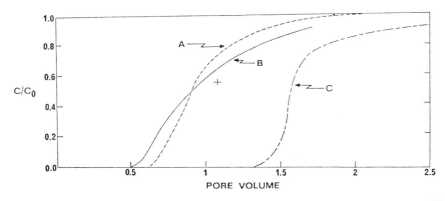

FIGURE 57 Examples of breakthrough curves which exhibit adsorption effects (after Elrick 1969).

	Medium	Tracer	Velocity (cm/hr)	Moisture content	Author
A	glass beads	Chloride	12.07	23%	Krupp and Elrick (1968)
B	1–2 mm aggregates in clay loam	Chloride	3.06	Saturated	Nielsen and Biggar (1962)
C	silt loam	2,4-D	0.39	Saturated	Elrick (1967)

extent by molecular diffusion. Hence unequal mobilities should not affect any of the predictions of the theory, except perhaps with regard to the residual molecular diffusion at extremely slow flow rates. This, of course, is only true as long as the displacement is not unstable (in the sense defined in sec. 9.4).

Some experiments on miscible displacement have been reported in the literature (e.g., Hall and Geffen 1957; Lacey et al. 1958; Slobod et al. 1959). The last of the papers mentioned presents some indications that the above predictions are correct, but, under adverse mobility conditions, instability (fingering) is observed which becomes the predominant phenomenon.

For the investigation of such instabilities, one can start with the general theory of section 9.4 of this book. However, the dispersion must now be superposed on the unstable displacement.

Just as in elementary displacement theory, there are several possibilities for trying to calculate the instability conditions.

First, one can introduce small perturbations in a stable displacement and investigate under what conditions they will grow. This approach was taken by Perrine (1961) and Wooding (1962). This corresponds to the usual procedure of hydrodynamic stability theory. The difficulty with this approach is that it only treats incipient fingers; the conditions for stabilization of developed fingers can, therefore, not be determined.

Second, a macrostatistical approach can be used corresponding to the method discussed in section 9.4.4, including, however, the effect of dispersion. This possibility was tried by Koval (1963), Dougherty (1963), and Perrine (1963). In all these cases one ends up with (non-linear) Buckley-Leverett type of equations, which have to be integrated numerically. This allows predictions of displacement behaviour.

Finally, direct attempts have been made to solve numerically the fundamental flow equations of miscible displacement for the floating boundary conditions obtaining at fingertips (Peaceman and Rachford 1962). The trouble with this approach is that, because of the non-linearity of the basic flow equations, no general statements can be made about the process. Each particular case that has been calculated represents a single experiment.

An evaluation of the various theories of instabilities in miscible displacement in porous media was made by Scheidegger (1965). Accordingly, it turns out that the macrostatistical theories leading to the Buckley–Leverett type equations are the most useful.

11.7
PRACTICAL APPLICATIONS OF DISPERSION THEORY

The dispersion theory has been applied to the explanation of various physical phenomena and to the prediction of events in practical engineering applications, as follows:

(a) To *the mixing of groundwater and salt water in permeable subsoils.* This is probably one of the oldest applications of miscible displacement theory. Originally, only the elementary displacement theory discussed in chapter 9 of this book was used (see sec. 9.2.3). An application of dispersion theory to this case has been discussed by Carrier (1958), who solved the diffusivity equation for the linear case and made some estimates and predictions with regard to the salinity distribution on Maui Island, Hawaii. Similar investigations have also been reported by Day (1956), Eriksson (1958), Cooper (1959), Rumer and Harleman (1963), Columbus (1965), and Pinder and Cooper (1970). Other applications to groundwater hydrology, viz. investigations on the mixing of waters in underground storage operations, were reported by Harpaz and Bear (1964).

(b) To *the hydrodynamic dispersion of solutes in the soil moisture stream,* which is a problem related to that of salt water intrusion. It has become of great importance in connection with the problem of the disposal of atomic wastes. It has been treated, for example, by Rifai et al. (1956) and Day and Forsythe (1957). The application of the linear case appears to yield a reasonable description of the observed phenomena. A related problem is that of the leaching of soil studied by Gardner and Brooks (1957). Much information on this problem will also be found in the textbooks on soil physics mentioned in the Introduction.

(c) To *the idea of recovery of oil by solvent floods* from underground reservoirs. This has been discussed in a great number of papers in the literature; reviews of the prevailing ideas have been given, for example, by Clark et al. (1957, 1958) and Craig (1970). As far as can be told, the dispersion theory seems to describe correctly the phenomena encountered.

Reviewing the above practical cases one may therefore say that the dispersion theory provides a satisfactory description of miscible displacement phenomena in porous media.

List of symbols*

a	radius of a tube; a constant; dispersivity
A	a constant; cross-sectional area; telegraph equation constant
b	a constant
B	a constant
c	a constant; velocity of shock; measure of correlation; normalized specific heat
C	concentration
C_p, C_v	specific heat at constant pressure or volume, respectively, of a gas
d	small distance; differential sign
D	coefficient of diffusion (and related entities); factor of dispersion
e	void ratio
E	modulus of elasticity
f	cumulative pore size distribution function; a force; a 'function of'; frequency
F	Helmholtz free energy; area in the Kozeny theory; porosity factor; formation factor
g	gravity acceleration; distribution function
$\mathbf{g}$	vector of gravity acceleration (pointing downward)
G	a constant; mass rate of flow (for gas)
h	height of free surface above datum; height, width of a system
H	hydraulic head; Hamiltonian function
$\mathfrak{H}$	first of Biot's constants
$J(s)$	Leverett function of saturation
k	permeability; Boltzmann's constant
k_r	relative permeability
K	a constant; modified Hankel function
L	mass rate of flow for liquid; length
lap	the Laplace operator
m	volume-to-surface ratio of capillary; a constant; Poisson's ratio; thermal gradient; viscosity ratio
$\mathbf{m}$	unit vector (components m_i)
M	molecular weight

* Note that vectors are printed in bold-face type.

n	denotes 'a number'; ratio defined in a packing of spheres
$\mathbf{n}$	unit vector
N	counting number
p	fluid pressure; elemental probability
p_c	capillary pressure
P	porosity; probability
Pe	Peclet number
$q, \mathbf{q}$	seepage velocity (vector)
Q	volume flow rate of a fluid
r	radial polar coordinate; fractional flow as a function of saturation
R	gas constant; radius; ratio
$\mathfrak{R}$	second of Biot's constants
Ra	Rayleigh number
Re	Reynolds number
s	saturation; length of flow path; arc length
S	specific internal area (per unit bulk volume); total internal area; entropy
S_0	internal area per unit solid volume
S_p	internal area per unit pore volume
t	time
T	tortuosity
$\mathscr{T}, \mathscr{T}_{ik}$	stress; stress tensor
$\mathfrak{T}$	absolute temperature
v	local fluid velocity
V	volume; local fluid velocity; pore velocity
w	a probability function
W	work, energy per unit surface; equivalent pressure
x	a length
x, y, z	Cartesian coordinates (z vertical, upward)
y	adsorbate surface density
Z	partition function
α	differential pore size distribution; a constant; storage coefficient
β	relative pressure in adsorption experiments; a constant; compressibility
γ	surface tension
δ	a microscopic diameter such as the pore diameter; Dirac's delta-function
δ_{ik}	Kronecker symbol ($=1$ for $i=k$, $=0$ for $i \neq k$)
Δ	symbol denoting finite difference
ε	proportionality factor; small quantity
ε_{ik}	strain tensor
ζ	zeta potential
η	mobility ratio

θ	contact angle; angle of spatial direction; volumetric liquid content
ϑ	liquid ratio
κ	sharpness factor; consolidation factor; coefficient of compaction; seepage coefficient
κ_m	thermal diffusivity
λ	mean free path length of molecules; friction factor; thermal coefficient of cubical expansion
λ'	the first constant of Lamé
μ	viscosity
μ'	the second constant of Lamé
ν	kinematic viscosity
ξ_α	parameters in finite strain theory
ρ	mass-density; resistivity
σ	a constant; standard deviation; solution for a slug
τ	shear stress; time interval
Σ	interfacial area; summation sign
φ	stream function; angular polar coordinate; phi-units
ϕ	force-potential in fluid flow $= Hg$ if H is the hydraulic head
Φ	various potential functions; 'a function of'
χ	$= \displaystyle\int_{p_0}^{p} \rho\,dp$
ψ	velocity potential in fluid flow; probability function
Ψ	capillary potential; moisture potential
ω	number of adsorbed molecules per unit area
Ω	solid angle; overburden potential

Bibliography*

INTRODUCTION

ARAVIN, V.I., and NUMEROV, S.N. 1965. Theory of fluid flow in undeformable porous media. Israel Program for Scientific Translations, Jerusalem.

BEAR, J., ZASLAVSKY, D., and IRMAY, S. 1968. Physical principles of water percolation and seepage. UNESCO, Paris.

CARMAN, P.C. 1956. The flow of gases through porous media. Academic Press, New York.

CEDERGREN, H.R. 1967. Seepage, drainage and flow nets. Wiley, New York.

CHILDS, E.C. 1969. An introduction to the physical basis of soil water phenomena. Wiley, New York.

COLLINS, R.E. 1961. Flow of fluids through porous materials. Reinhold, New York.

DE WIEST, R.J.M. 1965. Geohydrology. Wiley, New York.

– (ed.). 1969. Flow through porous media. Academic Press, New York.

ENGELUND, F. 1953. On the laminar and turbulent flows of ground water through homogeneous sand. Trans. Danish Acad. Tech. Sci., no. 3.

HAPPEL, J., and BRENNER, H. 1965. Low Reynolds-number hydrodynamics. Prentice-Hall, Englewood Cliffs, N.J.

HARR, M.E. 1962. Groundwater and seepage. McGraw-Hill, New York.

HOUPEURT, A. 1957. Eléments de mécanique des fluids dans les milieux poreux. Inst. Franç. de Pétrole, Paris.

KIRKHAM, D., and POWERS, W.L. 1969. Advanced soil physics. Wiley-Interscience, New York.

LEÍBENZON, L.S. 1947. Dvizhenie prirodnӯkh zhidkosteǐ i gazov v poristoi srede. Gosudarstv. Izdat. Tekh.-Teoret. Lit., Moscow and Leningrad.

LYKOW, A.W. 1958. Transporterscheinungen in kapillarporösen Körpern. Akademie-Verlag, Berlin.

MATHERON, G. 1967. Eléments pour une théorie des milieux poreux. Masson, Paris.

MUSKAT, M. 1937. The flow of homogeneous fluids through porous media. 1st ed. Edwards, Ann Arbor. (2nd prtg., 1946).

– 1949. Physical principles of oil production. McGraw-Hill, New York.

OROVEANU, T. 1963. Scurgerea fluidelor prin medii poroase neomogene. Acad. Rep. Pop. Romine, Bucharest.

POLUBARINOVA-KOCHINA, P.YA. 1952. Teoriya dvizheniya gruntovӯkh vod. Gosudarstv. Izdat. Tekh.-Teoret. Lit., Moscow. (English translation by R. de Wiest: Theory of groundwater movement. Princeton University Press, 1962).

* Slavonic names have been transliterated for this bibliography according to the British (Cambridge) system; this is the system used, for instance, for *Physics Abstracts*.

SCHEIDEGGER, A.E. 1960. Hydrodynamics in porous media. Encyclopedia of Physics, vol. 9. Springer, Berlin.

– 1966. Flow through porous media. *In* Applied mechanics surveys, ed. H.N. Abramson, H. Liebowitz, J.M. Crowley, and S. Juhasz, pp. 893–900. Spartan Books, Washington, D.C.

SCHOELLER, H. 1962. Les eaux souterraines. Masson, Paris.

TODD, D.K. 1959. Groundwater hydrology. Wiley, New York.

VON ENGELHARDT, W. 1960. Der Porenraum der Sedimente. Springer, Berlin.

YIH, C. 1965. Dynamics of nonhomogeneous fluids. Macmillan, New York.

ZIMENS, K.E. 1944. Poröse Stoffe, Kennzeichnung, Herstellung und Eigenschaften. Handbuch der Katalyse, vol. 4. Springer, Berlin.

CHAPTER 1

ACKERMAN, A.S.E. 1945. Nature *155*, 82.

ANDERSON, E.M. 1942. The dynamics of faulting and dyke formation with applications to Britain. Oliver & Boyd Ltd., Edinburgh.

ARCHIE, G.E. 1942. Trans. AIME *146*, 54.

ARTHUR, G. 1956. J. Inst. Metals *84*, 327.

ATHY, L.F. 1930. Bull. Amer. Assoc. Petrol. Geol. *14*, 1.

AVGUL, N.N. et al. 1951. Dokl. Akad. Nauk SSSR *76*, 855.

BABCOCK, A.B. 1945. Intern. Sugar J. *47*, 209.

BALLARD, J.H., and PIRET, E.L. 1950. Ind. Eng. Chem. *42*, 1088.

BEESON, C.M. 1950. Trans. AIME *189*, 313.

BELL, J.W. 1944. Trans. Can. Inst. Min. Met. *47*, 324.

BERNARD, R.A., and WILHELM, R.H. 1950. Chem. Eng. Progr. *46*, 223.

BIOT, M.A. 1941. J. Appl. Phys. *12*, 155.

BITTERLICH, W., and WÖBKING, H. 1970. Z. Geophys. *36*, 607

BOND, R.L., et al. 1950. Fuel *29*, (4), 83.

BRANDT, H. 1955. J. Appl. Mech. *22*, 479.

BRÖTZ, W., and SPENGLER, H. 1950. Brennstoff-Chemie *31*, 97.

BRUSSET, H. 1948. C.R. Acad. Paris *227*, 843.

BUSBY, T.S. 1950. J. Soc. Glass Technol. *34*, 10.

CADLE, R.D. 1955. Particle size determination. Interscience, New York.

CAQUOT, A., and KERISEL, J. 1956. Traité de mécanique des sols. 3me éd. Gauthier-Villars, Paris.

CARMAN, P.C. 1938. J. Soc. Chem. Ind. *57*, 225.

CATTANEO, C. 1938. Rend. Accad. Lincei (6) *27*, 342, 434, 474.

CHALKLEY, H.W., et al. 1949. Science *110*, 295.

CLARK, G.L., and LIU, C.H. 1957. Anal. Chem. *29*, 1539.

CLEARY, J.M. 1958. Ill. State Geol. Survey Circ. 252.

CLOUD, W.F. 1941. Oil Weekly *103*, (8), 26.

CORTE, H. 1955. Z. Erzeug. Holzst., Zellst., Papier und Pappe *9*, 289.

– 1962. Das Papier, *Oct. 1962*, 575.

CORTE, H., and KALLMES, O.J. 1962. Trans. Oxford Symposium on Form and Structure of Paper *2*, 13.

CROSS, A.H.B., and YOUNG, P.F. 1948. Trans. Brit. Ceram. Soc. *47*, 121.

DALLA VALLE, J.M. 1948. Micromeritics. Pitman, New York.

DALLMANN, H. 1941. Mühlenlab. *11*, 33.

DERESIEWICZ, H. 1958a. Appl. Mech. Revs. *11*, 259.

– 1958b. Adv. Appl. Mech. *5*, 233.

-- 1958c. Amer. Soc. Mech. Eng. Paper 57-A-90.
DONOGHUE, J.K. 1956. Brit. J. Appl. Phys. 7, 333.
DROTSCHMANN, C. 1943. Batterien 11, 207.
DUFFY, J.A. 1957. Tech. Rep. No. 3-Nonr 562(14), Brown Univ., Providence.
DUFFY, J.A., and MINDLIN, R.D. 1957. J. Appl. Mech. 24, 585.
DU MOND, J.W.M. 1947. Phys. Rev. 72, 83.
EICHLER, B. 1950. J. Soc. Glass. Technol. 34, 17.
EULER, F. 1957. J. Appl. Phys. 28, 1342.
FARA, H., and SCHEIDEGGER, A.E. 1961. J. Geophys. Res. 66, 3279.
FATT, I. 1957. J. Appl. Mech. 24, 148.
- 1958. J. Petrol. Tech. 10 (3), 64.
FERRANDON, J. 1950. Ann. Inst. Tech. Bat. Trav. Publics 145, 83.
FOORD, S.G. 1945. Nature 155, 427.
FRANKLIN, A.D., et al. 1953. J. Appl. Phys. 24, 1940.
FRICKE, H. 1931. Physics 1, 106.
FROMM, H. 1948. Optik 3, 137.
GAITHER, A. 1953. J. Sediment. Petrol. 23, 180.
GASSMANN, F. 1951. Viertelj. Natf. Gesellsch. Zürich 96, 1.
GEERTSMA, J. 1957. Trans. AIME 210, 331.
GERSEVANOV, N.M. 1933. Osnovy̆ dinamiki gruntovŏi massy̆. Gosstroĭizdat, Moscow.
GILCHRIST, J.D., and TAYLOR, J. 1951. J. Inst. Fuel 24, 207.
GORODETSKIĬ, O.S. 1940. Ogneupory̆ 8, 468.
GRACE, H.P. 1953. Chem. Eng. Progr. 49, 303.
GRATON, L.C., and FRASER, H.J. 1935. J. Geol. 43, 785.
GRIFFITHS, J.C. 1952. Bull. Amer. Assoc. Petrol. Geol. 36, 205.
GUMPRECHT, R.O., and SLIEPCEVICH, C.M. 1953. J. Phys. Chem. 57, 90.
HALL, H.N. 1954. Trans. AIME 198, 309.
HAWKSLEY, P.G.W. 1951. Bull. Brit. Coal Util. Res. Assoc. 15, 105.
HEDBERG, H.D. 1936. Amer. J. Sci. 31, 241.
HERTZ, H. 1895. Gesammelte Werke 1, 155 (Leipzig).
HEYWOOD, H. 1938. Inst. Mech. Eng., Nov. 1938.
- 1947. Nature 159, 717.
HIRSCH, P.B. 1954. Brit. J. Appl. Phys. 5, 257.
HOFSÄSS, M. 1948. Gas- und Wasserfach 89, 139.
HRUBÍŠEK, J. 1941. Kolloid-Beihefte 53, 385.
HUBBERT, M.K., and WILLIS, D.G. 1957. Trans. AIME 210, 153.
HUGHES, D.S., and COOKE, C.E. 1953. Geophysics 18, 298.
HUTTO, F.B. 1957. Chem. Eng. Progr. 53 (7), 328.
IMBT, W.C., and ELLISON, S.P. 1946. A.P.I. Drill. Prod. Pract. 1946, 364.
JARRETT, B.A., and HEYWOOD, H. 1954. Brit. J. Appl. Phys. Supp. 3, 521.
JEFFREYS, Sir H. 1931. Cartesian tensors. Cambridge University Press, London.
JOGLEKAR, G.D., and MARATHÉ, B.R. 1958. J. Sci. Ind. Res. 17B (6), 232.
KALLMES, O., and CORTE, H. 1960. Tappi 43 (9), 737.
KALLMES, O., CORTE, H., and BERNIER, G. 1961. Tappi 44 (7), 519.
KAYE, E., and FREEMAN, M.F. 1949. World Oil 128 (2), 24.
KIESSKALT, S., and MATZ, G. 1951. Z. Ver. deut. Ing. 93, (3), 58.
KING, J.G., and WILKINS, E.T. 1944. Proc. Conf. Ultra-fine Struct. Coals and Cokes, Brit. Coal Util. Res. Assoc., 46.
KISTLER, S.S. 1942. J. Phys. Chem. 46, 19.
KNÖLL, H. 1940. Kolloid-Z. 90, 189.
KRUMBEIN, W.C., and PETTIJOHN, F. 1938. Manual of sedimentary petrography. Appleton-Century, London.

LAKIN, J.R. 1951. Trans. Brit. Ceram. Soc. *50*, 208.

LEĬBENZON, L.S. 1947. See Introduction.

LEHMANN, H., and NÜDLING, H.D. 1954. Tonind.-Ztg. u. keram. Rdsch. *78*, 312.

LENZ, F. 1954. Optik *11*, 524.

LEPINGLE, M. 1945. Chaleur et ind. *26*, 101.

LIEBOWITZ, H. (ed.). 1969. Fracture, an advanced treatise in seven volumes. Academic Press, New York.

LOCKE, L.C., and BLISS, J.E. 1950. World Oil *131* (4), 206.

LOCKWOOD, W.N. 1950. Bull. Amer. Assoc. Petrol. Geol. *34*, 2061.

LUK'YANOVICH, V.M., and RADUSHKEVICH, L.V. 1953. Dokl. Akad. Nauk SSSR *91*, 585.

LYNOVSKIĬ, O.P., and POSTNIKOVA, E.N. 1940. Voenno-Sanit. 8–9, 99.

MAKSIMOVICH, G.K. 1957. Girdravlicheskiĭ razrȳv plastov. Gostoptekhizdat, Moscow.

MANEGOLD, E. 1937a. Kolloid-Z. *80*, 253.

– 1937b. Kolloid-Z. *81*, 19.

MANEGOLD, E., and SOLF, K. 1939. Kolloid-Z. *89*, 36.

MARSAL, D., and PHILIPP, W. 1970. Bull. Geol. Inst. Univ. Uppsala, N.S. *2* (7), 59.

MATHERON, G. 1967. See Introduction.

MC GEE, A.E. 1926. J. Amer. Ceram. Soc. *9*, 814.

MC LATCHIE, A.S., HEMSTOCK, R.A., and YOUNG, J.W. 1958. J. Petrol. Tech. *10* (6), 49.

MELCHER, A.F. 1921. Trans. AIME *65*, 469.

MILLIGAN, W.O., and ADAMS, C.R. 1954. J. Phys. Chem. *58*, 891.

MILLS, G.L. 1948. Nature *161*, 313.

MILNER, H.B. 1940. Sedimentary petrography. T. Murby & Co., London.

MINDLIN, R.D. 1949. J. Appl. Mech., Trans. ASME *71*, A259.

MINDLIN, R.D., and DERESIEWICZ, H. 1953. J. Appl. Mech. *20*, 327.

MITTON, R.G. 1945. J. Int. Soc. Leather Trades' Chem. *29*, 255.

MORAN, J.H. 1961. J. Geophys. Res. *67*, 2085.

MORGENSTERN, W. 1962. Geofis. Pura Appl. *52*, 104.

MÜLLER, H. 1947. Z. Naturf. *2a*, 473.

MURRELL, S.A.F. 1963. Proc. 5th Symposium on Rock Mechanics, Univ. Minnesota, p. 563.

MUSKAT, M. 1937. See Introduction.

NISHIDA, Y. 1956. Proc. Amer. Soc. Civ. Eng. *82*, SM3, Pap. 1027.

NISSAN, A.H. 1938. J. Inst. Petrol. Tech. *24*, 351.

NUSS, W.F., and WHITING, R.L. 1947. Bull. Amer. Assoc. Petrol. Geol. *31*, 2044.

OROWAN, E. 1949. Rep. Progr. Phys. *12*, 186.

OWEN, J.E. 1952. Trans. AIME *195*, 169.

PAGE, J.B. 1947. Proc. Soil Sci. Soc. Amer. *12*, 81.

PEERLKAMP, P.K. 1948. Landbouwkund. Tijdschr. *60*, 321.

PFANNKUCH, H.O. 1969. Trans. 1969 Haifa Symposium on Fundamentals of Transport Phenomena in Porous Media, p. 42. Ed. IAHR, Elsevier, Amsterdam, 1972.

POLAND, J.F., and DAVIS, G.H. 1969. Revs. Eng. Geol. *2*, 187.

POLLARD, T.A., and REICHERTZ, P.P. 1952. Bull. Amer. Assoc. Petrol. Geol. *36*, 230.

PRAMANIK, H.R. 1956. J. Instn. Eng. India *36* (8, Pt. 1), 1674.

PULPAN, H., and SCHEIDEGGER, A.E. 1965. J. Inst. Petrol. *51*, 169.

RALL, C.G., and TALIAFERRO, D.B. 1949. Producers Monthly *13* (11), 34.

RIM, M. 1952. Trans. Amer. Geophys. Un. *33*, 423.

RITTER, H.L., and ERICH, L.C. 1948. Anal. Chem. *20*, 665.

ROSENFELD, M.A. 1949. Producers Monthly *13* (7), 39.

RUSSELL, W.L. 1957. J. Petrol Tech. *9* (7), 51.

RYDER, H.M. 1948. World Oil *127* (13), 129.

SAUNDERS, H.L., and TRESS, H.J. 1945. J. Iron Steel Inst., advance copy, July 1945.
SCHEIDEGGER, A.E. 1956. Canad. J. Phys. *34*, 498.
‒ 1959. Canad. J. Phys. *37*, 276.
‒ 1960. Geologie und Bauw. *25*, 3.
SCHIFFMAN, R.L. 1970. J. Geophys. Res. *75* (20), 4035.
SCHMID, C. 1955. Producers Monthly *19* (8), 21.
SCHOFIELD, R.K., and TALIBUDDIN, O. 1948. Disc. Faraday Soc. *3*, 51.
SCHOPPER, J.R. 1966. Geophys. Prosp. *14*, 301.
SCHUMANN, H. 1944. Oel und Kohle *40*, 39.
SEDLÁČEK, B. 1956. Chem. Listy *50*, 304.
SEIL, G.E., et al. 1940. J. Amer. Ceram. Soc. *23*, 330.
SHAPIRO, I., and KILTHOFF, M. 1948. J. Phys. Coll. Chem. *52*, 1020.
SHULL, C.G., et al. 1948. J. Amer. Chem. Soc. *70*, 1410.
SKEMPTON, A.W. 1970. Quart J. Geol. Soc. London *125*, 373.
SLICHTER, C.S. 1899. U.S. Geol. Surv. 19th Ann. Rept. Pt. 2, 295.
SMITH, W.O., et al. 1929. Phys. Rev. *34*, 1271.
STAMM, A.J. 1931. Physics *1*, 116.
STEINBERG, A.R. 1946. J. Soc. Chem. Ind. *65*, 314.
STEVENS, A.B. 1939. AIME Tech. Paper 1061.
STULL, R.T., and JOHNSON, P.V. 1940. J. Res. Nat. Bur. Stand. *25*, 711.
TERZAGHI, K. 1945. Proc. Amer. Soc. Test. Mat. *45*, 777.
‒ 1951. Theoretical soil mechanics. Chapman and Hall, London.
THURSTON, C.W., and DERESIEWICZ, H. 1958. Tech. Rep. No. 27, Office of Naval
    Research, Nonr 266(09).
TICKELL, F.G., et al. 1933. Trans. AIME *103*, 254.
TILLER, F.M. 1953. Chem. Eng. Progr. *49*, 467.
TIMOSHENKO, S., and GOODIER, J.N. 1951. Theory of elasticity. McGraw-Hill, New York.
TOLLENAAR, D. and BLOCKHUIS, G. 1950. Appl. Sci. Res. *A2*, 125.
UREN, L.C. 1943. Petrol. Eng. *14* (9), 51.
VAN POOLLEN, H.K. 1957. Quart. Colo. School Mines *52* (3), 113.
VERBECK, G.J. 1947. J. Amer. Concrete Inst. *18*, 1025.
VON ENGELHARDT, W. 1960. See Introduction.
WALDO, A.W., and YUSTER, S.T. 1937. Bull. Amer. Assoc. Petrol. Geol. *21*, 259.
WASHBURN, W., and BUNTING, E.N. 1922. J. Amer. Ceram. Soc. *5*, 112.
WESTMAN, A.E.R. 1926. J. Amer. Chem. Soc. *48*, 311.
WIGGINS, E.J. et al. 1939. Canad. J. Res. *B17*, 318.
WISE, M.E. 1954. Phillips Res. Rep. *9*, 231.
WYLLIE, M.R.J., and ROSE, W.D. 1950. Trans. AIME *189*, 105.
WYLLIE, M.R.J., and SPANGLER, M.B. 1952. Bull. Amer. Assoc. Petrol. Geol. *36*, 359.
ZAVODOVSKAYA, E.K. 1937. Zavodsk. Lab. *6*, 1021.
ZIMENS, K.E. 1944. See Introduction.

CHAPTER 2

ADAM, N.K. 1941. The physics and chemistry of surfaces. 3rd ed., Oxford.
ADZUMI, H. 1937. Bull. Chem. Soc. Japan *12*, 292.
ARNELL, J.C. 1946. Canad. J. Res. *A24*, 103.
BARTELL, F.E., and SHEPARD, J.W. 1953. J. Phys. Chem. *57*, 211, 455, and 458.
BATCHELOR, G.K. 1953. The theory of homogeneous turbulence. Cambridge University
    Press, London.
BROWN, R.C. 1947. Proc. Phys. Soc. London *59*, 429.

BRUNAUER, S. 1943. The adsorption of gases and vapors. Princeton Univ. Press, Princeton, N.J.

BUCKINGHAM, E. 1921. Proc. Amer. Soc. Test. Mat. *21*, 1154.

BURDEN, R.S. 1949. Surface tension and the spreading of liquids. Cambridge Univ. Press, London.

CALDERWOOD, G.F.N., and MARDLES, E.W.G. 1955. Trans. Textile Inst. *46*, T161.

CASSIE, A.B.D. 1948. Disc. Faraday Soc. *3*, 11.

COMOLET, R. 1949. C.R. Acad. Paris *229*, 342.

DERYAGIN, B.V. et al. 1948. Dokl. Akad. Nauk SSSR *61*, 653.

DODGE, D.W., and METZNER, A.R. 1959. J. Amer. Inst. Chem. Eng. *5*, 189.

ELTON, G.A.H. 1951. J. Chem. Phys. *19*, 1066.

EMMETT, P.H. 1948. Adv. Catalysis and Rel. Subj. *1*, 65.

GURNEY, C. 1949. Proc. Phys. Soc. London *A62*, 639.

HARTMAN, R.J. 1947. Colloid chemistry. 2nd ed. Riverside Press, Cambridge, Mass.

JOOS, G. 1947. Theoretical physics. Blackie and Son, London.

KLOSE, W. 1931. Ann. Phys. *11*, 73.

KOENIG, F. 1950. J. Chem. Phys. *18*, 449.

– 1953. Z. Elektrochem. *57*, 361.

KONCAR-DJURDEVIC, S. 1953. Nature *172*, 858.

KNUDSEN, M. 1909. Ann. Phys. (4) *28*, 75.

KREITH, F., and EISENSTADT, R. 1956. Trans. ASME, Pap. 56–SA–15.

KUNDT, A., and WARBURG, E. 1875. Pogg. Ann. Phys. *195*, 337 and 625.

LAMB, H. 1932. Hydrodynamics. 6th ed. Cambridge Univ. Press, London.

LEVENSPIEL, O. 1958. Ind. Eng. Chem. *50*, 343.

LEVENSPIEL, O., and SMITH, W.K. 1957. Chem. Eng. Sci. *6*, 227.

MACLELLAN, A.G. 1952. Proc. Roy. Soc. *A213*, 274.

METZNER, A.B. 1956. Adv. Chem. Eng. *1*, 77.

MUSKAT, M. 1937. See Introduction.

OLDROYD, J.G. 1956. *In* Rheology, ed. O. Eirich, *1*, p. 653. Academic Press, New York.

PATTERSON, G.N. 1956. Molecular flow of gases. Wiley, New York.

POWELL, R.E., and EYRING, H. 1944. Nature *154*, 427.

PRIGOGINE, I. 1950. J. Chim. Phys. *47*, 33.

PRIGOGINE, I., and MARÉCHAL, A. 1952. J. Colloid Sci. *7*, 122.

PRIGOGINE, I., and SAROLÉA, L. 1952. J. Chim. Phys. *47*, 807.

RAY, B.R., and BARTELL, F.E. 1953. J. Coll. Sci. *8*, 214.

REINER, M. 1949. Twelve lectures on theoretical rheology. North-Holland, Amsterdam.

SCHAAF, S.A. 1956. Appl. Mech. Revs. *9*, 413.

SHUTTLEWORTH, R., and BAILEY, G.L. 1948. Disc Faraday Soc. *3*, 16.

STEVENS, W.E. 1953. Ph.D. Thesis in Chem. Eng., University of Utah, Salt Lake City.

TAYLOR, (Sir) G.I. 1953. Proc. Roy. Soc. *A219*, 186.

TOLLENAAR, D. 1954. Appl. Sci. Res. *A4*, 453.

TOPAKOGLU, C. 1951. Steady laminar flow of an incompressible fluid through a curved pipe of circular cross section (in Turkish). Pamphl. Istanbul Üniv.

VON ROSENBERG, D.U. 1956. J. Amer. Inst. Chem. Eng. *2*, 55.

CHAPTER 3

ADAM, N.K. 1948. Disc. Faraday Soc. *3*, 5.

ATTERBERG, A. 1918. Ber. Landw. Vers. Sta. *69*, 93.

BANGHAM, D.H. 1937. Trans. Faraday Soc. *33*, 805.

BARRETT, G.P., et al. 1951. J. Amer. Chem. Soc. *23*, 791.

BARTELL, F.E., and OSTERHOF, H.J. 1927. Ind. Eng. Chem. *19*, 1277.

BOBEK, J.E., MATTAX, C.C., and DENEKAS, M.O. 1958. J. Petrol. Tech. *10* (7), 155.

BOND, R.L., et al. 1948. Disc Faraday Soc. *3*, 29.

BOYER, R.L., et al. 1947. Trans. AIME *170*, 15.

BROAD, D.N., and FOSTER, A.G. 1945. J. Chem. Soc. 366.

BROWN, H.W. 1951. Trans. AIME *192*, 67.

BROWN, R.J., and FATT, I. 1956. Trans. AIME *201*, 262.

BRUCE, W.A., and WELGE, H.J. 1947. Oil and Gas J. *46* (12), 223.

BRUNAUER, S. 1943. See chapter 2.

BRUNAUER, S., et al. 1938. J. Amer. Chem. Soc. *60*, 309.

BUCKER, H.P., FELSENTHAL, M., and CONLEY, F.R. 1956. Trans. AIME *207*, 306.

CALHOUN, J.C., et al. 1949. Trans. AIME *186*, 189.

CAREY, J.W., and TAYLOR, S.A. 1967. Monogr. Amer. Soc. Agronomy *11*, Chap. 13.

CARMAN, P.C. 1941. Soil Sci. *52*, 1.

– 1951. Proc. Roy. Soc. *A209*, 69.

CARMAN, P.C., and RAAL, F.A. 1951. Proc. Roy. Soc. *A209*, 59.

CREMER, E. 1950. Z. Phys. Chem. *196*, 196.

DANEŠ, V. 1956. Collection Czech. Chem. Cummuns *21*, 1122.

DE BOER, J.H. 1953. The dynamical character of adsorption. Clarendon Press, Oxford.

– 1958. Angew. Chem. *70* (11), 346.

DERYAGIN, B.V. 1957a. Zhur. Fiz. Khim. *31*, 516.

– 1957b. Dokl. Akad. Nauk SSSR *113*, 842.

DONALDSON, E.C., THOMAS, R.D., and LORENZ, P.B. 1969. Soc. Petrol. Eng. J. *1969* (3), 13.

DRAKE, L.C. 1949. Ind. Eng. Chem. *41*, 780.

DUBININ, M.M. 1956. Zhur. Fiz. Khim. *30*, 1652.

DUBININ, M.M., and ZHUKOVSKAYA, E.G. 1956. Zhur. Fiz. Khim. *30*, 1840.

DUNNING, H.N., et al. 1954. Petrol. Eng. *26* (1), B82.

EMMETT, P.H. 1946. J. Amer. Chem. Soc. *68*, 1784.

– 1948. See chapter 2.

EMMETT, P.H., and BRUNAUER, S. 1937. Trans. Electrochem. Soc. *71*, 383.

EMMETT, P.H., and DE WITT, T.W. 1943. J. Amer. Chem. Soc. *65*, 1253.

EVERETT, D.H. 1950. Trans. Faraday Soc. *46*, 453.

FATT, I., and KLIKOFF, W.A. 1959. Preprint 62, A.I.Ch.E.-S.P.E. Symposium on Wetting and Capillarity in Fluid Displacement Processes, Kansas City.

FISHER, R.A. 1926. J. Agr. Sci. *16*, 492.

FOSTER, A.G. 1932. Trans. Faraday Soc. *18*, 645.

– 1934. Proc. Roy. Soc. *A147*, 128.

– 1945. J. Chem. Soc. *1945*, 769.

– 1948. Disc. Faraday Soc. *3*, 41.

FRASER, D.A.S. 1957. Nonparametric methods in statistics. Wiley, New York.

GATENBY, W.A., and MARSDEN, S.S. 1957. Producers Monthly *22* (1), 5.

GLASSTONE, S. 1946. Textbook of physical chemistry. 2nd ed. D. van Nostrand, New York.

GREGG, S.J. 1951. The surface chemistry of solids. Reinhold, New York.

HACKETT, F.E., and STRETTAN, J.S. 1928. J. Agr. Sci. *18*, 671.

HAINES, W.B. 1930. J. Agr. Sci. *20*, 97.

HALASZ, I. 1954. Magyar Kemikusok Lapja *9*, 93.

HALASZ, I., and SCHAY, G. 1956. Z. anorg. allg. Chem. *287*, 242.

HALASZ, I., SCHAY, G., and WENCKE, K. 1956. Z. anorg. allg. Chem. *287*, 253.

HARKINS, W.D., and JURA, G. 1943. J. Chem. Phys. *11*, 431.

– 1944a. J. Amer. Chem. Soc. *66*, 919.
– 1944b. J. Amer. Chem. Soc. *66*, 1362.
– 1944c. J. Amer. Chem. Soc. *66*, 1366.
– 1945. J. Chem. Phys. *13*, 449.
HARTMAN, R.J. 1947. See chapter 2.
HASSLER, G.L., and BRUNNER, E. 1945. AIME T.P. 1817.
HELLER, J.P. 1959. Rev. Sci. Instr. *30*, 1056.
– 1968. Proc. Soil Sci. Soc. Amer. *32*, 778.
HENDERSON, L.M., et al. 1940. Refiner *19*, 185.
HIRST, W. 1947. Nature *159*, 267.
HOLBROOK, O.C., and BERNARD, G.G. 1958. Trans. AIME *213*, 261.
IMELIK, B., and FRANÇOIS-ROSETTI, J. 1957. Bull. Soc. Chim. France *1957*, 153.
INNES, W.B. 1957. Anal. Chem. *29*, 1069.
JOYNER, L.G., et al. 1945. J. Amer. Chem. Soc. *67*, 2182.
JOYNER, L.G., et al. 1951. J. Amer. Chem. Soc. *73*, 3155.
KATZ, S.M. 1949. J. Phys. Coll. Chem. *53*, 1166.
KEENAN, A.G. 1948. J. Chem. Educ. *25*, 666.
LAIRD, A.D.K., and PUTNAM, J.A. 1951. Trans. AIME *192*, 275.
LANGMUIR, I. 1916. J. Amer. Chem. Soc. *38*, 2221.
LEVERETT, M.C. 1941. Trans. AIME *142*, 152.
LOISY, R. 1941. Bull. Soc. Chem. *8*, 589.
MARTIN, M., et al. 1938. Geophysics *3*, 258.
MATHEWS, D.H. 1957. Chem. & Ind. (London) *1957*, 465.
MCMILLAN, W.G., and TELLER, E. 1951. J. Phys. Coll. Chem. *55*, 17.
MEYER, H.I. 1953. J. Appl. Phys. *24*, 510.
MILLER, E.E., and MILLER, R.D. 1956. J. Appl. Phys. *27*, 324.
MOORE, T.F., and SLOBOD, R.L. 1956. Producers Monthly *20* (8), 20.
MORGAN, F., et al. 1950. Trans. AIME *189*, 183.
PECK, A.J., and RABBIDGE, R.M. 1966. Trans. UNESCO Symposium on Water in the Unsaturated Zone *1*, 165.
PHILIP, J.R. 1964. J. Geophys. Res. *69*, 1553.
– 1969a. Austral. J. Soil Res. *7*, 99.
– 1969b. Austral. J. Soil Res. *7*, 121.
– 1969c. Water Resources Res. *5* (5), 1070.
– 1970. Soil Sci. *109*, 294.
PLACHENOV, T.G., ALEKSANDROV, V.A., and BELOTSERKOVSKIĬ, G.M. 1953. Akad. Nauk SSSR, Trudȳ Soveshchaniya *1951*, 59.
POLANYI, M. 1920. Z. Elektrochem *26*, 370.
POLYAKOV, M.V. 1937. J. Phys. Chem. USSR *10*, 100.
POWERS, E.L., and BOTSET, H.G. 1949. Producers Monthly *13* (8), 15.
PURCELL, W.R. 1949. Trans. AIME *186*, 49.
RITTER, H.L., and DRAKE, L.C. 1945. Ind. Eng. Chem., Anal. Ed. *17*, 782.
ROSE, W. 1958. J. Appl. Phys. *29*, 687.
ROSE, W.D., and BRUCE, W.A. 1949. Trans. AIME *186*, 127.
RUBINSHTEĬN, A.M., and AFANAS'EV, V.A. 1956. Izv. Akad. Nauk SSSR, Otdel. Khim. Nauk *1956*, 1294.
SCHULTZE, K. 1925a. Kolloid-Z. *36*, 65.
– 1925b. Kolloid-Z. *37*, 10.
SHULL, C.G. 1948. J. Amer. Chem. Soc. *70*, 1405.
SIEGEL, S. 1956. Nonparametric statistics for the behavioral sciences. McGraw-Hill, New York.

SLOBOD, R.L., et al. 1951. Trans. AIME *192*, 127.

SMITH, W.O., et al. 1931. Physics *1*, 18.

STAHL, C.D., and NIELSEN, R.F. 1950. Producers Monthly *14* (3), 19.

STARKWEATHER, F.M., and PALUMBO, D.T. 1957. J. Electrochem. Soc. *104*, 287.

THORNTON, O.F., and MARSHALL, D.L. 1947. AIME T.P. 2126.

TOVBIN, M.V., and SAVINOVA, E.V. 1957. Ukrain. Khim. Zhur. *23*, 146.

VERSLUYS, J. 1917. Int. Mitt. Bodenk. *7*, 117.

– 1931. Bull. Amer. Assoc. Petrol. Geol. *15*, 189.

VON ENGELHARDT, W. 1955. Proc. Fourth World Petrol. Congr. *1*, 399.

WASHBURN, E.W. 1921. Proc. Nat. Acad. Sci. *7*, 115.

WATSON, A., MAY, J.O., and BUTTERWORTH, B. 1957. Trans. Brit. Ceram. Soc. *56*, 37.

WENZEL, R.N. 1938. Ind. Eng. Chem. *28*, 988.

WHALEN, J.W. 1954. Trans. AIME *201*, 203.

YUSTER, S.T., and STAHL, C.D. 1948. Producers Monthly *13* (2), 24.

ZSIGMONDY, R. 1911. Z. anorg. Chem. *71*, 356.

CHAPTER 4

ARAVIN, V.N., 1937. Trudy̅ Leningrad industr. in-ta; Rasdel. gidrotekh. *1937* (19), 2.

ARONOVICI, V.S., and DONNAN, W.W. 1946. Trans. Amer. Geophys. Un. *27*, 95.

BARENBLATT, G.I., and KRY̅LOV, A.I. 1955. Izv. Akad. Nauk SSSR, Otd. Tekh. Nauk *1955* (2), 5.

BIOT, M.A. 1941. See chapter 1.

– 1956. J. Appl. Phys. *27*, 459.

BODZIONY, J., and LITWINISZYN, J. 1962. Kolloid-Z. *185*, (2), 144.

BROWN, R.H. 1953. J. Am. Water Wks. Assoc. *45*, 844.

BROWN, R.L., and BOLT, R.H. 1942. J. Acoust. Soc. Amer. *13*, 337.

CARMAN, P.C. 1938. See chapter 1.

CHILDS, E.C., COLLIS-GEORGE, N., and HOLMES, J.W. 1957. J. Soil Sci. *8*, 27.

DARCY, H. 1856. Les fontaines publiques de la ville de Dijon. Dalmont, Paris.

DICKEY, G.D., and BRYDEN, C.L. 1946. Theory and practice of filtration. Reinhold, New York.

EASTMAN, J.E., and CARLSON, A.J. 1940. AIME T.P. 1196.

EMMERICH, A. 1954. Zucker *7*, 309.

EVANS, D.D., and KIRKHAM, D. 1950. Proc. Soil Sci. Soc. Amer. *14*, 65.

FATT, I. and DAVIS, D.H. 1952. Trans. AIME *195*, 329.

FERRANDON, J. 1948. Génie Civil *125*, 24.

FISHEL, V.C. 1935. Trans. Amer. Geophys. Un. *16*, 499.

FLORIN, V.A. 1948. Dokl. Akad. Nauk SSSR *59*, 219.

FROEHLICH, I.O.K. 1937. Mém. Ass. Int. Ponts Charpentes *5*, 133.

GARDNER, W., et al. 1934. Trans. Amer. Geophys. Un. *15*, 563.

GHIZETTI, A. 1949. Ann. Soc. Polon. Math. *22*, 195.

GRACE, H.P. 1953. Chem. Eng. Progr. *49*, 303.

GRIFFITHS, J.C. 1950. Producers Monthly *14* (8), 26.

HAZEBROOK, P., RAINBOW, H., and MATTHEWS, C.S. 1958. Trans. AIME *213*, 250.

HORNER, D.R. 1951. Proc. Third World Petrol. Congr. *2*, 504.

HUBBERT, M.K. 1940. J. Geol. *48*, 785.

HUBBERT, M.K., and WILLIS, D.G. See chapter 1.

HUTTA, J.J., and GRIFFITHS, J.C. 1955. Producers Monthly *19* (12), 24.

IRMAY, S. 1951. Proc. Assoc. Gén. Bruxelles; Assoc. Int. Hydrol. (UGGI) *2*, 179.

IWANAMI, I. 1940. Trans. Soc. Mech. Eng. Japan *6* (24), 4.

JACOB, C.E. 1940. Trans. Amer. Geophys. Un. *21*, 574.

JAEGER, J.C. 1959. J. Geophys. Res. *64*, 561.

JOHNSON, W.E., and BRESTON, J.N. 1951. Producers Monthly *15* (4), 10.

JOHNSON, W.E., and HUGHES, R.V. 1948. Producers Monthly *13* (1), 17.

KAWAKAMI, M. 1933. J. Chem. Soc. Japan *54*, 133.

KIRKHAM, D. 1955. Amer. Soc. Test. Mat. Spec. Publ. *163*, 80.

KOPPUIS, O.T., and HOLTON, W.G. 1938. Phys. Rev. *51*, 684.

LEÏBENZON, L.S. 1947. See Introduction.

LE ROSEN, A.L. 1942. J. Amer. Chem. Soc. *64*, 1905.

LITWINISZYN, J. 1950. Ann. Soc. Polon. Math. *22*, 185.

LOCKE, L.C., and BLISS, J.E. 1950. See chapter 1.

MAASLAND, M., and KIRKHAM, D. 1955. Proc. Soil Sci. Soc. Amer. *19*, 395.

MC LATCHIE, A.S., HEMSTOCK, R.A., and YOUNG, J.W. 1958. See chapter 1.

MC LAUGHLIN, J.F., and GOETZ, W.H. 1955. Highway Res. Bd. Proc. *34*, 274.

MANEGOLD, E. 1938. Kolloid-Z. *82*, 269; *83*, 146; *83*, 299.

MATHERON, G. 1966. Rev. Inst. Franç. Pétrole *21* (4), 564.

– 1967. Rev. Inst. Franç. Pétrole *22* (3), 443.

– 1968. Rev. Inst. Franç. Pétrole *23* (2), 201.

MEINZER, O.E. 1936. Bull. Amer. Assoc. Petrol. Geol. *20*, 704.

MEINZER, O.E., and FISHEL, V.C. 1934. Trans. Amer. Geophys. Un. *15*, 405.

MILLER, C.C., DYES, A.B., and HUTCHINSON, C.A. 1950. Trans. AIME *189*, 91.

MITTON, R.G. 1945. See chapter 1.

MUSKAT, M. 1937. See Introduction.

NUTTING, P.G. 1930. Bull. Amer. Assoc. Petrol. Geol. *14*, 1337.

ONODERA, T. 1958. Pub. Assoc. Int. Hydrologie (UGGI) *44*, 212.

PALLMANN, H., and DEUEL, H. 1945. Experientia *1*, 325.

PLUMMER, F.B., et al. 1934. AIME T.P. 578.

POLLARD, T.A., and REICHERTZ, P.P. 1952. See chapter 1.

PRESSLER, E.D. 1947. Bull. Amer. Assoc. Petrol. Geol. *31*, 1851.

RÜHL, W., and SCHMID, C. 1957. Geolog. Jahrb. *74*, 447.

RUTH, B.F. 1946. Ind. Eng. Chem. *38*, 564.

RUTH, B.F., et al. 1933. Ind. Eng. Chem. *25*, 153.

SCHAFFERNAK, F. 1933. Wasserwirtsch. *1933*, 399.

SCHEIDEGGER, A.E. 1954. Geofis. Pura Appl. *28*, 75.

– 1956. Geofis. Pura Appl. *33*, 111.

– 1959. Canad. J. Phys. *37*, 276.

SCHOELLER, H. 1955. Proc.-Verb. Soc. Sci. Phys. Nat. Bordeaux, Dec. 20, 1955.

– 1956. Publ. Assoc. Int. Hydrologie (UGGI) *41*, 67.

SCHWEIGL, K., and FRITSCH, V. 1942. Gas- und Wasserfach *83*, 481 and 501.

SECCHI, I.M. 1936. Chimica e industr. *18*, 514 and 563.

SHCHELKACHEV, V.N. 1946a. C.R. Acad. Sci. URSS *62*, 103.

– 1946b. C.R. Acad. Sci. URSS *52*, 203.

– 1946c. C.R. Acad. Sci. URSS *52*, 392.

– 1959. Razrabotka neftevodonosnỹkh plastov pri uprugom rezhime. Gostoptekhizdat, Moscow.

SHEKHTMAN, Y.M. 1961. Fil'tratsiya suspensii malykh kontsentratsii. USSR Academy of Sciences Press, Moscow.

SHESTAKOV, V.M. 1955. Razv. i Okhr. Nedr *1955* (6), 52.

SOUERS, R.C., and BINDER, R.C. 1952. Trans. Amer. Soc. Mech. Eng. *74*, 837.

STULL, R.T., and JOHNSON, P.V. 1940. See chapter 1.

SULLIVAN, R.R. 1941. J. Appl. Phys. *12*, 503.
TERLETSKAYA, M.N. 1954. Girdrotekh. Stroït. *23* (2), 37.
TERZAGHI, K. 1951. See chapter 1.
TILLER, F.M. 1953a. Pap. Sympos. Ind. Eng. Chem. Dec. 1953, Ann Arbor, Mich.
– 1953b. Chem. Eng. Progr. *49* (9), 467.
– 1955. Chem. Eng. Progr. *50* (6), 282.
TRZASKA, A. 1966. Bull. Acad. Polon. Sci., Ser. Sci. Techn. *14* (7), 433.
VIBERT, A. 1939. Génie Civil *115*, 84.
VREEDENBURGH, C.G., and STEVENS, O. 1936. Int. Conf. Soil Mech. Found. Eng.
WIGGINS, E.J., et al. 1939. See chapter 1.
WILKINSON, W.B., and SHIPLEY, E.L. 1969. Trans. 1969 Haifa Symposium on Fundamentals of Transport Phenomena in Porous Media, p. 285. Ed. IAHR, Elsevier, Amsterdam, 1972.
WYCKOFF, R.D., et al. 1933. Rev. Sci. Instr. *4*, 394.

CHAPTER 5

ALESKEROV, S.A., and MAKHMUDOV, A.YU. 1955. Izv. Akad. Nauk Azerb. SSR *1955* (8), 3.
ARAVIN, V.I., and NUMEROV, S.N. 1965. See Introduction.
ARONOFSKY, J.S., and FERRIS, O.D. 1954. J. Appl. Phys. *25*, 1289.
ARONOFSKY, J.S., and JENKINS, R. 1952. Proc. 1st Nat. Congr. Appl. Mech., 763.
– 1954. Trans. AIME *201*, 149.
ARONOFSKY, J.S., and PORTER, J.D. 1956. J. Appl. Mech. *23*, 128.
BAKER, W.J. 1955. Proc. Fourth World Petrol. Congr. *2*, 379.
BARENBLATT, G.I. 1952. Prikl. Mat. Mekh. *16* (1), 67.
– 1953. Prikl. Mat. Mekh. *17*, 739.
– 1954. Prikl. Mat. Mekh. *18*, 409.
– 1956a. Prikl. Mat. Mekh. *20*, 761.
– 1956b. Izv. Akad. Nauk SSSR, Otd. Tekh. Nauk *1956*, (11), 111.
BARENBLATT, G.I., and ZHELTOV, YU.P. 1960a. Dokl. Akad. Nauk. SSSR *132* (3), 545.
BARENBLATT, G.I., ZHELTOV, YU.P., and KOCHINA, I.N. 1960b. Prikl. Mat. Mekh. *24* (5), 1286.
BARRON, R.A. 1948. Proc. 2nd Int. Conf. Soil Mech. Found. Eng. *3*, 209.
BEAR, J., ZASLAVSKY, D., and IRMAY, S. 1968. See Introduction.
BIRKS, J. 1955. Proc. Fourth World Petrol. Congr. *2*, 425.
BLANCHARD, F.G., and BYERLY, P. 1935. Bull. Seism. Soc. Amer. *25*, 313.
BOUSSINESQ, J. 1904. J. Math. *10*, 5 and 363.
BOUWER, H. 1964. Ground Water *12* (3), 26.
BREDEHOEFT, J.D., and PINDER, G.F. 1970. Wat. Resources Res. *6*, 883.
BREITENÖDER, M. 1942. Ebene Grundwasserströmungen mit freier Oberfläche. Springer, Berlin.
BRUCE, G.H., et al. 1953. Trans. AIME *198*, 79.
BRUCE, W.A. 1943. Trans. AIME *151*, 112.
BUCKINGHAM, E. 1914. Phys. Rev. *4*, 345.
CARSLAW, H.S., and JAEGER, J.C. 1959. Conduction of heat in solids. 2nd ed. Clarendon Press, Oxford.
CASAGRANDE, A. 1940. Contributions to Soil Mechanics (Boston Soc. Civ. Engs.) 295.
CEDERGREN, H.R. 1967. See Introduction.
CHILDS, E.C. 1956. Science Progr. *174*, 208.
COURANT, R., and HILBERT, D. 1943. Methoden der mathematischen Physik (2 vols.). Interscience, New York.

CRAWFORD, P.B., and LANDRUM, B.L. 1955. J. Petrol. Tech. 7 (4), 47.

DAY, P.R., and LUTHIN, J.N. 1954. Proc. Soil Sci. Soc. Amer. 18, 133.

DE WIEST, R.J.M. 1965. See Introduction.

DODSON, C.R., and CARDWELL, W.T. 1945. Trans. AIME 160, 56.

DOLCETTA, A. 1948. Energia Elett. 26, 461.

DOUGLAS, J. 1961. Adv. Computers 2, 1.

DOUGLAS, J., PEACEMAN, D.W., and RACHFORD, H.H. 1955. Trans. AIME 204, 190.

DOUGLAS, J., and RACHFORD, H.H. 1956. Trans. Amer. Math. Soc. 32, 421.

DRACOS, T. 1962. Mitt. Vers. Anst. Wass. Erdb. E.T.H. Zürich 57.

DUPUIT, A.J.E.J. 1863. Etudes théoriques et pratiques sur le mouvement des eaux. Paris.

EATON, J.P., and TAKASAKI, K.J. 1959. Bull. Seism. Soc. Amer. 49, 227.

EMERY, K.O., and FOSTER, J.E. 1948. J. Marine Res. 7, 644.

EMSELEM, Y., and MARSILY, G.DE. 1969. Houille Blanche 1969 (8), 861.

– 1971. Water Resources Res. 7 (5), 1264.

FAN, S.S.T., and YEN, Y. 1968. U.S. Army Cold Regions Res. and Eng. Lab., Res. Rept. 256.

FIL'CHAKOV, P. 1949. Dokl. Akad. Nauk SSSR 66, 593.

FIL'CHAKOV, P.F., and PANCHISHIN, V.I. 1953. Gidrotekh. Stoĭtel. 22 (9), 39.

GEERTSMA, J., and SMIT, D.C. 1961. Geophysics 24, 169.

GHEORGHITZA, S.I. 1969. Trans. 1969 Haifa Symposium on Fundamentals of Transport Phenomena in Porous Media, p. 73. Ed. IAHR, Elsevier, Amsterdam, 1972.

GREEN, L., and WILTS, C.H. 1952. Proc. 1st U.S. Nat. Congr. Appl. Mech., 777.

GULNICK, M. 1956. Comm. Obs. Roy. Belg. 108 (Sér. Géophys. 37), 1.

GÜNTHER, E. 1940. Forsch. Geb. Ingenieurw. 11, 76.

HALEK, V., and NOVAK, M. 1969. Trans. Haifa 1969 Symposium on Fundamentals of Transport Phenomena in Porous Media, p. 103. Ed. IAHR, Elsevier, Amsterdam, 1972.

HAMEL, G. 1934. Z. ang. Math. Mech. 14, 129.

HANSEN, V.E. 1952. Trans. Amer. Geophys. Un. 33, 912.

HANTUSH, M. 1964. Adv. Hydroscience 1, 291.

HARR, M.E. 1962. See Introduction.

HELE-SHAW, H.S. 1897. Trans. Inst. Naval Arch. 39, 145.

– 1898. Trans. Inst. Naval Arch. 40, 21.

– 1899. Rep. 68th Meet. Brit. Assoc. Adv. Sci., 136.

HETHERINGTON, C.R., et al. 1942. Trans. AIME 146, 166.

HOPF, L., and TREFFTZ, E. 1921. Z. ang. Math. Mech. 1, 290.

JACOB, C.E. 1940. See chapter 4.

JEPPSON, R.W. 1968a. Proc. Amer. Soc. Civil Eng. 94 (HY 1), 259.

– 1968b. Proc. Amer. Soc. Civ. Eng. 94 (IR 1), 23.

– 1968c. Wat. Resources Res. 4, 435.

– 1968d. Wat. Resources Res. 4, 1277.

– 1969. Proc. Amer. Soc. Civ. Eng. 95 (HY 1), 363.

KALININ, N.K. 1950. Izv. Akad. Nauk SSSR, Odt. Tekh. Nauk, 1443.

KHRISTIANOVICH, S.A. 1941. Prikl. Mat. Mekh. 5, 277.

KIDDER, R.E. 1957. J. Appl. Mech. 24 (3), 329.

KIKKAWA, K. 1955. Japan. J. Limnology 17 (3), 91.

KIRKHAM, D. 1939. Trans. Amer. Geophys. Un. 20, 677.

– 1940. Trans. Amer. Geophys. Un. 21, 587.

– 1945. Trans. Amer. Geophys. Un. 26, 393.

– 1949. Trans. Amer. Geophys. Un. 30, 369.

– 1950a. Trans. Amer. Geophys. Un. 31, 425.

– 1950b. J. Appl. Phys. 21, 655.

– 1954. Trans. Am. Geophys. Un. *35*, 775.
– 1958. Trans. Am. Geophys. Un. *39*, 892.
KIRKHAM, D., and DE ZEEUW, J.W. 1952. Proc. Soil Sci. Soc. Amer. *16*, 286.
KOVACS, G. 1956. J. Hydrology *36* (3), 171.
LANDRUM, B.L., et al. 1959. Trans. AIME *216*, T.P. 8052.
LAURENT, J. 1949. Rev. Gen. Hyd. *15* (50), 79; *15* (51), 143.
LEE, B.D. 1948. Trans. AIME *174*, 41.
LEFRANC, E. 1938. Génie Civil *113*, 162.
LEÏBENZON, L.S. 1945. Dokl. Akad. Nauk SSSR *47*, 15.
– 1947. See Introduction.
LUTHIN, J.N., and SCOTT, V.H. 1952. Agr. Eng. *33*, 279.
MAC ROBERTS, D.T. 1949. Trans. AIME *186*, 36.
MARSAL, D. 1970. Erdöl und Kohle *23*, 10.
MATTA, G. 1957. Mém. Trav. Soc. Hydrotec. France *2*, 121.
MELCHIOR, P. 1956. Comm. Obs. Roy. Belg. *108* (Ser. Geophys. *37*), 7.
MORRIS, W.L. 1951. Ind. Eng. Chem. *43*, 2478.
MOSONYI, E., and KOVACS, G. 1956. Publ. Assoc. Int. Hydrologie (UGGI) *41*, 111.
MUSKAT, M. 1937. See Introduction.
– 1943. Trans. AIME *151*, 175.
MUSKAT, M., and BOTSET, H.G. 1931. Physics *1*, 27.
NAHRGANG, G. 1954. Zur Theorie des vollkommenen und unvollkommenen Brunnens. Springer, Berlin.
NEMETH, E. 1956. Publ. Assoc. Int. Hydrologie (UGGI) *41*, 116.
NEUMAN, S.P., and WITHERSPOON, P.A. 1969. Water Resources Res. *5* (4), 817.
– 1970. Water Resources Res. *6*, 889.
– 1971. Water Resources Res. *7* (3), 611.
ÖLLÖS, G. 1958. Hidrol. Közl. *38*, 1.
OPSAL, F.W. 1955. Trend. Eng. Univ. Wash. *7*, (2), 15.
OROVEANU, T. 1961a. Rev. Méc. Appl. Roum. *6*, 283.
– 1961b. Rev. Méc. Appl. Roum. *6*, 669.
– 1962. Stud. Cerc. Mec. Apl. Rom. *13*, 363.
– 1963. See Introduction.
OROVEANU, T., and PASCAL, H. 1959 Rev. Méc. Appl. Roum. *4*, 445.
PASCAL, H. 1964. Rev. Méc. Appl. Roum. *9*, 747.
PATTERSON, O.L., et al. 1951. A.P.I. Drill. Prod. Pract. *1951*, 47.
PISKUNOV, N.S. 1951. Dokl. Akad. Nauk SSSR *76*, 505.
PLUMMER, F.B., and WOODWARD, J.S. 1936. Trans AIME *123*, 120.
POLUBARINOVA-KOCHINA, P.YA. 1949. Dokl. Akad. Nauk SSSR *62*, 623.
– 1952. See Introduction.
RAM, G., et al. 1935. Proc. Ind. Acad. Sci. *2*, 22.
REINIUS, E. 1947. Tekn. Tidskr. *77*, 83.
ROBERTS, R.C. 1952. Proc. 1st U.S. Nat. Congr. Appl. Mech.
SANTING, G. 1958. Publ. Assoc. Int. Hydrologie (UGGI) *44*, 105.
SCHEIDEGGER, A.E. 1953. Petrol. Eng. *25* (5), B121.
– 1962. Proc. N.R.C. (Canada) Hydrology Symposium *3*, 107.
– 1963. Canad. J. Phys. *41*, 90.
SHCHELKACHEV, V.N. 1946a. See chapter 4.
– 1946b. See chapter 4.
– 1946c. See chapter 4.
STALLMAN, R.W. 1956. Pub. Assoc. Int. Hydrologie (UGGI) *41*, 227.
THEIS, C.V. 1935. Trans. Amer. Geophys. Un. *16*, 519.

TODD, D.K. 1954. Trans. Amer. Geophys. Un. *35*, 905.
– 1955. Civil Engineering *85*, 51.
VAN DER PLOEG, R.R., KIRKHAM, D., and BOAST, C.W. 1971. Water Resources Res. *7* (4), 942.
VAN EVERDINGEN, A.F., and HURST, W. 1949. Trans. AIME *186*, 302.
WERNER, P.W. 1946a. Trans. Amer. Geophys. Un. *27*, 687.
– 1946b. J. Inst. Eng. Australia *18* (6).
WERNER, P.W., and NORÉN, D. 1951. Trans. Amer. Geophys. Un. *32*, 238.
WILSON, L.H., and MILES, A.J. 1950. J. Appl. Phys. *21*, 532.
WYCKOFF, R.D., and REED, D.W. 1935. Physics *6*, 395.
YOUNGS, E.G., and SMILES, D.E. 1963. J. Geophys. Res. *68*, 5905.
ZELLER, J. 1957. Schweiz. Z. Hydrologie *19*, 164.

CHAPTER 6

ADAMS, J.T., et al. 1949. Chem. Eng. Progr. *45*, 665.
ADAMSON, J.F. 1950. Nature *166*, 314.
ARTHUR, J.R. et al. 1950. Trans. Faraday Soc. *46*, 270.
BACKER, S. 1951. Textile Res. J. *21*, 703.
BAKHMETEFF, B.A., and FEODOROFF, N.V. 1937. J. Appl. Mech. *4A*, 97.
BARTELL, F.E., and OSTERHOF, H.J. 1928. J. Phys. Chem. *32*, 1553.
BAVER, L.D. 1949. *In* O.E. Meinzer, Hydrology, p. 364. Dover, New York.
BENNER, F.C., RICHES, W.W., and BARTELL F.E. 1943. *In* Fundamental research on occurrence and recovery of petroleum, p. 74. Publ. by Am. Petrol. Inst.
BJERRUM, N., and MANEGOLD, E. 1927. Kolloid-Z. *43*, 514.
BLAINE, R.L. 1941. Amer. Soc. Test. Mat. Bull. *108*, 17.
BLAINE, R.L., and VALIS, H.J. 1949. J. Res. Nat. Bur. Stand. *42*, 257.
BREVDY, J. 1948. M.Sc. Thesis, Univ. of Minnesota, Minneapolis.
BRINKMAN, H.C. 1947. Appl. Sci. Res. *A1*, 27.
– 1948. Appl. Sci. Res. *A1*, 81.
– 1949. Research, Lond. *2*, 190.
BROOKS, C.S., and PURCELL, W.R. 1952. Trans. AIME *195*, 289.
BUCHANAN, A.S. and HEYMANN, E. 1948. Trans. Faraday Soc. *44*, 318.
BÜCHE, W. 1937. Z. Ver. deuts. Ing., Beih. Verfahrenstech. *1937*, 155.
BULNES, A.C., and FITTING, R.V. 1945. Trans. AIME *160*, 179.
CARMAN, P.C. 1937. Trans. Inst. Chem. Eng. Lond. *15*, 150.
– 1938a. Trans. Inst. Chem. Eng. Lond. *16*, 168.
– 1938b. J. Soc. Chem. Ind. *57*, 225.
– 1939a. J. Agr. Sci. *29*, 262.
– 1939b. J. Soc. Chem. Ind. *58*, 1.
– 1939c. J. Chem. Met. Mining Soc. S. Africa *39*, 26.
– 1941. See chapter 3.
– 1948. Disc. Faraday Soc. *3*, 72.
CHILDS, E.C., and COLLIS-GEORGE, N. 1950. Proc. Roy. Soc. *A201*, 392.
CLOUD, W.F. 1941. See chapter 1.
CORNELL, D. and KATZ, D.L. 1953. Ind. Eng. Chem. *45*, 2145.
COULSON, J.M. 1949. Trans. Inst. Chem. Eng. *27*, 237.
DONAT, J. 1929. Wasserkr. u. Wasserw. *24*, 225.
DUBININ, M.M. 1941. J. Appl. Chem. USSR *14*, 906.
DUPUIT, A.J.E.J. 1863. See chapter 5.

EMERSLEBEN, O. 1924. Bautechnik 2, 73.
- 1925. Physikal. Z. 26, 601.
- 1964. Z. angew. Math. Mech. 44, 23.
EYRAUD, C., BRICOUT, J., and GRILLET, G. 1963. C.R. Acad. Paris 257 (17), 2460.
FAIR, G.M., and HATCH, L.P. 1933. J. Amer. Water Wks. Assoc. 25, 1551.
FARA, H., and SCHEIDEGGER, A.E. 1961. See chapter 1.
FATT, I. 1956. Trans. AIME 207, 144.
FOWLER, J.L., and HERTEL, K.L. 1940. J. Appl. Phys. 11, 496.
FRANZINI, J.B., 1951. Trans. Amer. Geophys. Un. 32, 443.
GOODEN, E.L., and SMITH, C.M. 1940. Ind. Eng. Chem., An. Ed. 12, 479.
GORING, D.A.I., and MASON, S.G. 1950. Canad. J. Res. B28, 307.
GRATON, L.C., and FRASER, H.J. 1935. See chapter 1.
GRIFFITHS, J.C. 1950. See chapter 4.
- 1952 a. Bull. Amer. Assoc. Petrol. Geol. 36, 205.
- 1952b. Penn. State Coll. M.I. Exp Sta. Bull. 60, 47.
GRIFFITHS, J.C., and ROSENFELD, M.A. 1953. Amer. J. Sci. 251, 192.
HAPPEL, J. 1959. J. Amer. Inst. Chem. Eng. 5, 174.
HAPPEL, J., and BRENNER, H. 1965. See Introduction.
HAPPEL, J., and BYRNE, B.J. 1954. Ind. Eng. Chem. 46, 1181.
HAPPEL, J. and EPSTEIN, N. 1954. Ind. Eng. Chem. 46, 1187.
HEISS, J.F., and COULL, J. 1952. Chem. Eng. Progr. 48, 133.
HENDERSON, J.H. 1949. Producers Monthly 14 (1), 32.
HOFFING, E.H., and LOCKHART, F.J. 1951. Chem. Eng. Progr. 47 (1), 3.
HUDSON, H.E., and ROBERTS, R.E. 1952. Proc. 2nd Midw. Conf. Fluid Mech. Eng. Series, 31 (3), 105.
IBERALL, A.S. 1950. J. Res. Nat. Bur. Stand. 45, 398.
JACOB, C.E. 1946. Trans. Amer. Geophys. Un. 27, 245.
KLYACHKO, V.A. 1948. Dokl. Akad. Nauk SSSR 60, 1329.
KOLB, H. 1937. Zement 26, 93.
KOZENY, J. 1927a. S.-Ber. Wiener Akad., Abt. IIa, 136, 271.
- 1927b. Wasserkr. u. Wasserw. 22, 86.
- 1932. Kulturtechniker 35, 478.
KRAUS, G., and ROSS, J.W. 1953. J. Phys. Chem. 57, 334.
KRAUS, G. et al. 1953. J. Phys. Chem. 57, 330.
KRÜGER, E. 1918. Int. Mitt. Bodenk. 8, 105.
KRUMBEIN, W.C., and MONK, G.D. 1942. Trans. AIME 151, 153.
KUHN, H. 1946. Experientia 2, 64.
KWONG, J.N.S., et al. 1949. Chem. Eng. Progr. 45, 508.
LAMB, H. 1932. See chapter 2.
LEA, F.M., and NURSE, R.W. 1939. J. Soc. Chem. Ind. Lond. 58, T277.
LEÏBENZON, L.S. 1947. See Introduction.
LOUDON, A.G. 1952. Géotechnique Lond. 3, 165.
MACEY, H.H. 1940. Proc. Phys. Soc. 52, 625.
MAKHL, R.T. 1939. Keram. Sbornik 5, 45.
MARTIN, J.J., et al. 1951. Chem. Eng. Progr. 47 (2), 91.
MATHERON, G. 1966. Rev. Inst. Franç. Pétrole 21 (11), 1697.
MAVIS, F.T., and WILSEY, E.F. 1937. Eng. News Rec. 118, 299.
MISSBACH, A. 1938. Z. Zuckerind, Československ. Rep. 62.
MITTON, R.G. 1945. See chapter 1.
MOTT, R.A. 1951. In Some aspects of fluid flow, p. 242. Edward Arnold, London.
MROSOWSKI, S. 1958. Bull. Amer. Phys. Soc. Ser. II, 3, 98.

NELSON, W.R., and BAVER, L.D. 1940. Proc. Soil. Sci. Soc. Amer. 5, 69.

O'NEAL, A.M. 1949. Soil Sci. 67, 403.

PHILIP, J.R. 1969. Trans. Ciba Foundation Symposium on Circulatory and Respiratory Mass Transport, ed. G.E.W. Wolstenhome and J. Knight, p. 25. Churchill Ltd., London.

PILLSBURY, A.F. 1950. Soil Sci. 70, 299.

PROCKAT, F. 1940. Chem. App. 27, 129.

PURCELL, W.R. 1949. Trans. AIME 186, 39.

RAYLEIGH, LORD. 1893. Phil. Mag. (5) 36, 354.

RIGDEN, P.J. 1943. J. Soc. Chem. Ind. 62, 1.

RINK, M., and SCHOPPER, J.R. 1968. Geophys. Prospect. 16 (2), 277.

ROSE, H.E. 1945. Proc. Inst. Mech. Eng. Appl. Mech. 153 (5), 141.

ROSE, W.D., and BRUCE, W.A. 1949. See chapter 3.

ROSE, W.D., and WITHERSPOON, P.A. 1956. Div. Ill. State Geol. Survey Circ. 224.

SATHAPATHY, R.R., and RAO, C.V. 1954. Trans. Ind. Inst. Chem. Eng. 7, 85.

SCHEIDEGGER, A.E. 1953. Producers Monthly 17 (10), 17.

– 1962. Trans. Oxford Symposium of British Paper Board Makers Assoc. 829.

SCHOPPER, J.R. 1966. See chapter 1.

SEN-GUPTA, N.C., and NYUN, M.G.T. 1943. Indian J. Phys. 17, 39.

SHUSTER, W.W. 1952. Ph.D. Thesis, Rensselaer Polytech. Inst.

SLICHTER, C.S. 1899. See chapter 1.

SULLIVAN, R.R. 1941. See chapter 4.

– 1942. J. Appl. Phys. 13, 725.

SULLIVAN, R.R., and HERTEL, K.L. 1940. J. Appl. Phys. 11, 761.

SVENSSON, J. 1949. Jernkontorets Annaler 133 (2), 33.

THORNTON, O.F. 1949. Trans. AIME 186, 328.

TICKELL, F.G. 1935. Bull. Amer. Assoc., Petrol. Geol. 19, 1233.

TICKELL, F.G., and HIATT, W.N. 1938. Bull. Amer. Assoc. Petrol. Geol. 22, 1272.

TICKELL, F.G., et al. 1933. See chapter 1.

TRAXLER, R.N., and BAUM, L.A.H. 1936. Physics 7, 9.

WALAS, S.M. 1946. Trans. Amer. Inst. Chem. Eng. 42, 783.

WHITAKER, S. 1969. Ind. Eng. Chem. 61 (12), 14.

WIGGINS, E.J., et al. 1939. See chapter 1.

WILSON, B.W. 1953. Austral. J. Appl. Sci. 4, 300.

WYLLIE, M.R.J., 1951. Trans. AIME 192, 1.

WYLLIE, M.R.J., and GARDNER, G.H.F. 1958a. World Oil 146 (4), 121.

– 1958b. Nature 181, 477.

WYLLIE, M.R.J., and GREGORY, A.R. 1955. Ind. Eng. Chem. 47, 1379.

WYLLIE, M.R.J., and ROSE, W.D. 1950. Nature 165, 972.

WYLLIE, M.R.J., and SPANGLER, M.B. 1952. See chapter 1.

ZHURAVLEVI, V.F., and SYCHEV, M.M. 1947. Zhur. Prikl. Khim. 20 (3), 171.

ZIMENS, K.E. 1944. See Introduction.

ZUNKER, F. 1932. Z. pfl. Ernähr. Düng. A25, 1.

CHAPTER 7

ABD-EL-AZIZ, M.H., and TAYLOR, S.A. 1965. Proc. Soil Sci. Soc. Amer. 29, 141.

ADAM, N.K. 1941. See chapter 2.

ADZUMI, H. 1937a. See chapter 2.

– 1937b. Bull. Chem. Soc. Japan 12, 304.

ALLEN, H.V. 1944. Petrol. Refiner *23* (7), 247.

ARNELL, J.C. 1946. See chapter 2.

− 1947. Canad. J. Res. *A25*, 191.

ARNELL, J.C., and HENNEBERRY, G. 1948. Canad. J. Res. *A26*, 29.

ARONOFSKY, J.S. 1954. J. Appl. Phys. *25*, 48.

ARONOFSKY, J.S., STEIN, N., and WALLICK, G.C. 1955. Soil Sci. *79*, 49.

BAKHMETEFF, B.A., and FEODOROFF, N.V. 1937. See chapter 6.

− 1938. Proc. 5th Int. Congr. Appl. Mech., 555.

BAPTIST, O.C., and SWEENEY, S.A. 1955. U.S. Bureau of Mines, R.I. 5180.

− 1957. U.S. Bureau of Mines R.I. 5331.

BAPTIST, O.C., and WHITE, E.J. 1957. Trans. AIME *210*, 414.

BARRER, R.M. 1939. Phil. Mag. *28*, 148.

− 1941. Diffusion in and through solids. Cambridge Univ. Press.

− 1948. Disc. Faraday Soc. *3*, 61.

BARRER, R.M., and BARRIE, J.A. 1952. Proc. Roy. Soc. *A213*, 250.

BARRER, R.M., and GROVE, D.M. 1951a. Trans. Faraday Soc. *47*, 826.

− 1951b. Trans. Faraday Soc. *47*, 837.

BARTH, W., and ESSER, W. 1933. Forschung *4*, 82.

BEUTNER, R. 1933. Physical chemistry of living tissues. Baltimore.

BIESEL, F. 1950. Houille blanche *5*, 2, 157.

BJERRUM, N., and MANEGOLD, E. 1928. Kolloid-Z. *43*, 5.

BLAKE, F.E. 1922. Trans. Amer. Inst. Chem. Eng. *14*, 415.

BODMAN, G.B. 1937. Proc. Soil Sci. Soc. Amer. *2*, 45.

BRESTON, J.N., and JOHNSON, W.E. 1945. Producers Monthly *9* (12), 19.

BRIDGWATER, A.B. 1950. Civ. Eng. Lond. *45*, 234, 313, 385, 451.

BRIEGHEL-MÜLLER, A. 1940. Kolloid-Z. *92*, 285.

BROWN, G.P. et al. 1946. J. Appl. Phys. *17*, 802.

BROWN, R.L., and BOLT, R.H. 1942. See chapter 4.

BROWNELL, L.E., and KATZ, D.L. 1947. Chem. Eng. Progr. *43*, 537.

BROWNELL, L.E. et al. 1950. Chem. Eng. Progr. *46*, 415.

BULKLEY, R. 1931. J. Res. Nat. Bur. Stand. *6*, 88.

BULL, H.B., and WRONSKY, J.P. 1937. J. Phys. Chem. *41*, 463.

BURKE, S.P., and PARRY, V.F. 1935. AIME T.P. 607.

BURKE, S.P., and PLUMMER, W.B. 1928. Ind. Eng. Chem. *20*, 1196.

CALHOUN, J.C. 1946. Ph.D. Thesis, Penn State Coll.

CALHOUN, J.C., and YUSTER, S.T. 1946. A.P.I. Drill. Prod. Pract., 335.

CAMBEFORT, H. 1951. Wwys. Exp. Sta. Transl. 51–3.

CARMAN, P.C. 1937. See chapter 6.

− 1947. Nature *160*, 301.

− 1949. Nature *163*, 684.

− 1950. Proc. Roy. Soc. *A203*, 55.

− 1952. Proc. Roy. Soc. *A211*, 526.

CARMAN, P.C., and ARNELL, J.C. 1948. Canad. J. Res. *A26*, 128.

CARMAN, P.C., and MALHERBE, P.LE R. 1950. Proc. Roy. Soc. *A203*, 165.

CARMAN, P.C., and RAAL, F.A. 1951. Proc. Roy. Soc. *A209*, 38.

CASSIE, A.B.D. 1945. Trans. Faraday Soc. *41*, 458.

CHALMERS, J. et al. 1932. Trans. AIME *98*, 375.

CHARDABELLAS, P.E. 1940. Mitt. preuss. Vers. Anst. Wasser-, Erd- u. Schiffbau H40.

CHILTON, T.H. 1938. The science of petroleum. Oxford Univ. Press.

CHILTON, T.H., and COLBURN, A.P. 1931. Ind. Eng. Chem. *23*, 313.

− 1931b. Ind. Eng. Chem. *23*, 913.

COLLINS, R.E., and CRAWFORD, P.B. 1953. Trans. AIME *198*, 339.
CORNELL, D., and KATZ, D.L. 1951. Ind. Eng. Chem. *43*, 992.
– 1953. See chapter 6.
DANIELLI, J.F. 1937. Trans. Faraday Soc. *33*, 1139.
DERYAGIN, B.V., et al. 1948. See chapter 2.
DURIEZ, M. 1952. Ann. Trav. Pub. Belg. *104* (2), 201.
EHRENBERGER, R. 1928. Z. Oester. Ing. Arch. Ver. *9/10*, 71.
EKEDAHL, E., and SILLÉN, L.G. 1947. Ark. Kemi. Min. Geol. *25A*, pap. 4.
ENGELUND, F. 1953. Trans. Dan. Acad. Techn. Sci., 3.
ERGUN, S. 1952a. Chem. Eng. Progr. *48*, 89.
– 1952b. Analyt. Chem. *24*, 388.
– 1953. Ind. Eng. Chem. *45*, 477.
ERGUN, S., and ORNING, A.A. 1949. Ind. Eng. Chem. *41*, 1179.
FAIR, G.M., and HATCH, L.P. 1933. See chapter 6.
FANCHER, G.H., and LEWIS, J.A. 1933. Ind. Eng. Chem. *25*, 1139.
FANCHER, G.H., et al. 1933. Penn. State M.I. Exp. Sta. Bull. *12*, 65.
– 1934. Proc. World Petrol. Congr. *1*, 322.
FERGUSON, A.F.D. et al. 1956. Chem. & Ind. *1956*, 1523.
FLOOD, E.A. et al. 1952a. Canad. J. Chem. *30*, 348.
– 1952b. Canad. J. Chem. *30*, 372.
– 1952c. Canad. J. Chem. *30*, 389.
FORCHHEIMER, P. 1901. Z. Ver. deuts. Ing. *45*, 1782.
FOX, J.W. 1949. Proc. Phys. Soc. *B62*, 829.
FUJITA, S., and UCHIDA, S. 1934. Soc. Chem. Ind. Japan *37*, 791B.
FURNAS, C.C. 1931. Ind. Eng. Chem. *23*, 1052.
GAMSON, B.W., THODOS, G., and HOUGEN, O.A. 1943. Trans. Amer. Inst. Chem. Eng. *39*, 1.
GIVAN, C.V. 1934. Trans. Amer. Geophys. Un. *15*, 572.
GLÜCKAUF, E. 1944. Nature *154*, 831.
GRACE, H.P. 1953. Chem. Eng. Progr. *49*, 303 and 367.
GRIFFITHS, J.C. 1946. J. Inst. Petrol. *32*, 18.
GRIMLEY, S.S. 1945. Trans. Inst. Chem. Eng. *23*, 228.
GRISEL, F. 1936. C.R. Acad. Paris. *203*, 1351.
GRUNBERG, L., and NISSAN, A.H. 1943. J. Inst. Petrol. *29*, 193.
GUSTAFSON, Y. 1940. Lantbruks-Högskol. Ann. *8*, 425.
HAPPEL, J. 1949. Ind. Eng. Chem. *41*, 1161.
HATFIELD, M.L. 1939. Ind. Eng. Chem. *31*, 1419.
HEID, J.G., et al. 1950. A.P.I. Drill. Prod. Pract., 230.
HICKOX, G.H. 1934. Trans. Amer. Geophys. Un. *15*, 567.
HILES, J., and MOTT, R.A. 1945. Fuel *24*, 135.
HODGINS, J.W. et al. 1946. Canad. J. Res. *B24*, 167.
HOLLER, H. 1943. Deutsche Wasserw. *36*, 9.
HOLMES, W.R. 1946. Nature, *157*, 694.
HOOGSCHAGEN, J. 1953. J. Chem. Phys. *21*, 2097.
HUDSON, H.E., and ROBERTS, R.E. 1952. See chapter 6.
IBERALL, A.S. See chapter 6.
IRMAY, S. 1958. Trans. Amer. Geophys. Un. *39*, 702.
ISHIKAWA, H. 1942. Waseda Appl. Chem. Soc. Bull. *19*, 51.
IWANAMI, S. 1940. Trans. Soc. Mech. Eng. Japan *6* (24), 18.
JOHNSON, W., and TALIAFERRO, D.B. 1938. U.S. Bur. Mines T.P. 592.
JONES, W.M. 1951. Trans. Faraday Soc. *47*, 381.
– 1952. Trans. Faraday Soc. *48*, 562.

KEYES, W.F. 1946. Ind. Eng. Chem., An. Ed. *18*, 33.

KHANIN, A.A. 1948. Neftyanoe Khoz. *26* (5), 32.

KING, G. 1940. Z. Ver. deuts. Ing. *84*, 85.

KLINKENBERG, L.J. 1941. A.P.I. Drill. Prod. Pract., 200.

KRISCHER, O. 1963. Die wissenschaftlichen Grundlagen der Trocknungstechnik. Springer, Berlin.

KRUTTER, H., and DAY, R.J. 1941. Oil Weekly *104* (4), 24.

KRUYT, H.R. 1952. Colloid science, vol. I. Elsevier, New York.

KUTÍLEK, M. 1969. Trans. 1969 Haifa Symposium on Fundamentals of Transport Phenomena in Porous Media, p. 327. Ed. IAHR, Elsevier, Amsterdam, 1972.

LAPUK, B.B., and EVDOKIMOVA, V.A. 1951. Dokl. Akad. Nauk SSSR *76*, 509.

LEA, F.M., and NURSE, R.W. 1947. Sympos. Particle Size Anal., Suppl. Trans. Inst. Chem. Eng. *25*, 54.

LEĬBENZON, L.S. 1954a. Izv. Akad. Nauk SSSR, Ser. Geog. Geofis. *9*, 3.

– 1945b. Izv. Akad. Nauk SSSR, Ser. Geog. Geofiz. *9*, 7.

LEVA, M. 1947. Chem. Eng. Progr. *43*, 549.

– 1949. Chem. Eng. *56* (5), 115.

LEVA, M., and GRUMMER, M. 1947. Chem. Eng. Progr. *43* (11).

LEVA, M., et al. 1951. U.S. Bur. Mines Bull. 504.

LINDQUIST, E. 1933. Proc. 1ᵉʳ Congr. Grands Barr. Stockholm, *5*, 81.

LINN, H.A.D. 1950. Water (Holland) *34*, 19.

LYKOW, A.W. 1958. See Introduction.

MACH, E. 1935. Forschungsh. Ver. deuts. Ing. 375.

– 1939. Dechema, Monographieen *6*, 38.

MANEGOLD, E. 1937. Kolloid-Z. *81*, 269.

MEADLEY, C.K. 1962. Canad. J. Chem. Eng. *1962*, 256.

MEINZER, O.E., and WENZEL, L.K. 1940. Econom. Geol. *35* (8), 915.

MEYER, G., and WORK, L.T. 1937. Trans. Amer. Inst. Chem. Eng. *33*, 13.

MEYER, K.H., and SIEVERS, J.F. 1936. Helv. chim. Acta *19*, 649.

MICHAELS, A.S., and LIN, C.S. 1954. Ind. Eng. Chem. *46*, 1239.

– 1955. Ind. Eng. Chem. *47*, 1249.

MILLER, L.E. 1946. Producers Monthly *11* (1), 35.

MILLER, K.T., et al. 1946. Producers Monthly *11* (1), 31.

MISSBACH, A. 1937. Listy Cukrovar. *55*, 293.

MOODY, L.F. 1944. Trans. Amer. Soc. Mech. Eng. *66*, 671.

NAYAR, M.R., and SHUKLA, K.P. 1943a. Current Sci. *12* (5), 156.

– 1943b. Current Sci. *12* (6), 183.

– 1943c. Current Sci. *12* (7), 206.

– 1949. J. Sci. Ind. Res. *8B* (8), 137.

NEMENYI, P. 1934. Wasserbauliche Strömungslehre. Springer, Berlin.

NIELSEN, R.F. 1951. World Oil *132* (6), 188.

NISSAN, A.H. 1942. J. Inst. Petroleum *28*, 257.

OMAN, A.O., and WATSON, K.M. 1944. Natl. Petrol. News *36*, R795.

PENMAN, H.L. 1940a. J. Agr. Sci. *30*, 437.

– 1940b. J. Agr. Sci. *30*, 570.

PHILIP, J.R., and DE VRIES, D.A. 1957. Trans. Amer. Geophys. Un. *38*, 222.

PLAIN, G.J. and MORRISON, H.L. 1954. Amer. J. Phys. *22*, 143.

POLUBARINOVA-KOCHINA, P.YA. 1952. See Introduction.

REYNOLDS, O. 1900. Papers on mechanical and physical subjects. Cambridge Univ. Press, London.

RIGDEN, P.J. 1946. Nature *157*, 268.

– 1947. See chapter 6.
ROMITA, P.L. 1951. Ricer. Scient. *21*, 1978.
ROSE, H.E. 1945a. See chapter 6.
– 1945b. Proc. Inst. Mech. Eng. *153*, 148.
– 1945c. Proc. Inst. Mech. Eng. *153*, 154.
– 1949. Proc. Inst. Mech. Eng. *160*, 492.
– 1951. *In* Some aspects of fluid flow, p. 136. Edward Arnold, London.
ROSE, H.E., and RIZK, A.M.A. 1949. Proc. Inst. Mech. Eng. *160*, 493.
ROSE, W.D. 1948. A.P.I. Drill. Prod. Pract., 209.
RUTH, B.F. 1946. See chapter 4.
SAMESHIMA, J. 1926. Bull. Chem. Soc. Japan *1*, 5.
ŠANDERA, K., and MIRČEV, A. 1938. Listy Cukrovar *57*, 51.
SCHAFFERNAK, F., and DACHLER, R. 1934. Wasserwirtsch. *1*, 145.
SCHEIDEGGER, A.E. 1953. See Chapter 6.
SCHLÖGL, R. 1964. Stofftransport durch Membranen. Fortschr. physikal. Chemie, vol. 9. Steinkopff, Darmstadt.
SHOUMATOFF, N. 1952. Amer. Soc. Mech. Eng. Pap. 52, SA-42.
SHUKLA, K.P. 1944. Current Sci. *13*, 45.
SHUKLA, K.P., and NAYAR, M.R. 1943. Current Sci. *12*, 155.
SILLÉN, L.G. 1946. Ark. Kemi. Min. Geol. *22A* (5), Pap. 15.
– 1950a. Ark. Kemi *2*, Pap. 34, 477.
– 1950b. Ark. Kemi *2*, Pap. 35, 499.
SILLÉN, L.G., and EKEDAHL, E. 1946. Ark. Kemi Min. Geol. *22A* (5), Pap. 16.
SOKOLOVSKIĬ, V.V. 1949a. Dokl. Akad. Nauk SSSR *65*, 617.
– 1949b. Prikl. Mat. Mekh. *13*, 525.
SPAUGH, O.H. 1948. Food Technol. *2*, 33.
SPIEGLER, K.S. 1958. Trans. Faraday Soc. *54*, 1408.
STARK, K.P. 1969. Trans. 1969 Haifa Symposium on Fundamentals of Transport Phenomena in Porous Media, p. 86. Ed. IAHR, Elsevier, Amsterdam, 1972.
STEWART, C.R., and OWENS, W.W. 1958. Trans. AIME *213*, 121.
SWARTZENDRUBER, D. 1966. Advances in Agronomy *18*, 327.
– 1967. Trans. Int. Soil Water Sympos. 207.
TAKAGI, M., and ISHIKAWA, H. 1942. Waseda Appl. Chem. Soc. Bull. *19*, 59.
TEORELL, T. 1935. Proc. Soc. Exp. Biol. Med. *33*, 282.
– 1937. Trans. Faraday Soc. *33*, 1053.
TOMLINSON, R.H., and FLOOD, E.A. 1948. Canad. J. Res. *B26*, 38.
UCHIDA, S. 1952. Proc. First Jap. Nat. Congr. Appl. Mech., 1951, Nat. Comm. Theor. Appl. Mech. *1952*, 437.
UCHIDA, S., and FUJITA, S.J. 1934. Soc. Chem. Ind. Japan (Suppl. Binding) *37*, 724B.
UGUET, D. 1951. C.R. Acad. Paris *232*, 383.
URBAIN, P. 1941. C.R. Soc. Géol. France *1941*, 106.
VAN DER KOOI, J. 1971. Moisture transport in cellular concrete roofs. Diss. Tech. Hoogeschool, Eindhoven.
VERONESE, A. 1941. L'ingenere *15*, 463.
VERSCHOOR, K. 1950. Ingenieur (Utrecht) *62*, 29.
VON ENGELHARDT, W., and TUNN, W.L.M. 1954. Heidelberger Beiträge Min. Petrogr. *2*, 12.
WADELL, H.J. 1934. J. Franklin Inst. *217*, 459.
WALLICK, G.C., and ARONOFSKY, J.S. 1954. Trans. AIME *201*, 322.
WARD, W.H. 1939. Engineering *148*, 435.
WEINTRAUB, M., and LEVA, M. 1948. Chem. Eng. Progr. *44*, 801.

WENTWORTH, C.K. 1944. Amer. J. Sci. *242*, 478.
- 1946. Trans. Amer. Geophys. Un. *27*, 540.
WHITE, A.M. 1935. Trans. Amer. Inst. Chem. Eng. *31*, 390.
WICKE, E. 1938. Z. Elektrochem. *44*, 587.
- 1939a. Kolloid-Z. *86*, 167.
- 1939b. Kolloid-Z. *86*, 295.
WICKE, E., and KALLENBACH, R. 1941. Kolloid-Z. *97*, 135.
WILSON, L.H. et al. 1951. J. Appl. Phys. *22*, 1027.
WODNYANSZKY, G. 1938. Magyar Timár *2*, 33.
YUSTER, S.T. 1946. A.P.I. Drill. Prod. Pract., 356.
ZABEZHINSKIĬ, YA.L. 1939. J. Phys. Chem. USSR *13*, 1858.
ZEISBERG, F.C. 1919. Trans. Amer. Inst. Chem. Eng. *12*, 231.
ZUNKER, F. 1920. J. Gasbel. Wasservers.

CHAPTER 8

BACHMAT, Y., and BEAR, J. 1964. J. Geophys. Res. *69*, 2561.
BERGE, C. 1958. Théorie des graphes et ses applications. Dunod, Paris.
BLACKWELL, R.J. 1962. J. Soc. Petrol. Eng. *2*, 1.
BROADBENT, S.R., and HAMMERSLEY, J.M. 1957. Proc. Cambridge Phil. Soc. *53*, 629.
BRUCH, J.C., and STREET, R.L. 1967. J. Sanit. Eng. Div., Amer. Soc. Civ. Engr. *SA6*, 17.
CAYLEY, A. 1859. Phil. Mag. *18*, 374.
CHAUDHARI, N., and SCHEIDEGGER, A.E. 1964. Pure Appl. Geophys. *59*, 45.
- 1965. Canad. J. Phys. *43*, 1776.
DE GROOT, S.R. 1961. Thermodynamics of irreversible processes. North-Holland, Amsterdam.
DE JOSSELIN DE JONG, G. 1969. Trans. 1969 Haifa Symposium on Fundamentals of Transport Phenomena in Porous Media, p. 459. Ed. IAHR, Elsevier, Amsterdam, 1972.
DE WIEST, R.J.M. 1965. See Introduction.
GOLDSTEIN, S. 1951. Quart. Appl. Math. *4*, 129.
HARLEMAN, D.R.F., and RUMER, R.R. 1963. J. Fluid Mech. *16* (3), 385.
KAMPÉ DE FÉRIET, J. 1939. Ann. Soc. Sci. de Bruxelles *59*, Sér. 1, 145.
LIAO, K.H., and SCHEIDEGGER, A.E. 1968. Bull. Internat. Assoc. Sci. Hydrol. *13* (1), 5.
- 1969. Bull. Internat. Assoc. Sci. Hydrol. *14* (4), 137.
NYE, J.F. 1957. Physical properties of crystals. Clarendon Press, Oxford.
PERKINS, T.K., and JOHNSTON, O.C. 1963. J. Soc. Petrol. Eng. *3*, 70.
SCHEIDEGGER, A.E. 1954. J. Appl. Phys. *25*, 994.
- 1955. Geofis. Pura Appl. *30*, 17.
- 1956. Canad. J. Phys. *34*, 692.
- 1958. Canad. J. Phys. *36*, 649.
- 1961a. J. Geophys. Res. *66*, 3273.
- 1961b. Canad. J. Phys. *39*, 1573.
- 1968. Canad. J. Phys. *47*, 209.
TODOROVIĆ, P. 1970. Water Resources Res. *6* (1), 211.
TOMKORIA, B.N., and SCHEIDEGGER, A.E. 1967. Canad. J. Phys. *45*, 3569.
TORELLI, L., and SCHEIDEGGER, A.E. 1971. Pure Appl. Geophys. *89*, 32.
- 1972. J. Hydrol. *15*, 23.
VON MISES, R. 1945. Wahrscheinlichkeitsrechnung. Rosenberg, New York.
YUHARA, K. 1954. Tikyubuturi Geophys. Inst. Kyoto Univ. *9* (2), 127.

CHAPTER 9

ADIVARAHAN, P., et al. 1962. J. Soc. Pet. Eng. 2, 290.

ANZELIUS, A. 1926. Z. angew. Math. Mech. 6, 291.

ARONOFSKY, J.S. 1952. Trans. AIME 195, 15.

ARONOFSKY, J.S., and RAMEY, H.J. 1956. Trans. AIME 207, 205.

ARTHUR, K.B., and METANOMSKI, Z.G. 1966. C.R. 2me Colloq. Assoc. Rech. Tech. Forage et Production, p. 23. Ed. Technip, Paris.

BACHMAT, Y., and ELRICK, D.E. 1970. Water Resources Res. 6, 156.

BÉNARD, H. 1901. Ann. Chim. Phys. (7) 23, 62.

BERNHARD, H., LINDEMANN, N., and MEDER, H. 1958. Erdöl u. Kohle 11, 231.

BLAND, D.R. 1954. Proc. Roy. Soc. (London) A221, 1.

BORIES, S. 1970a. C.R. Acad. Sci. Paris. 270, 66.

– 1970b. C.R. Acad. Sci. Paris 271, 269.

– 1970c. Revue Gén. Thermique 108, 1377.

BRESTON, J.N., and PEARMAN, B.R. 1953. Producers Monthly 18 (1), 15.

BURTON, M.B., and CRAWFORD, P.B. 1956. Trans. AIME 207, 333.

CHOW, C.C., and SCHEIDEGGER, A.E. 1972. J. Hydrol. 15, 1.

CHUOKE, R.L., VAN MEURS, P., and VAN DER POEL, C. 1959. Trans. AIME 216, T.P. 8073.

COLLINS, M.A., and GELHAR, L.W. 1961. Water Resources Res. 7, 971.

CRAUSSE, E., and POIRIER, Y. 1957. C.R. Acad. Sci. Paris 243, 475.

CRAWFORD, P.B., and COLLINS, R.E. 1955. Producers Monthly 20 (1), 26.

DAGAN, G. 1966. Rept. A.E.C. Proj. No. 401-25-07, Iowa State Univ., Ames, Iowa.

DEJAK, C., and TREVISSOI, C. 1954. Boll. sci. fac. chim. ind. (Bologna) 12 (2), 21.

ELDER, J.W. 1965. In Terrestrial heat flow, ed. W.H.K. Lee, Chap. 8. Amer. Geophys. Un. Monogr. 8. Washington.

– 1967. J. Fluid Mech. 27, 29.

ENGELBERTS, W.F., and KLINKENBERG, L.J. 1951. Proc. Third World Petrol. Congr. 2, 544.

FURNAS, C.C. 1930. Trans. Amer. Inst. Chem. Eng. 24, 142.

GAMSON, B.W., THODOS, G., and HOUGEN, O.A. 1943. See chapter 7.

GARBUS, R. 1956. Petrol. Eng. 28 (4), B83.

GOGUEL, J. 1953. Ann. Mines (Paris) 10, 3.

GOSWAMI, A.B. 1968a. Trans. Min. Geol. Met. Inst. India 65 (2), 19.

– 1968b. Bull. Internat. Assoc. Sci. Hydrol. 13 (3), 77.

HADIDI, T.A.R., NIELSEN, R.F., and CALHOUN, J.C. 1956. Producers Monthly 20 (10), 38.

HAUSEN, H. 1931. Z. angew. Math. Mech. 11, 105.

HORTON, C.W., and ROGERS, F.T. 1945. J. Appl. Phys. 16, 367.

HUBBERT, M.K. 1940. See chapter 4.

JEFFREYS, H. 1926. Phil. Mag. (7) 2, 833.

– 1928. Proc. Roy. Soc. (London) A118, 195.

JENKINS, R., and ARONOFSKY, J.S. 1955. Producers Monthly 19 (5), 37.

KARPLUS, W.J. 1956. Trans. AIME 207, 240.

KATTO, M. 1967. J. Heat and Mass Transfer 10, 297.

KAWABATA, H. 1965. Bull. Kyoto Gakugei Univ. B27, 19.

KLINKENBERG, A. 1948. Ind. Eng. Chem. 40, 1992.

LANDRUM, B.L., et al. 1958. Producers Monthly 22 (5), 40.

LAPWOOD, E.R. 1948. Proc. Camb. Phil. Soc. 44, 508.

LUSCZINSKY, N.J. 1961. J. Geophys. Res. 66, 4247.

MEYER, H.I., and GARDER, A.O. 1954. J. Appl. Phys. 25, 1400.

MEYER, H.I., and SEARCY, D.F. 1956. Trans. AIME 207, 302.

MORRISON, H.L., and ROGERS, F.T. 1952. J. Appl. Phys. 23, 1058.

MORRISON, H.L., ROGERS, F.T., and HORTON, C.W. 1949. J. Appl. Phys. *20*, 1027.
MUSKAT, M. 1934. Physics *5*, 250.
− 1937. See Introduction.
NOBLES, M.A., and JANZEN, H.B. 1958. J. Petrol. Tech. *10* (2), 60.
NUSSELT, W. 1927. Z. Ver. deutsch. Ing. *71*, 85.
ODEH, A.S., BRADLEY, H.P., and HELLER, J.P. 1956. Trans. AIME *207*, 200.
PERKINS, T.K., and JOHNSTON, O.C. 1969. J. Soc. Petrol. Eng. *9*, 39.
PERLMUTTER, N.M., GERAGHTY, J.J., and UPSON, J.E. 1959. Econom. Geol. *54*, 416.
POLUBARINOVA-KOCHINA, P.YA. 1952. See Introduction.
PRESTON, F.W. 1955. Mechanism of heat transfer in unconsolidated porous·media. Ph.D. Thesis, Pennsylvania State Univ.
− 1956. Amer. J. Sci. *254*, 754.
PRESTON, F.W., and HAZEN, R.D. 1954. Producers Monthly *18* (4), 24.
RAYLEIGH, LORD. 1916. Phil. Mag. *32*, 529.
REILLY, P.M. 1957. J. Am. Inst. Chem. Eng. *3*, 513.
ROGERS, F.T. 1953. J. Appl. Phys. *24*, 877.
ROGERS, F.T., and MORRISON, H.L. 1950. J. Appl. Phys. *21*, 1177.
ROGERS, F.T., and SCHILBERG, L.E. 1951. J. Appl. Phys. *22*, 233.
ROGERS, F.T., SCHILBERG, L.E., and MORRISON, L.E. 1951. J. Appl. Phys. *22*, 1476.
SCHEIDEGGER, A.E. 1960a. Canad. J. Phys. *38*, 153.
− 1960b. Physics of Fluids *3*, 94.
− 1967. Canad. J. Phys. *45*, 1783.
− 1969. Canad. J. Phys. *47*, 209.
SCHEIDEGGER, A.E., and JOHNSON, E.F. 1961. Canad. J. Phys. *39*, 326.
SCHILD, A.E. 1957. Trans. AIME *210*, 1.
SCHUMANN, T.E.W. 1929. J. Franklin Inst. *208*, 405.
SINEL'NIKOVA, O.L. 1957. Neft. Khoz. *35* (3), 40.
TERRY, W.M., BLACKWELL, R.J., and RAYNE, J.R. 1958. AIME paper, No. 1131-G.
TODD, D.K. 1959. See Introduction.
TOMKORIA, B.N., and SCHEIDEGGER, A.E. 1967. See chapter 8.
VAN MEURS, P. 1957. Trans. AIME *210*, 295.
VERMA, A.P. 1969. Canad. J. Phys. *47*, 319.
VERRUIJT, A. 1969. Trans. 1969 Haifa Symposium on Fundamentals of Transport Phenomena in Porous Media, p. 25. Ed. IAHR, Elsevier, Amsterdam, 1972.
WENTWORTH, C.K. 1947. Pacific Sci. *1*, 172.
− 1951. C.R. Assoc. Gén. Bruxelles, Assoc. Int. Hydrologie (UGGI) *2*, 238.
WILHELM, R.C., et al. 1948. Chem. Eng. Progr. *44*, 105.
WOODING, R.A. 1957. J. Fluid Mech. *2* (3), 273.
− 1958. J. Fluid Mech. *3*, 582.
− 1959. Proc. Roy. Soc. *A252*, 120.
− 1962. Z. angew. Math. Phys. *13*, 255.
− 1963. J. Fluid Mech. *15*, 527.
− 1964. J. Fluid Mech. *19*, 103.

CHAPTER 10

ATKINSON, D.I.W. 1948. J. Imp. Coll. Chem. Eng. Soc. *4*, 78.
BABBITT, J.D. 1939. Canad. J. Res. *A17*, 15.
− 1940. Canad. J. Res. *A18*, 105.
− 1948. Pulp Paper Mag. Canada *49*, 83,

BAKER, T., et al. 1935. Trans. Amer. Inst. Chem. Eng. *31*, 296.

BANKOFF, S.G. 1969. Trans. 1969 Haifa Symposium on Fundamentals of Transport Phenomena in Porous Media, p. 166. Ed. IAHR, Elsevier, Amsterdam, 1972.

BARTELL, F.E., and MILLER, F.L. 1932. Ind. Eng. Chem. *24*, 335.

BEAR, J., and BRAESTER, C. 1969. Trans. 1969 Haifa Symposium on Fundamentals of Transport Phenomena in Porous Media, p. 177. Ed. IAHR, Elsevier, Amsterdam, 1972.

BERGELIN, O.P. 1949. Chem. Eng. *56* (5), 104.

BERTETTI, J.W. 1942. Trans. Amer. Inst. Chem. Eng. *38*, 1023.

BLAIR, P.M., and PEACEMAN, D.W. 1963. Soc. Petrol. Eng. J. *1963* (3), 19.

BOELTER, L.M.K., and KEPNER, R.H. 1939. Ind. Eng. Chem. *31*, 426.

BOKSERMAN, A.A., ZHELTOV, YU.P., and KOCHESHKOV, A.A. 1964. Dokl. Akad. Nauk SSSR *9* (4), 285.

BOTSET, H.G. 1940. Trans. AIME *136*, 91.

BRANSON, U.S. 1951. World Oil *133* (1), 184.

BRINKMAN, H.C. 1948. Appl. Sci. Res. *A1*, 333.

BROWNELL, L.E., and KATZ, D.L. 1947. Chem. Eng. Progr. *43*, 601.

BROWNSCOMBE, E.R., and DYES, A.B. 1952. A.P.I. Drill. Prod. Pract., 383.

BROWNSCOMBE, E.R., et al. 1949. A.P.I. Drill. Prod. Pract., 302.

– 1950a. Oil and Gas J. *48* (40), 68.

– 1950b. Oil and Gas J. *48* (41), 98.

BRUCE, R.R., and KLUTE, A. 1963. Proc. Soil Sci. Soc. Amer. *27*, 18.

BUCKLEY, S.E., and LEVERETT, M.C. 1942. Trans. AIME *146*, 107.

BURDINE, N.T. 1953. Trans. AIME *198*, 71.

BUTIJN, J., and WESSELING, J. 1959. Netherl. J. Agric. Sci. *7*, 155.

CALHOUN, J.C. 1951. Oil and Gas J. *50* (21), 117.

– 1951b. Oil and Gas J. *50* (20), 308.

CARMAN, P.C. 1941. Soil Sci. *52*, 1.

CAUDLE, B.H., et al. 1951. Trans. AIME *192*, 145.

CEAGLSKE, N.H., and KIESLING, F.C. 1940. Trans. AIME *36*, 211.

CHATENEVER, A. 1952. Oil and Gas J. *51* (3), 174.

CHATENEVER, A., and CALHOUN, J.C. 1952. Trans. AIME *195*, 149.

CHATENEVER, A., INDRA, M.K., and KYTE, J.R. 1959. J. Petrol. Tech. *9* (6), 13.

CHILDS, E.C., and COLLIS-GEORGE, N. 1948. Disc. Faraday Soc. *3*, 78.

CHRISTENSEN, H.R. 1944. Soil Sci. *57*, 381.

CHRISTIANSEN, J.E. 1944. Soil Sci. *58*, 355.

CLOUD, W.F. 1930. Trans. AIME *86*, 337.

COLLIS-GEORGE, N. 1953. Trans. Amer. Geophys. Un. *34*, 589.

COOPER, C.M., et al. 1941. Trans. Amer. Inst. Chem. Eng. *37*, 979.

COREY, A.T., and RATHJENS, C.H. 1957. Trans. AIME *207*, 358.

COREY, A.T., et al. 1956. Trans. AIME *207*, 349.

DOMBROWSKI, H.S., and BROWNELL, L.E. 1954. Ind. Eng. Chem. *46*, 1207.

DOUGLAS, J., PEACEMAN, D.W., and RACHFORD, H.H. 1959. Trans. AIME *216*, 297.

DUNLAP, E.N. 1938. Trans. AIME *127*, 215.

ELGIN, J.C., and WEISS, F.B. 1939. Ind. Eng. Chem. *31*, 435.

ENRIGHT, R.J. 1954. Oil and Gas J. *53* (2), 104.

ESTES, R.K., and FULTON, P.F. 1956. Trans. AIME *207*, 338.

EVINGER, H.H., and MUSKAT, M. 1942. Trans. AIME *146*, 194.

FATT, I. 1953. J. Petrol. Technol. *5* (10), 15.

– 1956. Trans. AIME *207*, 160.

FATT, I., and DYKSTRA, H. 1951. Trans. AIME *192*, 249.

FERGUSON, H., and GARDNER, W.R. 1963. Proc. Soil Sci. Soc. Amer. *27*, 243.

FLETCHER, J.E. 1949. Trans. Amer. Geophys. Un. *30*, 548.

FREEZE, R.A. 1971. Water Resources Res. *7*, 347.

FULTON, P.F. 1951. Producers Monthly *15* (12), 14.

FURNAS, C.C., and BELLINGER, F. 1938. Trans. Amer. Chem. Eng. *34*, 251.

GARDESCU, I.I. 1930. Trans. AIME *86*, 351.

GARDNER, W.R. 1956. Proc. Soil Sci. Soc. Amer. *20*, 317.

GARDNER, W.R., and MAYHUGH, M.S. 1958. Proc. Soil Sci. Soc. Amer. *22*, 187.

GARRISON, A.D. 1934. A.P.I. Drill. Prod. Pract., 130.

GATES, J.I., and LIETZ, W.T. 1950. A.P.I. Drill. Prod. Pract., 285.

GAZLEY, C. 1949. Ph.D. Thesis, Univ. of Delaware.

GEFFEN, T.M., et al. 1951. Trans. AIME *192*, 99.

GRAHAM, J.W., and RICHARDSON, J.G. 1959. J. Petrol. Tech. *11* (2), 65.

HANKS, R.J., and BOWER, S.A. 1962. Proc. Soil. Sci. Soc. Amer. *26*, 530.

HARRINGTON, J.D. 1949. M.Sc. Thesis, Univ. of Oklahoma.

HASSAN, M.E., and NIELSEN, R.F. 1953. Petrol. Eng. *25* (3), B61.

HASSLER, G.L. 1944. U.S. Patent 2,345,935.

HASSLER, G.L., et al. 1936. Trans. AIME *118*, 116.

HAUSER, P.M., and MCLAREN, A.D. 1948. Ind. Eng. Chem. *40*, 112.

HENDERSON, J.H., and MELDRUM, A.H. 1949. Producers Monthly *13* (5), 12.

HENDERSON, J.H., and YUSTER, S.T. 1948a. Producers Monthly *12* (3), 13.

– 1948b. World Oil *127* (12), 139.

HILL, S. 1952. Chem. Eng. Sci. *1*, 247.

HOLMGREN, C.R., and MORSE, R.A. 1951. Trans. AIME *192*, 135.

HOUGEN, O.A., and MARSHALL, W.R. 1947. Chem. Eng. Progr. *43* (4).

HOUGHTON, F.C., et al. (1924). J. Amer. Soc. Heating Vent. Eng. *30*, 139.

HOUPEURT, A. 1949. Rév. Inst. franç. pétrole *4*, 107.

IRMAY, S. 1954. Trans. Amer. Geophys. Un. *35*, 463.

IVAKIN, V.V. 1951. Izv. Akad. Nauk SSSR, Otd. Tekh. Nauk *1951*, 1874.

JAMIN, J.C. 1860. C.R. Acad. Paris *50*, 172.

JESSER, B.W., and ELGIN, J.C. 1943. Trans. Amer. Inst. Chem. Eng. *39*, 277.

JOHANSSON, C.H., et al. 1949. Acta Polytech. No. 29 (Chem. Met. Ser. No. 7), 5.

JONES, P.J. 1946. Petroleum production. Reinhold, New York.

– 1949. World Oil *129* (2), 170.

KERN, L.R. 1952. Trans. AIME *195*, 39.

KIRKHAM, D., and FENG., C.L. 1949. Soil Sci. *67*, 29.

KLUTE, A. 1952. Soil Sci. *73*, 105.

KRUTTER, H., and DAY, R.J. 1943. Pet. Chem. Techn. *6*, 1.

KRYNINE, D.P. 1950. Highway Res. Board, Proc. 29th Ann. Meet., 520.

KUNZE, R.J., and KIRKHAM, D. 1962. Proc. Soil Sci. Soc. Amer. *26*, 421.

LEAS, W.J., et al. 1950. Trans. AIME *189*, 65.

LERNER, B.J., and GROVE, C.S. 1951. Ind. Eng. Chem. *43*, 216.

LEVERETT, M.C. 1939. Trans. AIME *142*, 152.

LEVERETT, M.C., and LEWIS, W.B. 1941. Trans. AIME *142*, 107.

LEWIS, J.O. 1944. Trans. AIME *155*, 131.

LOBO, W.E., et al. 1949. Trans. Amer. Inst. Chem. Eng. *1949*, 693.

LYKOW, A.W. See Introduction.

MACH, E. 1935. See chapter 7.

MAHONEY, C.F. 1947. Thesis. Univ. of Oklahoma.

MARTIN, J.C. 1958. Producers Monthly *22* (6), 22.

MARTINELLI, R.C., et al. 1944. Trans. Amer. Inst. Mech. Eng. *66*, 139.

– 1946. Trans. Amer. Inst. Chem. Eng. *42*, 681.

MATTAX, C.C., and KYTE, J.R. 1962. J. Soc. Petrol. Eng. *2*, 177.

MAYO, F., et al. 1935. J. Soc. Chem. Ind. *54*, 373T.

MILLER, E.E., and ELRICK, D.E. 1958. Proc. Soil Sci. Soc. Amer. *22*, 483.

MILLER, E.E., and MILLER, R.D. 1956. See chapter 3.

MORSE, R.A., et al. 1947a. Producers Monthly *11* (10), 19.

– 1947b. Oil and Gas J. *46* (16), 109.

MUSKAT, M. 1949. See Introduction.

– 1950. Trans. AIME *189*, 349.

MUSKAT, M., and MERES, M.W. 1936. Physics *7*, 346.

MUSKAT, M., et al. 1937. Trans. AIME 123, 69.

– 1953. Oil and Gas J. *52* (28), 238.

NIELSEN, R.F. 1949. Producers Monthly *14* (2), 29.

NIELSON, D.R., et al. 1962. Proc. Soil Sci. Soc. Amer. *26*, 107.

O'BANNON, L.S. 1924. J. Amer. Soc. Heating Vent. Eng. *30*, 157.

O'CONNOR, G.V. 1946. Chem. Eng. *53* (11), 162.

OROVEANU, T., and PASCAL, H. 1956. Stud. si. Cercet. Mecan. Appl. *7* (2), 387.

OSOBA, J.S., et al. 1951. Trans. AIME *192*, 47.

PECK, A.J. 1966. Trans. UNESCO Syposium on Water in the Unsaturated Zone *1*, 191.

– 1969. Proc. Soil Sci. Soc. Amer. *33*, 980.

PEIRCE, F.T., et al. 1945. J. Textile Inst. *36*, T169.

PETERS, W.A. 1922. Ind. Eng. Chem. *14*, 476.

PFALZNER, P.M. 1950. Canad. J. Res. *A28*, 389.

PHILIP, J.R. 1954. J. Inst. Eng. Austral. *26*, 255.

– 1955a. Trans. Faraday Soc. *51*, 885.

– 1955b. Proc. Nat. Acad. Sci. India (Allahabad) *A24*, 93.

– 1957a. Austral. J. Phys. *10*, 29.

– 1957b. Soil Sci. *83*, 163, 257, 329, 345, 435.

– 1958. Soil Sci. *85*, 278.

– 1966a. Trans. UNESCO Symposium on Water in the Unsaturated Zone *2*, 559.

– 1966b. *Ibid. 1*, 503.

– 1969a. Austral. J. Soil Res. *7*, 213.

– 1969b. Adv. Hydroscience *5*, 215.

– 1970. Annual Rev. Fluid Mech. *2*, 177.

PILLSBURY, A.F., and APPLEMAN, D. 1945. Soil Sci. *59*, 115.

PIRET, E.L., et al. 1940. Ind. Eng. Chem. *32*, 861.

PIRVERDYAN, A.M. 1952. Prikl. Mat. Mekh. *16*, 711.

PLUMMER, F.B., et al. 1937. A.P.I. Drill. Prod. Pract., 417.

PORKHAEV, A.P. 1949. Kolloid Zhur. *11*, 346.

PURCELL, W.R. 1950. Trans. AIME *189*, 369.

RAATS, P.A.C., and GARDNER, W.R. 1971. Water Resources Res. *7*, 921.

RACHFORD, H.H. 1964. Soc. Petrol. Eng. J. *1964* (6), 134.

RAPOPORT, L.A. 1955. Trans. AIME *204*, 143.

RAPOPORT, L.A., and LEAS, W.J. 1951. Trans. AIME *192*, 83.

– 1953. Trans. AIME *198*, 139.

RAWLINGS, S.L., and GARDNER, W.R. 1963. Proc. Soil Sci. Soc. Amer. *27*, 507.

REID, L.S., and HUNTINGTON, R.L. 1938. AIME T.P. 873.

RICHARDS, L.A. 1931. Physics *1*, 318.

RICHARDS, L.A., and MOORE, D.C. 1952. Trans. Amer. Geophys. Un. *33*, 531.

RICHARDSON, J.G., et al. 1952. Trans. AIME *195*, 187.

RIJTEMA, P.E. 1959. Netherland J. Agric. Sci. *7*, 209.

ROSE, W. 1948. See chapter 7.
- 1949. Trans. AIME *186*, 111.
- 1951a. Proc. 3rd. World Petrol. Congr. *2*, 446.
- 1951b. Trans. AIME *192*, 373.
- 1969. Trans. 1969 Haifa Symposium on Fundamentals of Transport Phenomena in Porous Media, p. 229. Ed. IAHR, Elsevier, Amsterdam, 1972.
ROSE, W., et al. 1962. Z. Phys. Chem. *34*, 182.
ROSE, W., and BRUCE, W.A. 1949. See chapter 3.
ROSE, W., and WITHERSPOON, P.A. 1956. Producers Monthly *21* (2), 32.
ROSE, W., and WYLLIE, M.R.J. 1949. Trans. AIME *186*, 329.
RYDER, H.M. 1948. World Oil *128* (2), 142.
SARCHET, B. 1942. Trans. Amer. Inst. Chem. Eng. *38*, 283.
SCOTT, P.H., and ROSE, W. 1953. Trans. AIME *198*, 323.
SEN-GUPTA, N.C. 1943. Indian J. Phys. *17*, 338.
SHERWOOD, T.K. 1937. Adsorption and extraction. McGraw-Hill, New York.
SIMMONS, C.W., and OSBORN, H.B. 1934. Ind. Eng. Chem. *26*, 529.
TALSMA, T. 1970a. Water Resources Res. *6*, 964.
- 1970b. Austral. J. Soil Res. *8*, 179.
TARNER, J. 1944. Oil Weekly *114* (2), 32.
TEMPLETON, C.C. 1953. AIME T.P. 307-G; Trans. AIME *201*, 162.
- 1954. Bull. Amer. Phys. Soc. *29* (2), 16.
TERWILLIGER, P.L., et al. 1951. Trans. AIME *192*, 285.
TERWILLIGER, P.L., and YUSTER, S.T. 1946. Producers Monthly *11* (1), 42.
- 1947. Oil Weekly *126* (1), 54.
THIRROT, C., and ARIBERT, J.M. 1969. Trans. 1969 Haifa Symposium on Fundamentals of Transport Phenomena in Porous Media, p. 371. Ed. IAHR, Elsevier, Amsterdam, 1972.
UCHIDA, S., and FUJITA, S. 1936. J. Soc. Chem. Ind. Japan *39*, 886.
- 1937. J. Soc. Chem. Ind. Japan *40*, 238.
UREN, L.C., and BRADSHAW, E.J. 1932. Trans. AIME *98*, 438.
UREN, L.C., and DOMERECQ, M. 1937. Trans. AIME *114*, 25.
VAN WINGEN, H. 1938. Oil Weekly, Oct. 10.
VERMA, A.P. 1968. Rev. Roum. Sci. Tech., Méc. Appl. *13* (2), 277.
- 1969. Canad. J. Phys. *47*, 2519.
VERSCHOOR, H. 1938. Trans. Inst. Chem. Eng. *16*, 66.
VERSLUYS, J. 1917. See chapter 3.
- 1931. See chapter 3.
VILBRANDT, F.C., et al. 1938. Trans. Amer. Inst. Chem. Eng. *34*, 51.
WELGE, H. 1952. Trans. AIME *195*, 91.
WHISLER, F.D., and KLUTE, A. 1967. Trans. Amer. Soc. Agric. Eng. *10*, 391.
WHITE, A.M. 1935. See chapter 7.
WHITING, L.L., and GUERRERO, E.T. 1951. Oil and Gas J. *50* (12), 272.
WILSON, D.A., and CALHOUN, J.C. 1952. Oil and Gas J. *51* (3), 175.
WILSON, J.W. 1956. J. Amer. Inst. Chem. Eng. *2*, 94.
WINK, W.A., and DEARTH, L.R. 1949. TAPPI *32*, 232.
WYCKOFF, R.D., and BOTSET, H.G. 1936. Physics *7*, 325.
WYLLIE, M.R.J. 1951. Trans. AIME *192*, 381.
WYLLIE, M.R.J., and SPANGLER, M.B. 1952. See chapter 1.
YOUNGS, E.G., and TOWNER, G.D. 1963. Soil Sci. *95*, 369.
YUSTER, S.T. 1951. Proc. 3rd World Petrol. Congr. *2*, 436.
ZENZ, F.A. 1947. Trans. Amer. Inst. Chem. Eng. *43*, 415.

CHAPTER 11

ARIS, R. 1956. Proc. Roy. Soc. *A235*, 67.
ARIS, R., and ADMUNDSON, N.R. 1957. J. Am. Inst. Chem. Eng. *3*, 280.
ARONOFSKY, J.S., and HELLER, J.P. 1957. Trans. AIME *210*, 345.
BACHMAT, Y. 1967. Water Resources Res. *3* (4), 1079.
BERNARD, R.A., and WILHELM, R.H. 1950. See chapter 1.
BRUCH, J.C. 1970. Water Resources Res. *6* (3), 791.
BRUCH, J.C., and STREET, R.L. 1967. See chapter 8.
CARBERRY, J.J. 1958. Canad. J. Chem. Eng. *36* (5), 207.
CARBERRY, J.J., and BRETTON, R.H. 1958. J. Am. Inst. Chem. Eng. *4* (3), 367.
CARRIER, G.F. 1958. J. Fluid Mech. *4*, 479.
CLARK, N.J., et al. 1957. Petrol. Eng. *29* (10), B21.
− 1958. J. Petrol. Tech. *10* (6), 11.
COLUMBUS, N. 1965. Water Resources Res. *1* (2), 313.
COOPER, H.H. 1959. J. Geophys. Res. *64*, 461.
CRAIG, F.F. 1970. J. Petrol. Technol. *22* (5), 529.
CRANE, F.E., and GARDNER, G.H.F. 1961. Chem. Eng. Data *6*, 283.
CRANK, J. 1956. The mathematics of diffusion. University Press, Oxford.
DAY, P.R. 1956. Trans. Am. Geophys. Un. *37*, 595.
DAY, P.R., and FORSYTHE, W.M. 1957. Proc. Soil Sci. Soc. Amer. *21*, 477.
DE JOSSELIN DE JONG, G. 1956. Pub. Assoc. Int. Hydrologie (UGGI) *41*, 139.
− 1958. Trans. Am. Geophys. Un. *39*, 67.
DOUGHERTY, E.L. 1963. J. Soc. Petrol. Eng. *3*, 155.
EBACH, E.A., and WHITE, R.R. 1958. J. Am. Inst. Chem. Eng. *4*, 161.
ELRICK, D.E. 1967. Proc. 1st Canad. Conf. Micromet. *2*, 477.
ERIKSSON, E. 1958. Trans. Am. Geophys. Un. *39*, 937.
GARDER, A.E., PEACEMAN, D.W., and POZZI, A.L. 1964. J. Soc. Petrol. Eng. *4* (1), 26.
GARDNER, W.R., and BROOKS, R.H. 1957. Soil Sci. *83*, 295.
GOLDSTEIN, S. 1951. Quart. J. Mech. Appl. Math. *4*, 129.
HALL, H.N., and GEFFEN, T.M. 1957. Trans. AIME *210*, 48.
HANDY, L.L. 1959. J. Petrol. Tech. *11* (3), 61.
HARLEMAN, D.R.F., and RUMER, R.R. 1963. See chapter 8.
HARPAZ, Y., and BEAR, J. 1964. Pub. Internat. Assoc. Sci. Hydrol. *64*, 132.
HENRY, H.R. 1959. J. Geophys. Res. *64*, 1911.
HOOPES, J.A., and HARLEMAN, D.R.F. 1965. MIT Dept. Civil Eng., Hydrod. Lab. Rept. 75.
JAHNKE, E., and EMDE, F. 1945. Tables of functions with formulae and curves. Dover, New York.
KLINKENBERG, A., and SJENITZER, F. 1956. Chem. Eng. Sci. *5*, 258.
KOVAL, D.J. 1953. J. Soc. Petrol. Eng. *3*, 145.
KRAMERS, H. and ALBERDA, G. 1953. Chem. Eng. Sci. *2*, 173.
KRUPP, H.K., and ELRICK, D.E. 1968. Water Resources Res. *4*, 809.
LACEY, J.W., DRAPER, A.L., and BINDER, G.G. 1958. Trans. AIME *213*, 76.
LAPIDUS, L., and AMUNDSON, N.R. 1952. J. Phys. Chem. *56*, 984.
LI, W.H., and LAI, F.H. 1966. Proc. Amer. Soc. Civ. Eng. *92*, HY6, 141.
LIST, E.J., and BROOKS, N.H. 1967. J. Geophys. Res. *72*, 2531.
MCHENRY, K.W., and WILHELM, R.H. 1957. J. Amer. Inst. Chem. Eng. *3*, 83.
NIELSEN, D.R., and BIGGAR, J.W. 1962. Proc. Soil Sci. Soc. Amer. *26*, 216.
OFFERINGA, J., and VAN DER POEL, C. 1954. Trans. AIME *201*, 310.
OGATA, A. 1958. Dispersion in porous media. Ph.D. Thesis, Northwestern Univ., Evanston, Illinios.

– 1961. U.S. Geol. Surv. Prof. Pap. 411-B.

OGATA, A., and BANKS, R.B. 1961. U.S. Geol. Surv. Prof. Pap. 411-A.

PEACEMAN, D.W., and RACHFORD, H.H. 1962. J. Soc. Petrol. Eng. *2* (4), 327.

PERKINS, T.K., and JOHNSTON, O.C. 1963. See chapter 8.

PERRINE, R.L. 1961. J. Soc. Petrol. Eng. *1* (1), 17.

– 1963. J. Soc. Petrol. Eng. *3*, 205.

PINDER, G.F., and COOPER, H.H. 1970. Water Resources Res. *6* (3), 875.

RAIMONDI, P., GARDNER, G.H.G., and PETRICK, C.B. 1959. A.I.Ch.E. & S.P.E. Joint Symposium, San Francisco *2*, Paper 43.

RIFAI, M.N.E., KAUFMAN, W.J., and TODD, D.K. 1956. Dispersion phenomena in laminar flow through porous media. San. Eng. Res. Lab., Div. Div. Eng. Univ. Calif. Berkeley; Rep. I.E.B. Series No. 93, Issue No. 2.

RUMER, R.R., and HARLEMAN, D.R.F. 1963. Proc. Amer. Soc. Civ. Eng. *HY6*, 193.

SCHEIDEGGER, A.E. 1955. See chapter 8.

– 1958a. Trans. Amer. Geophys. Un. *39*, 929.

– 1958b. C. R. Assoc. Toronto. Assoc. Int. Hydrologie (UGGI) *2*, 236.

– 1959. Proc. Theory of Fluid Flow in Porous Media Conf., Univ. Oklahoma, 101.

– 1965. Rev. Inst. Franç. Pétrole *20* (6), 879.

SCHEIDEGGER, A.E., and LARSON, V.C. 1958. Canad. J. Phys. *36*, 1476.

SHAMIR, U.Y., and HARLEMAN, D.R.F. 1967. Water Resources Res. *3* (2), 557.

SIMPSON, E.S. 1962. U.S. Geol. Surv. Prof. Paper 411-C.

SLOBOD, R.L., BURCIK, E.J., and CASHDOLLAR, B.H. 1959. Producers Monthly *23* (8), 11.

STONE, H.L., and BRIAN, P.L.T. 1963. J. Amer. Inst. Chem. Eng. *9* (5), 681.

VON ROSENBERG, D.U. 1956. See chapter 2.

WOODING, R.A. 1962. See chapter 9.

YAGI, S., and MIYAUCHI, T. 1955. Chem. Eng. Japan *19*, 507.

Index

RETURN P